SAXON ALGEBRA 1

Practice Workbook

HOUGHTON MIFFLIN HARCOURT
Supplemental Publishers

www.SaxonPublishers.com
800-531-5015

ISBN 13: 978-1-6027-7504-6
ISBN 10: 1-6027-7504-4

© 2009 Saxon®, an imprint of HMH Supplemental Publishers Inc.

All rights reserved. No part of this material protected by this copyright may be reproduced or utilized in any form or by any means, in whole or in part, without permission in writing from the copyright owner. Requests for permission should be mailed to: Paralegal Department, 6277 Sea Harbor Drive, Orlando, FL 32887.

Saxon® is a registered trademark of HMH Supplemental Publishers Inc.

Printed in the United States of America

If you have received these materials as examination copies free of charge, HMH Supplemental Publishers Inc. retains title to the materials and they may not be resold. Resale of examination copies is strictly prohibited and is illegal.

Possession of this publication in print format does not entitle users to convert this publication, or any portion of it, into electronic format.

1 2 3 4 5 6 7 8 170 15 14 13 12 11 10 09 08

Table of Contents

Lesson 1 1
Lesson 2 3
Lesson 3 5
Lesson 4 7
Lesson 5 9
Lesson 6 11
Lesson 7 15
Lesson 8 17
Lesson 9 21
Lesson 10 23
Lesson 11 27
Lesson 12 31
Lesson 13 35
Lesson 14 39
Lesson 15 43
Lesson 16 47
Lesson 17 51
Lesson 18 55
Lesson 19 59
Lesson 20 63
Lesson 21 67
Lesson 22 71
Lesson 23 75
Lesson 24 79
Lesson 25 83
Lesson 26 87
Lesson 27 91
Lesson 28 95
Lesson 29 99
Lesson 30 103
Lesson 31 107
Lesson 32 111
Lesson 33 115
Lesson 34 119
Lesson 35 123
Lesson 36 127
Lesson 37 131
Lesson 38 135
Lesson 39 139
Lesson 40 143
Lesson 41 147
Lesson 42 151
Lesson 43 155
Lesson 44 159

Saxon Algebra 1

Table of Contents

Lesson 45 163
Lesson 46 167
Lesson 47 171
Lesson 48 175
Lesson 49 179
Lesson 50 183
Lesson 51 187
Lesson 52 191
Lesson 53 195
Lesson 54 199
Lesson 55 203
Lesson 56 207
Lesson 57 211
Lesson 58 215
Lesson 59 219
Lesson 60 223
Lesson 61 227
Lesson 62 231
Lesson 63 235
Lesson 64 239
Lesson 65 243
Lesson 66 247
Lesson 67 251
Lesson 68 255
Lesson 69 259
Lesson 70 263
Lesson 71 267
Lesson 72 271
Lesson 73 275
Lesson 74 279
Lesson 75 283
Lesson 76 287
Lesson 77 291
Lesson 78 295
Lesson 79 299
Lesson 80 303
Lesson 81 307
Lesson 82 311
Lesson 83 315
Lesson 84 319
Lesson 85 323
Lesson 86 327
Lesson 87 331
Lesson 88 335

Saxon Algebra 1

Table of Contents

Lesson 89 339
Lesson 90 343
Lesson 91 347
Lesson 92 351
Lesson 93 355
Lesson 94 359
Lesson 95 363
Lesson 96 367
Lesson 97 371
Lesson 98 375
Lesson 99 379
Lesson 100 383
Lesson 101 387
Lesson 102 391
Lesson 103 395
Lesson 104 399
Lesson 105 403
Lesson 106 407
Lesson 107 411
Lesson 108 415
Lesson 109 419
Lesson 110 423
Lesson 111 427
Lesson 112 431
Lesson 113 435
Lesson 114 439
Lesson 115 443
Lesson 116 447
Lesson 117 451
Lesson 118 455
Lesson 119 459
Lesson 120 463

Name _____ Date _____ Class _____

Classifying Real Numbers **LESSON 1**

1. (SB 2) Multiply 26.1×6.15.

2. (SB 3) Add $\frac{4}{7} + \frac{1}{8} + \frac{1}{2}$.

3. (SB 2) Divide $954 \div 0.9$.

4. (SB 3) Add $\frac{3}{5} + \frac{1}{8} + \frac{1}{8}$.

5. (SB 5) Write $\frac{3}{8}$ as a decimal.

6. (SB 6) Write $0.\overline{666}$ as a fraction.

7. (SB 3) Add $2\frac{1}{2} + 3\frac{1}{5}$.

8. (SB 7) Name a fraction equivalent to $\frac{2}{5}$.

9. (SB 12) **Error Analysis** Two students determine the prime factorization of 72. Which student is correct? Explain the error.

Student A	Student B
72	72
$= 9 \cdot 8$	$= 9 \cdot 8$
$= 9 \cdot 4 \cdot 2$	$= 9 \cdot 4 \cdot 2$
$= 9 \cdot 2 \cdot 2 \cdot 2$	$= 3 \cdot 3 \cdot 2 \cdot 2 \cdot 2$

10. (SB 12) Find the prime factorization of 144.

11. (SB 5) Write 0.15 as a percent. If necessary, round to the nearest tenth.

12. (SB 5) Write 7.2 as a percent. If necessary, round to the nearest tenth.

*13. (1) Use braces and digits to designate the set of natural numbers.

*14. (1) The set $\{0, 1, 2, 3, \ldots\}$ represents what set of numbers?

*15. (1) Represent the following numbers as being members of set K: 2, 4, 2, 0, 6, 0, 10, 8.

*16. (1) **Multiple Choice** Which of the following numbers is an irrational number?

　　A 15　　　　B $\sqrt{15}$　　　　C 15.15151515…　　D $-\frac{15}{3}$

*17. (1) **Measurement** The surface area of a cube is defined as $6s^2$, where s is the length of the side of the cube. If s is an integer, then would the surface area of a cube be a rational or irrational number?

18. (SB 1) **Verify** True or False: A right triangle can have an obtuse angle. Explain your answer.

19. (SB 3) (**Anatomy**) A baby's head is approximately one fourth of its total body length. If the baby's body measures 19 inches, what does the baby's head measure?

20. (SB 13) True or False: An acute triangle has 3 acute angles. Explain your answer.

21. (SB 14) True or False: A trapezoid has two pairs of parallel sides. Explain your answer.

© Saxon. All rights reserved.　　　　　　　　Saxon Algebra 1

Classifying Real Numbers

LESSON 1

*22. **(Track Practice)** Tyrone ran 7 laps on the quarter-mile track during practice. Which subset of real numbers would include the distance Tyrone ran at practice?
(1)

23. True or False: A parallelogram has two pairs of parallel sides. Explain your answer.
(SB 14)

24. **Write** Use the divisibility test to determine if 1248 is divisible by 2. Explain your answer.
(SB 4)

*25. **Geometry** The diagram shows a right triangle. The length of the hypotenuse is a member of which subset(s) of real numbers?
(1)

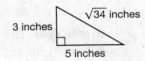

*26. **Multi-Step** The diagram shows a rectangle.
(1)
  a. Find the area of the rectangle.
  b. The number of square feet is a member of which subset(s) of real numbers?

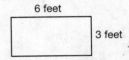

27. **(Lunar Rover)** The surface-speed record set by the lunar rover on the moon is 10.56 miles per hour. At that speed, how far would the rover travel in 3.5 hours?
(SB 2)

28. **Write** Use the divisibility test to determine if 207 is divisible by 3. Explain your answer.
(SB 4)

29. **(Swimming)** Vidiana and Jaime went swimming before school. Vidiana swam $\frac{3}{5}$ mile and Jaime swam $\frac{4}{7}$ mile. Write a comparison to show who swam farther. Use <, >, or =.
(SB 1)

*30. **(Banking)** Shayla is balancing her checkbook. Which subset of real numbers best describes her balance?
(1)

Name _____ Date _____ Class _____

Understanding Variables and Expressions LESSON 2

Find the GCF of each pair of numbers.

1. 24, 32 (SB 9)

2. 28, 42 (SB 9)

Find the LCM of each group of numbers.

3. 9, 12 (SB 10)

4. 3, 5, 6 (SB 10)

Multiply or divide.

5. $\dfrac{3}{4} \cdot \dfrac{8}{15}$ (SB 3)

6. $\dfrac{7}{15} \div \dfrac{21}{25}$ (SB 3)

Identify the coefficients and variables in each expression.

*7. $rst - 12v$ (2)

*8. $2xy + 7w - 8$ (2)

*9. $47s + \dfrac{2}{5}t$ (2)

Identify the following statements as true or false. Explain your choice.

*10. Verify All whole numbers are natural numbers. (1)

11. Verify All integers are real numbers. (1)

12. Verify A number can be a member of the set of rational numbers and the set of irrational numbers. (1)

13. Multi-Step Use the following set of data. (SB 29)

3, 6, 4, 3, 6, 5, 6, 7, 4, 3, 2, 4, 6

  a. What is the frequency of each number?

  b. Display the set of data in a line plot.

14. All natural numbers are members of which other subsets of real numbers? (1)

15. Measurement Add $7\dfrac{3}{8}$ meters + $6\dfrac{1}{3}$ meters. Does the sum belong to the set of rational numbers, integers, or whole numbers? (1, SB 3)

16. Find the prime factorization of 153. (SB 12)

17. Verify True or False: An obtuse triangle can have more than one obtuse angle. Explain your choice. (SB 13)

18. Geometry A line can be classified as a _____ angle. (SB 13)

19. Write Use the divisibility test to determine if 2345 is divisible by 4. Explain your answer. (SB 4)

Saxon Algebra 1

Understanding Variables and Expressions

LESSON 2

20. Write 0.003 as a percent. If necessary, round to the nearest tenth.
(SB 5)

21. Use braces and digits to designate the set of whole numbers.
(1)

22. The set {1, 2, 3,...} represents what set of numbers?
(1)

*23. **Multiple Choice** What is the second term in the expression
(2) $\sqrt{8} + \frac{gh}{5} + (3x + y) + 15gh$?

    A $(3x + y)$ **B** $15gh$ **C** $\sqrt{8}$ **D** $\frac{gh}{5}$

*24. (**Astronomy**) To calculate the amount of time it takes for a planet to travel around the
(2) sun, you use the following expression: $\frac{2\pi r}{v}$. Which values are constants, which are variables, and which are coefficients?

*25. (**Entertainment**) Admission price for a matinee movie is $5.75 for children and $6.25
(2) for adults. Brad uses the expression $5.75c + $6.25a to calculate the cost for his family. What are the variables in the expression?

*26. **Error Analysis** The surface area of a rectangular prism is $2lw + 2lh + 2wh$. Two
(2) students determined the variables in the formula. Which student is correct? What was the error of the other student?

Student A	Student B
variables: $2lw$, $2lh$	variables: l, w, h

*27. (**Cost Analysis**) A large medical organization wants to put two cylindrical aquariums in
(2) the pharmacy area. It will cost the pharmacy 53 cents per cubic inch of aquarium. This is the formula for figuring out the cost: $P = (\pi r^2 h)(\$0.53)$.
   a. Find the coefficients of the expression.
   b. Find the variables of the expression.

28. **Multiple Choice** Which shape is not a parallelogram?
(SB 14)
    A square   **B** rectangle   **C** trapezoid   **D** rhombus

*29. (**Cycling**) A bicycle shop uses the expression $5 + $2.25h to determine the charges for
(2) bike rentals. How many terms are in the expression?

30. (**Attendance**) The attendance clerk keeps records of students' attendance. Which
(1) subset of real numbers would include the number of students in attendance each school day?

Name _____ Date _____ Class _____

Simplifying Expressions Using the Product Property of Exponents
LESSON 3

Find the GCF for each pair of numbers.

1. 15, 35 (SB 9)
2. 32, 48 (SB 9)

Find the LCM for each group of numbers.

3. 8, 12 (SB 10)
4. 2, 4, 7 (SB 10)

Multiply or divide.

5. $\dfrac{9}{16} \cdot \dfrac{12}{15}$ (SB 3)
6. $\dfrac{6}{15} \div \dfrac{24}{30}$ (SB 3)

Identify the coefficients and variables in each expression.

7. $6mn + 4b$ (2)

8. $5j - 9cd + 2$ (2)

9. $23t + \dfrac{4}{7}w$ (2)

Identify the following statements as true or false. Explain your choice.

*10. **Verify** All real numbers are integers. (1)

11. **Verify** All natural numbers are whole numbers. (2)

12. **Verify** All irrational numbers are real numbers. (2)

Complete the comparisons. Use <, >, or =.

13. 42.53 ◯ 42.35 (SB 1)
14. $\dfrac{5}{9}$ ◯ $\dfrac{7}{12}$ (SB 1)

15. Add $1\dfrac{1}{8} + 7\dfrac{2}{5}$. (SB 3)

16. **Measurement** Use braces and digits to designate the set of integers. Which measurement can be described by the set of integers: temperature or volume? (1)

17. Find the prime factorization of 98. (SB 12)

*18. **Error Analysis** Two students are trying to simplify the expression $x^2 \cdot x^5$. Which student is correct? Explain the error. (3)

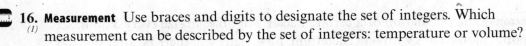

Student A	Student B
$x^2 \cdot x^5$	$x^2 \cdot x^5$
$x^{2 \cdot 5} = x^{10}$	$x^{2+5} = x^7$

19. **Verify** True or False: A rhombus is always a square. Explain your choice. (SB 14)

Saxon Algebra 1

Simplifying Expressions Using the Product Property of Exponents

LESSON 3

20. Write Use the divisibility test to determine if 306 is divisible by 6. Explain your answer.
(SB 4)

*21. The expression 3^6 indicates the number of times 3 is used as a factor.
(3)
  a. Which number in the expression is the base?
  b. Which number is the exponent?
  c. What is the simplified value of this expression?

*22. Multiple Choice MFLOPS, TFLOPS, and PFLOPS are used to measure the speed
(3) of a computer. One PFLOP is equal to 10^3 TFLOPS. Each TFLOP is equal to 10^6 MFLOPS. How many MFLOPS are in a PFLOP?
  A 10^{18} **B** 10^9 **C** 10^6 **D** 10^3

*23. (Cooking) A cooking magazine advertises 4^4 recipes in every issue. How many recipes
(3) are in 4^2 issues?

*24. Multi-Step A business is worth 10^6 dollars this year. The business expects to be 10^3
(3) more valuable in five years.
  a. Simplify 10^3 to determine how many times more valuable the business will be.
  b. What will the business be worth in five years? Express your answer in exponential form, then simplify your answer.

*25. (Population) The population of Bridgetown triples every decade. If the population in
(3) the year 2000 was 25,000, how many people will be living in Bridgetown in 2030?

26. Multiple Choice Which triangle is a right triangle?
(SB 13)
  A a triangle with angle measures of 45°, 45°, and 90°
  B a triangle with angle measures of 40°, 110°, and 30°
  C a triangle with angle measures of 55°, 45°, and 80°
  D a triangle with angle measures of 60°, 60°, and 60°

*27. (Bacteria) The population of a certain bacteria doubles in size every 3 hours. If
(3) a population begins with one bacterium, how many will there be after one day? Simplify the expression $1 \cdot (2)^8$ to determine the population after one day.

28. Geometry You can calculate the area of a trapezoid using the following equation:
(2) $A = h \times \frac{b_1 + b_2}{2}$. Identify the constant(s) in the equation.

*29. (Aquarium) A fish tank is in the shape of a cube. Each side measures 3 feet. What is the
(SB 26) volume of the fish tank?

*30. (Remodeling) Vanessa is remodeling her bathroom. She uses the expression $2l + 2w$ to
(2) determine the amount of wallpaper border she needs.
  a. How many terms are in the expression?
  b. What are the variables?

Name _____ Date _____ Class _____

Using Order of Operations

LESSON 4

Add, subtract, multiply, or divide.

1. (SB 3) $2\frac{1}{4} + 4\frac{1}{2}$

2. (SB 3) $5\frac{2}{5} - 3\frac{1}{4}$

3. (SB 3) $1\frac{3}{4} + 4\frac{1}{8} - 2\frac{1}{2}$

4. (SB 3) $4\frac{1}{3} \div 2\frac{1}{6}$

5. (SB 2) $3.519 \div 0.3$

6. (SB 2) $4.16 \cdot 2.3$

7. (2) How many terms are in the algebraic expression $14x^2 + 7x + \frac{x}{4}$?

8. (SB 12) Find the prime factorization of 225.

9. Write (SB 4) Use the divisibility test to determine if 124,302 is divisible by 3. Explain your answer.

10. (1) Represent the following numbers as being members of set L: −15, 1, 7, 3, −8, 7, 0, 12, 6, 12

*11. Verify (1) True or False: All whole numbers are integers. Explain your answer.

*12. (1) To which set(s) of numbers does $\sqrt{5}$ belong?

13. (SB 5) Write $\frac{1}{6}$ as a percent. If necessary, round to the nearest tenth.

14. (SB 5) Write $\frac{5}{9}$ as a percent. If necessary, round to the nearest tenth.

*15. (4) Compare $3 \cdot 4^2 + 4^2 \bigcirc 3 \cdot (16 + 16)$ using <, >, or =. Explain.

16. Multiple Choice (SB 13) Which triangle is an obtuse triangle?

A a triangle with angle measures of 45°, 45°, and 90°

B a triangle with angle measures of 40°, 120°, and 20°

C a triangle with angle measures of 55°, 45°, and 80°

D a triangle with angle measures of 60°, 60°, and 60°

17. (SB 29) Display the following set of data in a line plot:

6, 7, 8, 4, 5, 4, 3, 4, 5, 3, 2, 6, 2, 7

18. Verify (SB 14) True or False: A square is a rectangle. Explain your choice.

19. Measurement (SB 3) Subtract $15\frac{1}{3}$ yards − $7\frac{4}{5}$ yards.

20. Error Analysis (SB 12) Two students determine the prime factorization of 108. Which student is correct? Explain the error.

Student A	Student B
$108 = 2 \cdot 2 \cdot 3 \cdot 3 \cdot 3 \cdot 3$	$108 = 2^2 \cdot 3^3$

Saxon Algebra 1

Using Order of Operations

LESSON 4

21. Write Use the divisibility test to determine if 1116 is divisible by 9. Explain your answer.
(SB 4)

22. $\frac{n}{6} + 3xy - 19$
(2)

a. Find the variables of the expression.

b. Find the terms of the expression.

*23. (Biology) A survey found that there were 1100 gray wolves in Minnesota in 1976. By 2003, the number of gray wolves had increased to 2300. What was the average growth of the wolf population in one year? (Round to the nearest whole number.)
(4)

*24. Multiple Choice A bouquet is made from nine red roses that cost $1.75 each and five white roses that cost $1.50 each. Use the expression $9 \cdot (\$1.75) + 5 \cdot (\$1.50)$ to find the cost of the bouquet.
(4)

A $31.00

B $23.25

C $25.25

D $21.75

*25. A can of soup in the shape of a cylinder has a radius of 3.8 cm and a height of 11 cm. What is the surface area of the can to the nearest tenth? Use 3.14 for π.
(4)

*26. Multi-Step Two friends compare the amount of change they have in their pockets. Ashley has 12 nickels, 2 dimes, and 4 quarters. Beto has 10 nickels, 4 dimes, and 3 quarters. Who has more money?
(4)

a. Write an expression to represent the value of Ashley's money. (Hint: Use 10¢ to represent the value of each dime, 5¢ for each nickel, and so on). Simplify the expression.

b. Write an expression to represent the value of Beto's money. Simplify the expression.

c. Compare the value of money that each friend has. Who has more?

*27. (School Supplies) Anthony had 10 packages of markers. Each package contained 8 markers. He gave 2 packages to each of the other 3 people in his group. Use the expression $8(10 - 3 \cdot 2)$ to determine how many markers Anthony kept for himself.
(4)

28. Geometry Use the cube shown to write a formula for the volume of any cube.
(3)

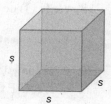

*29. (Temperature) The hottest day in Florida's history was 109°F, which occurred on June 29, 1931 in Monticello. Use the expression $\frac{5}{9}(F - 32)$ to convert this temperature to degrees Celsius. Round your answer to the nearest tenth of a degree.
(4)

*30. (Billing) Each month Mrs. Li pays her phone company $28 for phone service and $0.07 per minute for long-distance calls. Use the expression $28 + 0.07m$ to find the amount she was billed if her long-distance calls totaled 223 minutes.
(4)

Name _____ Date _____ Class _____

Finding Absolute Value and Adding Real Numbers
LESSON 5

Add, subtract, multiply, or divide.

1. (SB 3) $1\frac{1}{6} + 3\frac{1}{3}$
2. (SB 3) $2\frac{3}{8} - 1\frac{1}{4}$
3. (SB 3) $3\frac{2}{3} + 1\frac{5}{8} - 1\frac{3}{4}$
4. (SB 3) $3\frac{1}{3} \div 1\frac{3}{5}$
5. (SB 2) $1.506 \div 0.2$
6. (SB 2) $2.89 \cdot 1.2$

7. (2) How many terms are in the algebraic expression $2x^2 + 3x + 7$?

8. (SB 12) Find the prime factorization of 150.

9. (SB 4) **Write** Use the divisibility test to determine if 125,000 is divisible by 10. Explain your answer.

10. (1) **Model** Represent the following numbers as being members of set L: $-12, 0, -8, 4, -4, 4, 0, 8, 8, 12$.

11. (1) **Verify** True or False: All integers are rational numbers. Explain your answer.

12. (1) **Error Analysis** Student A said that $\frac{\sqrt{2}}{1}$ is a rational number. Student B said that it is an irrational number. Which student is correct? Explain your answer.

13. (SB 5) Write $\frac{5}{8}$ as a decimal and a percent.

14. (SB 1) **Measurement** Order the lengths 1.25 yards, 3 feet, $1\frac{1}{3}$ yards from least to greatest.

15. (SB 5) Write 7% as a fraction in simplest form and as a decimal.

16. (5) **Formulate** Write an equation using absolute values to represent the sentence. "The distance from -11 to 0 is 11."

17. (SB 13) **Multiple Choice** Which angle measures form an acute triangle?
    A 45°, 45°, and 90°
    B 40°, 110°, and 20°
    C 55°, 45°, and 80°
    D 30°, 30°, and 120°

18. (SB 8) Estimate: $1.48 + $0.12 - $0.27.

19. (5) **Write** Use the definition of absolute value to write $|-5| = 5$ in words.

20. (SB 14) **Verify** True or False: A rectangle is a parallelogram. Explain your choice

*21. (4) **Geometry** The hypotenuse squared (c^2) can be determined by solving for $a^2 + b^2$ in the Pythagorean Theorem. Using the order of operations, decide if the expression $(a + b)$ should be determined before a^2?

*22. (5) **Weather** One winter day the temperature rose 29°F from a low of -3°F in the morning. What was the day's high temperature?

© Saxon. All rights reserved.   9   Saxon Algebra 1

Finding Absolute Value and Adding Real Numbers

LESSON 5

*23. **(Football)** On the first down, the Tigers gained 8 yards. Then they were pushed back for a loss of $13\frac{1}{2}$ yards on the second down. Write and solve an addition problem to find the total number of yards lost or gained on the first two downs.

*24. **Multiple Choice** Which of these sets of numbers is closed under addition?
 A integers **B** rational numbers
 C real numbers **D** all of these

*25. **Multi-Step** Airplane A took off from an airport that is 43 feet below sea level, and then climbed 20,512 feet to its cruising altitude. Airplane B took off at the same time from an airport that was 1924 feet above sea level, and then climbed 18,527 feet to its cruising altitude. Which airplane is currently cruising at a higher altitude?

*26. **(Banking)** Martha had $500 in her checking account. She made a withdrawal of $34.65. Write and solve an addition problem to find Martha's balance after the withdrawal.

*27. **Multi-Step** A china cup-and-saucer set sells for $15.25 and a plate sells for $25. A woman buys 3 cup-and-saucer sets and 4 plates. If she pays a 5% sales tax, how much does she pay for her purchase?
 a. Determine how much the woman spends before sales tax. Use the expression $3 \cdot (\$15.25) + 4 \cdot (\$25)$ to solve.
 b. How much does she pay with sales tax included? Round your answer to the nearest hundredth. Use the expression $\$145.75 + (0.05) \cdot (\$145.75)$ to solve.

*28. **(Stocks)** Stock in the ABC Company fell 12.67 points on Monday and 31.51 points on Tuesday. Determine the total change in the stock for the two days.

*29. **Multiple Choice** Which expression correctly represents 1.6^5?
 A $1.6 \times 1.6 \times 1.6 \times 1.6 \times 1.6$ **B** $1.6 + 1.6 + 1.6 + 1.6 + 1.6$
 C $0.6 \times 0.6 \times 0.6 \times 0.6 \times 0.6 + 1$ **D** $1 \times 1 \times 1 \times 1 \times 1 + 0.6$

*30. **(Temperature)** At midnight the temperature was $-7°F$. By noon the temperature had risen $23°F$. What was the temperature at noon?

Name _____ Date _____ Class _____

Subtracting Real Numbers — LESSON 6

Add, subtract, multiply, or divide.

1. (SB 3) $5\frac{1}{3} \div 2\frac{1}{3}$

2. (SB 3) $40\frac{1}{8} - 21\frac{1}{4}$

3. (5) $5\frac{2}{3} + 2\frac{5}{6} + (-2\frac{1}{6})$

4. (SB 3) $1\frac{2}{3} \div 1\frac{1}{4} \cdot 1\frac{1}{2}$

5. (SB 2) $0.74 \div 0.2 \cdot 0.3$

6. (SB 2) $5.4 \cdot 0.3 \div 0.4$

7. (SB 2) $1.24 \cdot 0.2 \div 0.1$

8. (SB 2) $112.4 \div 3.2$

9. (SB 12) Find the prime factorization of 592.

10. (SB 12) Find the prime factorization of 168.

11. (SB 29) **Model** Display the following set of data in a line plot.

    8, 6, 9, 7, 5, 4, 6, 7, 9, 8, 5, 6, 6, 8

12. (SB 4) **Write** Use the divisibility test to determine if 2326 is divisible by 3. Explain your answer.

13. (SB 5) Write 6% as a fraction in simplest form and as a decimal.

14. (SB 5) **Measurement** Write 1.25 feet as a fraction in simplest form and compare it to $\frac{5}{3}$ feet. Which is greater?

15. (SB 5) Write $\frac{3}{5}$ as a decimal and as a percent.

16. (4) **Multiple Choice** What is the value of the expression below?

    $$\frac{(3 \cdot 20 + 2 \cdot 20) \cdot 6 - 20}{10^2}$$

    A 2.8   **B** 58   **C** −14   **D** 5.8

17. (4) Simplify $\dfrac{(45 + 39 + 47 + 40 + 33 + 39 + 41)}{(2 \cdot 2)^2 - 12}$.

*18. (6) **Multiple Choice** Which of these differences will be negative?

   A −4.8 − (−5.2)   **B** 4.8 − 5.2

   C 4.8 − 3.2   **D** 6.7 − (−7.8)

*19. (6) **Football** Ryan's varsity football team is on its own 25-yard line. The quarterback stumbles for a loss of 15 yards. What line is Ryan's varsity football team on now?

*20. (6) **Geometry** If one angle in a triangle measures 105.5° and another measures 38.2°, what is the measurement of the third angle? Use the expression 180 − 105.5 − 38.2 to solve.

© Saxon. All rights reserved.   Saxon Algebra 1

Subtracting Real Numbers
LESSON 6

*21. **Temperature** On a winter day, a wind gust makes the temperature in Antarctica feel sixteen degrees colder than the actual temperature. If the temperature is −5°C, how cold did it feel?

*22. **Consumer Math** Leila issued a check for $149.99 and deposited $84.50 in her account. What is the net change in her account?

23. **Multiple Choice** Which triangle is an equiangular triangle?
   A a triangle with angle measures of 45°, 45°, and 90°
   B a triangle with angle measures of 60°, 60°, and 60°
   C a triangle with angle measures of 55°, 35°, and 90°
   D a triangle with angle measures of 30°, 30°, and 120°

24. **Boating** The tour boat can leave the dock only if the level of the lake is no more than 2 feet below normal. Before the recent rainfall, the level of the lake was $5\frac{1}{3}$ feet below normal. After the recent rainfall, the level of the lake rose $3\frac{1}{4}$ feet. Can the tour boat leave the dock? Explain.

*25. **Geometry** The triangle inequality is a theorem from geometry stating that for any two real numbers a and b, $|a + b| \le |a| + |b|$. Verify the triangle inequality by simplifying $|-18.5 + 4.75| \le |-18.5| + |4.75|$.

*26. **Error Analysis** Two students solved this problem. Which student is correct? Explain the error.

The elevator started on the second floor and went up 8 floors, then down 11 floors to the garage level, and then up 6 floors. Which floor is the elevator on now?

Student A	Student B
$2 + 8 + (-11) + 6 = 5$ The 5th floor	$2 + 8 - (-11) + 6 = 27$ The 27th floor

27. **Multi-Step** A bit is a binary digit and can have a value of either 0 or 1. A byte is a string of 8 bits.
   a. Write the number of bits in one byte as a power of 2.
   b. Write 32 as a power of 2.
   c. Write the number of bits in 32 bytes as a power of 2.

*28. $16c + (-4d) + \frac{8\pi}{15} + 21efg$
   a. Find the coefficients of the expression.
   b. Find the number of terms in the expression.
   c. **Justify** Rewrite the expression so that there are no parentheses. Justify your change.

Subtracting Real Numbers

*29. Multiple Choice What subset of numbers does the number $-9.0909090909\overline{09}$ belong to?
(1)
   A integers   **B** irrational numbers

   C natural numbers   **D** rational numbers

*30. (**Oceanography**) The Pacific Ocean has an average depth of 12,925 feet, while the
(6) Atlantic Ocean has as average depth of 11,730 feet. Find the difference in average depths.

Name _____ Date _____ Class _____

Simplifying and Comparing Expressions with Symbols of Inclusion

LESSON 7

Add, subtract, multiply, or divide.

1. $(5 + 2)^2 - 50$
2. $(3 - 5) + 7^2$
3. $3\frac{1}{3} - 1\frac{1}{6} - 5\frac{1}{4}$
4. $2\frac{1}{3} \cdot 3\frac{1}{4} \cdot 1\frac{1}{2}$
5. $(0.56 + 0.3) \cdot 0.2$
6. $3.25 \cdot 0.4 + 0.1$
7. $1.2 \div 0.1 \div 0.1$
8. $20.2 \cdot 0.1 \cdot 0.1$

9. **Verify** True or False: All whole numbers are counting numbers. If true, explain your answer. If false, give a counterexample.

10. The set $\{\ldots, -3, -2, -1, 0, 1, 2, 3, \ldots\}$ represents which set of numbers?

11. **Justify** True or False: An obtuse triangle has two obtuse angles. Explain your choice.

12. Find the prime factorization of 207.

13. Find the prime factorization of 37.

14. **Write** Use the divisibility test to determine if 10,048 is divisible by 8. Explain.

15. Write 0.345 as a fraction in simplest form and as a percent.

16. Write 0.07% as a fraction in simplest form and as a decimal.

*17. Evaluate $(|-3| \cdot 4) + \left[\left(\frac{1}{2} + \frac{1}{4}\right) \div \frac{1}{3}\right]$.

*18. Compare: $\frac{1}{3} + \frac{1}{5} \cdot \frac{2}{15}$ ◯ $\left(\frac{1}{3} + \frac{1}{5}\right) \cdot \frac{2}{15}$.

*19. **Temperature** The following two formulas are used to convert degrees Celsius (°C) to degrees Fahrenheit (°F) and vise versa: $C = \frac{5}{9}(F - 32)$ and $F = \frac{9}{5}C + 32$. Explain how the equations are different.

*20. **Fencing** The diagram represents the fencing around a backyard. The fence is formed with parallel lines and a half-circle. Write and solve an equation to determine how many feet of fencing are needed. Round the answer to the nearest tenth.

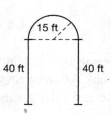

*21. **Multiple Choice** Simplify $[(10 - 8)^2 - (-1)] + (5 - 3)$.
   A -38 B 7 C -80 D 37

Saxon Algebra 1

Simplifying and Comparing Expressions with Symbols of Inclusion

LESSON 7

*22. (Manufacturing) A company produces two different types of 6-sided boxes. Box A is 12 inches long, 12 inches wide, and 12 inches tall. Box B is 16 inches long, 16 inches wide, and 6.75 inches tall. Both boxes have the same volume, but the company wants to know which box uses less material to produce.

  a. Write and solve an expression to find the surface area of Box A.
  b. Write and solve an expression to find the surface area of Box B.
  c. Compare the box sizes. Which box uses less material?

*23. Multi-Step A ball is dropped from a height of 25.6 feet. After it hits the ground, it bounces to 12.8 feet and falls back to the ground. Next it bounces to 6.4 feet and falls back to the ground. Then it bounces to 3.2 feet and falls back to the ground.

  a. Find the difference in heights between each consecutive bounce.
  b. If the pattern continues, will the ball ever stop bouncing? Explain.

24. Geometry What is the perimeter of the rectangle?

8.42 units

22.312 units

25. Measurement A valley is 250 below sea level and a small hill is 78 feet above sea level. Solve $|-250| + 78$ to determine the distance from the bottom of the valley to the top of the hill.

*26. (Transportation) In the last hour, 7 planes have landed at the airport and 11 planes have taken off. Use addition to find the change in the total number of planes at the airport in the last hour.

27. Error Analysis The temperature in the morning was $-18°F$. It increased by $5°$ by noon and dropped $10°$ in the evening. Two students determined the temperature in the evening. Which student is correct? Explain the error.

Student A	Student B
$-18 + 5 + 10 = -3$	$-18 + 5 + (-10) = -23$

*28. (Meteorology) The water level of the reservoir in Purcellville, Virginia was 2 feet below normal. After a heavy rainstorm, the water level increased to 5 feet above normal. Write and solve a subtraction problem to find the change in the water level caused by the rainstorm.

29. Multiple Choice Which term in the expression $\frac{\sqrt{9}ny}{nx} + a^2 - \frac{n}{4} + \frac{3\pi}{8}$ contains an irrational constant?

  A $\frac{\sqrt{9}ny}{nx}$ B $\frac{3\pi}{8}$ C $\frac{n}{4}$ D a^2

*30. Geometry The measure of each interior angle of a hexagon is given by the expression $\frac{180(6-2)°}{6}$. What is the measure of an interior angle of a hexagon?

Using Unit Analysis to Convert Measures — LESSON 8

Add, subtract, multiply, or divide.

1. $4\frac{1}{3} \div 1\frac{1}{3} + 3\frac{1}{3}$

2. $2\frac{3}{8} - 1\frac{3}{4} \div 1\frac{1}{2}$

3. $2\frac{2}{3} + 1\frac{5}{6} - 6\frac{3}{4}$

4. $3\frac{1}{3} \div 1\frac{1}{4} \cdot \frac{1}{2}$

5. $0.37 \div 0.2 \cdot 0.1$

6. $1.74 \cdot 0.3 \div 0.2$

7. Given the sets $A = \{1, 3, 5\}$, $B = \{0, 2, 4, 6\}$, and $C = \{1, 2, 3, 4\}$, are the following statements true or false?
   a. $A \cup B = \{0, 1, 2, 3, 4, 5, 6\}$
   b. $A \cap B = \{0, 1, 2, 3, 4, 5, 6\}$
   c. $B \cup C = \{2, 4\}$
   d. $A \cap C = \{1, 3\}$

8. Compare the expressions using $<$, $>$, or $=$. Explain.
   $$8^2 \div 4 - 6^2 \bigcirc (6 \cdot 7 \cdot 5) \div 6 - 15$$

9. Draw a line plot for the frequency table.

Number	2	3	4	5	6	7
Frequency	4	3	2	1	4	3

10. Subtract $78\frac{2}{5} - 14\frac{7}{10}$.

11. Find the prime factorization of 484.

12. **Write** Use the divisibility test to determine if 22,993 is divisible by 5. Explain your answer.

13. Write 125% as a fraction in simplest form and as a decimal.

14. Convert 105 kilometers per hour to kilometers per minute.

*15. Convert 74 square meters to square centimeters.

*16. (Camping) Norman's camping tent has a volume of 72,576 cubic inches. What is the volume of the tent in cubic feet?

17. **Multiple Choice** Which of these differences will be positive?

   A $-\frac{1}{2} - \frac{1}{8}$

   B $\frac{9}{12} - 1$

   C $\frac{5}{7} - \frac{3}{10}$

   D $-\frac{14}{15} - \left(\frac{4}{15}\right)$

Using Unit Analysis to Convert Measures

LESSON 8

18. Error Analysis Two students used unit analysis to convert a measurement of length. Which student is correct? Explain the error.

Student A
1 cm = 10 mm
5540 mm = 5540 mm × $\dfrac{1 \text{ cm}}{10 \text{ mm}}$
5540 mm = 554 cm

Student B
1 cm = 10 mm
5540 mm = 5540 mm × $\dfrac{10 \text{ mm}}{1 \text{ cm}}$
5540 mm = 55400 cm

*19. **Multiple Choice** Which one of the following ratios can be used to convert 120 cm into an equivalent measure in inches? (Hint: There are 2.54 cm in one inch.)

A $\dfrac{2.54 \text{ cm}}{1 \text{ in.}}$

B $\dfrac{1 \text{ in.}}{2.54 \text{ cm}}$

C $\dfrac{2.54 \text{ cm}}{1 \text{ in.}} \cdot \dfrac{2.54 \text{ cm}}{1 \text{ in.}}$

D $\dfrac{1 \text{ in.}}{2.54 \text{ cm}} \cdot \dfrac{1 \text{ in.}}{2.54 \text{ in.}}$

*20. (**Weather Forecasting**) One knot is exactly 1.852 kilometers per hour. The highest wind gust for the day was measured at 38 knots.

a. How many km/hr was the recorded wind gust? (Hint: 1 knot = 1.852 km/hr)

b. How many mph was the recorded wind gust? (Hint: 1 mi = 1.609 km)

*21. **Multi-Step** How can you find the area of the triangle in square inches?

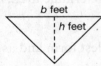

a. Write a formula for computing the area of the triangle in units of square feet.

b. Write a new formula that computes the area of the triangle in units of square inches.

c. What is the area of the triangle in square inches if $b = 3$ ft and $h = 2.2$ ft?

*22. (**Chemistry**) Water has a density of 1 gram per cubic centimeter. What is the density of water in grams per cubic inch?

*23. **Justify** True or False: A right triangle has one right angle, one obtuse angle, and one acute angle. If false, explain why.

Using Unit Analysis to Convert Measures

LESSON 8

24. Error Analysis Student A and Student B simplified the expression $\frac{24}{8} + (2+4)^2$. Which student is correct? Explain the error.
(7)

Student A	Student B
$\frac{24}{8} + (2+4)^2$	$\frac{24}{8} + (2+4)^2$
$3 + 6^2$	$3 + (2+16)$
$3 + 36 = 39$	$3 + 18 = 21$

25. Geometry A square pyramid has a base with edges of 8 inches and a height of 12 inches.
(7)

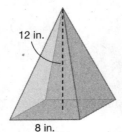

a. Use the following formula to find the volume: $V = \frac{1}{3}s^2h$.

b. **Analyze** Which term did you simplify first? Why?

*26. (Economics) Use the table of the Profit and Loss Report of ABC Company. What was the total profit or loss for the year? [( ) indicates a loss.]
(6)

	1st Quarter	2nd Quarter	3rd Quarter	4th Quarter
Profit or Loss	$6 million	($3.5 million)	($2 million)	$5 million

*27. Multi-Step Two groups of students measured the length of a tabletop. Group A's measurement was $56\frac{3}{4}$ inches. Group B's measurement was $57\frac{3}{4}$ inches. If the actual length of the tabletop was $57\frac{1}{2}$ inches, which group's measurement had the smaller error?
(5)

28. Error Analysis Two students simplified an expression containing an exponent. Which student is correct? Explain the error.
(3)

Student A	Student B
$b^2 \cdot c^2 \cdot c \cdot b^2 \cdot b =$	$b^2 \cdot c^2 \cdot c \cdot b^2 \cdot b =$
$b^2 \cdot b^2 \cdot b \cdot c^2 \cdot c = b^5c^3$	$b^2 \cdot b^2 \cdot b \cdot c^2 \cdot c = b^4c^2$

29. (Speed) A giraffe can run 32 miles per hour. What is the speed in feet per hour?
(8)

Using Unit Analysis to Convert Measures — LESSON 8

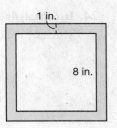

30. Geometry A square photograph measuring 8 inches by 8 inches is positioned within a 1-inch-wide picture frame as shown.
(3)
  a. What is the area of the photograph?
  b. What is the combined area of the photograph and frame?
  c. What is the area of the frame alone?
  d. If the 1-inch-wide frame is replaced with a 2-inch-wide frame, how much more wall space will be needed to hang the framed photograph?

Name _____ Date _____ Class _____

Evaluating and Comparing Algebraic Expressions

LESSON 9

Add, subtract, multiply, or divide.

1. (SB 3) $4\frac{1}{3} \div 2\frac{1}{3}$

2. (SB 3) $42\frac{3}{8} - 21\frac{3}{4}$

3. (SB 3) $1\frac{2}{3} + 2\frac{5}{6}$

4. (SB 3) $2\frac{2}{3} \div 1\frac{3}{4}$

5. (SB 2) $0.75 \div 0.2$

6. (SB 2) $1.74 \div 0.3$

7. (SB 2) $1.25 \cdot 0.2$

8. (SB 2) 12.2×3.2

9. (SB 14) **Verify** True or False: A square is a rhombus. Explain your choice.

10. (7) Simplify $4[(6-4)^3 - 5]$.

11. (8) Convert 1.86 km² to m².

*12. (9) Evaluate the expression $14c + 28 - 12cd$ for the given values $c = 4$ and $d = 5$.

13. (SB 13) A straight angle measures _____.

14. (SB 12) Find the prime factorization of 125.

*15. (9) Find the value of the expression $\frac{t-36}{36} + l$ if $t = 72$ and $l = 1$.

16. (7) **Multiple Choice** Evaluate the expression $14 + \frac{36}{9} \cdot (2 + 5)$.

    A 126   **B** 21   **C** 42   **D** 140

17. (7) Simplify $(3 + 12) + (|-4| - 2)^3 + 1$.

*18. (9) **(Flight)** A rocket is fired upward at an initial speed of 112 feet per second (ft/sec). It travels at a speed of $112 - 32t$ ft/sec, where t is the flight time in seconds.
    a. What is the rocket's speed after 1 second?
    b. What is the rocket's speed after 2 seconds?

*19. (9) **(Canoeing)** Rachel wants to rent a canoe for 3 hours. Use the expression $\$6.50 + \$1.75h$, where h represents the number of hours, to calculate the cost of renting the canoe.

*20. (9) **Error Analysis** Two students were asked to evaluate $\frac{y^2}{-x}$ when x is 5 and y is -5. Which student is correct? Explain the error.

Student A	Student B
$\frac{y^2}{-x} = \frac{-5^2}{-(5)} = \frac{-25}{-5} = 5$	$\frac{y^2}{-x} = \frac{(-5)^2}{-(5)} = \frac{25}{-5} = -5$

*21. (9) **Data Analysis** The variance of a set of data can be found with the expression $\frac{s}{n}$, where s is the sum of the squared deviation and n is the total number of data items in the set. What is the variance for a set of data with 12 items and a sum of the squared deviations equal to 30?

Evaluating and Comparing Algebraic Expressions

LESSON 9

*22. (Sports) In volleyball, the statistic for total blocks at the net is calculated with the expression $s + 0.5a$, where s is the number of solo blocks and a is the number of assisted blocks. What is the total-blocks statistic for a player who has 80 solo blocks and 53 assisted blocks?

23. Write Use the divisibility test to determine if 224 is divisible by 6. Explain your answer.

24. Write 35.2% as a fraction in simplest form and as a decimal.

*25. Multi-Step The rectangle has dimensions measured in centimeters. What is the ratio of the area of the rectangle in centimeters to the area of the rectangle in millimeters?

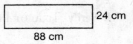

a. Calculate the area of the rectangle.

b. Find the area of the rectangle in square millimeters.

c. Find the ratio of square centimeters to square millimeters for the rectangle.

*26. Geometry A right circular cylinder has a base radius of 56 mm and a height of 128 mm. What is its volume in cubic centimeters? Round the answer to the nearest hundredth. Use 3.14 for π.

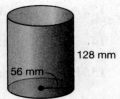

27. (Loans) The formula for long-term loans is $F = P(1 + i)^{n \div 12}$, where F is the future value of money, P is the present value, i is the interest rate, and n is the length of time the money is borrowed in months. When solving this equation for F, what step would you perform after adding 1 and i?

28. (Golf) Below is Rickie's golf score for two golf tournaments. What is the difference in his final score for the 1st and 2nd tournament?

1st Tournament	1	−2	−3	2
2nd Tournament	−2	−1	1	−1

29. Error Analysis Two students were asked to evaluate $(30 - 10)^2$. Student A answered 400, and Student B answered −70. Which student is correct? Explain the error.

Student A	Student B
$(30 - 10)^2$ $= 20^2$ $= 400$	$(30 - 10)^2$ $= (30 - 10 \cdot 10)$ $= (30 - 100)$ $= -70$

*30. (Typing) Jared can type 35 words per minute. Use the expression $35m$ to find the number of words he can type in 15 minutes.

Adding and Subtracting Real Numbers

LESSON 10

Add, subtract, multiply, or divide.

1. (SB 3) $\frac{1}{2} + \frac{3}{5}$

2. (SB 3) $15\frac{1}{3} - 7\frac{4}{5}$

3. (SB 3) $3\frac{2}{3} \cdot 2\frac{1}{4}$

4. (SB 3) $3\frac{2}{5} \div 1\frac{2}{3}$

5. (SB 3) $78\frac{2}{5} - 14\frac{7}{10}$

6. (SB 3) $2\frac{1}{3} \cdot 1\frac{1}{4}$

7. (SB 2) $10.2 \cdot 3.15$

8. (SB 2) $20.46 \div 2.2$

9. (SB 2) $12.3 \cdot 2.02$

10. (SB 2) $0.8 \div 0.25$

*11. (10) Order from greatest to least: $\frac{6}{7}, \frac{3}{5}, \frac{1}{7}, -\frac{4}{3}$.

12. (SB 1) A(n) _____ angle measures less than 90°.

13. (8) Convert 8673 g to kg.

14. (8) Convert 26 mi to km. Round your answer to the nearest tenth.

15. (4) True or False: $(2 + 5) - (3 \cdot 4) = 2 + 5 - 3 \cdot 4$. Explain.

*16. (10) **Multiple Choice** Simplify $1.29 + 3.9 - 4.2 - 9.99 + 6.1$.
    A −2.9   **B** −1   **C** 1   **D** 2.9

*17. (10) **Error Analysis** Which student is correct? Explain the error.

Student A	Student B
$1 - \left(\frac{1}{5} - \frac{2}{10} - \frac{1}{10}\right)$	$1 - \left(\frac{1}{5} - \frac{2}{10} - \frac{1}{10}\right)$
$= 1 - \left(-\frac{1}{10}\right)$	$= 1 - \frac{1}{5} - \frac{2}{10} - \frac{1}{10}$
$= 1\frac{1}{10}$	$= \frac{1}{2}$

*18. (10) **Time** A ship sailed northeast for $2\frac{1}{4}$ hours. It then sailed east for $1\frac{1}{3}$ hours. How much longer did it sail northeast than east?

19. (SB 29) **Model** Draw a line plot for the frequency table.

Number	9	10	11	12	13	14
Frequency	4	3	2	1	0	4

*20. (9) **Multi-Step** A map shows streets $\frac{1}{1000}$ of their size.
    a. Write an expression that represents the real length of a block if the length of the block on the map is b.
    b. Find the actual length of a block that is 0.4 feet on the map.

Adding and Subtracting Real Numbers

LESSON 10

*21. Geometry A parallelogram has a base of z and a height of $2z$. Write an expression to find the area of the parallelogram. If z is equal to 12 cm, what is the area of the parallelogram?

22. Error Analysis Two students used unit analysis to convert a measurement of area to a different unit. Which student is correct? Explain the error.

Student A	Student B
1 ft = 12 in.	1 ft = 12 in.
$9 \text{ ft}^2 = 9 \text{ ft}^2 \times \frac{12 \text{ in.}}{1 \text{ ft}}$	$9 \text{ ft}^2 = 9 \text{ ft}^2 \times \frac{12 \text{ in.}}{1 \text{ ft}} \times \frac{12 \text{ in.}}{1 \text{ ft}}$
$9 \text{ ft}^2 = 108 \text{ in}^2$	$9 \text{ ft}^2 = 1{,}296 \text{ in}^2$

23. Meteorology When a weather system passes through, a barometer can be used to measure the change in atmospheric pressure in millimeters of mercury. What is the pressure difference in inches of mercury for a measured change of +4.5 mm of mercury? Round the answer to the nearest thousandth.

*24. Multi-Step A plot of land contains a rectangular building that is 9 yards long and 6 yards wide and a circular building with a diameter of 4 yards.

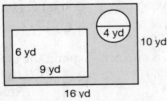

a. Write an expression for the area of the two buildings.

b. Write an expression and solve for how much area on the plot of land is not being taken up by the buildings. Round the answer to the nearest hundredth yard.

*25. Error Analysis Student A and Student B each added the numbers −4.8 and 3.6 as shown below. Which student is correct? Explain the error.

Student A	Student B
−4.8 + 3.6	−4.8 + 3.6
|−4.8| = 4.8	|−4.8| = 4.8
|+3.6| = 3.6	|+3.6| = 3.6
4.8	4.8
+3.6	−3.6
8.4	1.2
8.4	−1.2

26. Banking Raul had $500 in his checking account. He wrote checks for $157.62 and $43.96. Then he deposited $225. Find Raul's balance after these three transactions.

Saxon Algebra 1

Adding and Subtracting Real Numbers

LESSON 10

27. Write Mutually exclusive means that two sets of numbers have no numbers in
(1) common. Name two subsets of real numbers that are mutually exclusive. Explain.

*28. (Stocks) Stock in the 123 Company fell 8.2 points on Monday and 5.3 points on
(5) Tuesday. On Wednesday the stock rose 9.1 points. Determine the total change in the stock for the three days.

*29. (Football) The Rams gained 4 yards on the first down, lost 6 yards on the second
(5) down, and gained 14 yards on the third down. How many total yards did the Rams gain on the three downs?

30. Measurement A kite flies 74 feet above the ground. The person flying the kite is 5 feet
(10) 6 inches tall. How far above the person is the kite?

Name _____ Date _____ Class _____

Multiplying and Dividing Real Numbers

LESSON 11

*1. **Verify** True or False: The product of a number and its reciprocal is equal to one. Verify your answer.
(11)

*2. Simplify $-(-4)^2$.
(11)

3. **Error Analysis** Which student is correct? Explain the error.
(10)

Student A	Student B
$\left(\frac{11}{12} - \frac{2}{4} - \frac{1}{3}\right) + \frac{11}{12}$	$\left(\frac{11}{12} - \frac{2}{4} - \frac{1}{3}\right) + \frac{11}{12}$
$= \frac{1}{12} + \frac{11}{12}$	$= \frac{8}{12} + \frac{11}{12}$
$= 1$	$= \frac{19}{12}$

4. Simplify $\frac{2 \cdot 14 + 3 \cdot 7}{71 - 15}$.
(4)

5. Draw a line plot for the frequency table.
(SB 29)

Number	5	6	7	8	9	10
Frequency	4	2	0	1	0	3

6. A(n) _____ angle measures more than 90° and less than 180°.
(SB 13)

7. Evaluate $3(x + 4) + y$ when $x = 8$ and $y = 7$.
(9)

8. Evaluate the expression $3x^2 + 2(x - 1)^3$ for the given value $x = 6$.
(9)

9. **Multiple Choice** Which rate is the fastest?
(8)
 A 660 ft/15 s

 B 645 ft/11 s

 C 616 ft/12 s

 D 1100 ft/30 s

10. **Justify** Simplify $5 + \frac{9}{3}[4(\frac{1}{2} + 4)]$. Justify each step.
(7)

*11. **Multiple Choice** The temperature at noon was 20°C. The temperature fell 2 degrees every hour until 3 a.m. the next day. What was the temperature at 11 p.m. that evening?
(11)
 A 22°C

 B −2°C

 C −30°C

 D −22°C

© Saxon. All rights reserved.   27   Saxon Algebra 1

Multiplying and Dividing Real Numbers — LESSON 11

*12. **(Physics)** The magnitude of the instant acceleration of an object in uniform circular motion is found using the formula $a = \frac{v^2}{r}$, where r is the radius of the circle and v is the constant speed. Evaluate $a = \frac{v^2}{r}$ if $v = 35$ cm/s and $r = 200$ cm.
(9)

*13. **(Retailing)** A grocery store is having a sale on strawberries. Suppose 560 pints of strawberries are sold at a loss of $0.16 for each pint. How much money does the store lose on the sale of the strawberries?
(11)

*14. **(Ocean Travel)** The deepest point of the Kermandec trench in the Pacific Ocean is 10,047 meters below sea level. A submarine made two dives from above the deepest point of the trench at a rate of 400 meters per minute. The first of the two dives was 10 minutes long and the second was 4 minutes. How far did the submarine travel in each dive?
(11)

15. Add $-1.06 + 2.01 + 4.13$.
(10)

16. Multi-Step A purple string is 0.99 m long. A green string is 0.23 m long. What is the difference in length of the two pieces?
(10)
   a. Estimate Estimate the difference using fractions.
   b. Find the exact value of the difference using fractions.

17. Error Analysis Two students solved a homework problem as shown below. Which student is correct? Explain the error.
(9)

Student A	Student B
Evaluate $3g - 4(g + 2b)$; $g = 9$ and $b = 4$.	Evaluate $3g - 4(g + 2b)$; $g = 9$ and $b = 4$.
$3(9) - 4(9 + 2(4))$	$3(9) - 4(9 + 2(4))$
$27 - 4(17)$	$27 - 4(17)$
$23(17)$	$27 - 68$
391	-41

18. (Science) Scientists can use the expression $2.6f + 65$ to estimate the height of a person if they know the length of the femur bone, f. What is the approximate height of a person if the femur bone is 40 centimeters long?
(9)

19. Error Analysis The highest point in North America, Mount McKinley, in the Alaska Range, is 20,320 feet above sea level. The lowest point in North America is 282 feet below sea level and is in Death Valley in California. Which student correctly calculated the difference in elevations? Explain the error.
(6)

Student A	Student B
$20,320 - 282$	$20,320 - (-282)$
$20,320 + (-282)$	$20,320 + (+282)$
$20,038$ feet	$20,602$ feet

Multiplying and Dividing Real Numbers

LESSON 11

 20. Probability Describe each of the events below as impossible, unlikely, as likely as not, likely, or certain.
  a. Joshua rolls an odd number on a standard number cube.
  b. Maria's birthday is September 31st.
  c. The basketball team has won 11 of their last 12 games. The team will win the next game.

Simplify each expression.

*21. $5(-2)$
(11)

*22. $(-3)(-5)$
(11)

23. $-|-15 + 5|$
(5)

*24. $(-3)(-6)(-2)(5)$
(11)

*25. $(3)(5)$
(11)

 26. Geometry Can the perimeter of a rectangle be any integer value?
(1)

27. **Model** Mary is playing a board game using a number cube to decide the number of spaces she moves. She moves forward on an even number and backward on an odd number. Her first 5 rolls were 4, 2, 3, 6, 1.
(10)
  a. Model her moves on a number line with zero being the starting point.
  b. Using addition and subtraction, write an expression showing her moves.
  c. At a the end of 5 rolls, how many spaces is she away from the starting point?

*28. **Analyze** Jan bought 2 yards of ribbon. She needs 64 inches of ribbon to make a bow. Does she have enough ribbon? Explain your answer.
(8)

29. **Meteorology** A meteorologist reported the day's low temperature of $-5°F$ at 7 a.m. and the day's high temperature of $20°F$ at 5 p.m. How much did the temperature rise from 7 a.m. to 5 p.m.?
(6)

30. **Phone Charges** Fast Talk Phone Company charges an initial fee of $20 plus 10¢ per minute used. The total bill is expressed as $20 + 0.10m$, where m is the minutes used. If 200 minutes are used, what is the amount of the bill?
(9)

Name _____ Date _____ Class _____

Using the Properties of Real Numbers to Simplify Expressions
LESSON 12

***1.** Identify the property illustrated in the equation $100 \cdot 1 = 100$.
(12)

Simplify each expression.

2. $-18 \div 3$
(11)

3. $|12 - 30|$
(5)

4. $(-3)(-2)(-1)(-8)$
(11)

*5. True or False: $p(q + r) = (p + q)r$. Justify your answer using the properties.
(12)

6. Write a fraction equivalent to $\frac{2}{3}$.
(SB 7)

7. True or False: The sum of the measures of complementary angles is 90°.
(SB 15)

*8. Multiple Choice Which equation demonstrates the Identity Property of Addition?
(12)

  A $a \cdot 0 = 0$

  B $a + 0 = a$

  C $a \cdot \frac{1}{a} = 1$

  D $a + 1 = 1 + a$

9. Add $\frac{11}{15} + \frac{1}{30} + \frac{3}{60}$.
(10)

10. Error Analysis Students were asked to simplify $\frac{5}{6} \div \left(-\frac{3}{2}\right)$. Which student is correct? Explain the error.
(11)

Student A	Student B
$\frac{5}{6} \div \left(-\frac{3}{2}\right)$	$\frac{5}{6} \div \left(-\frac{3}{2}\right)$
$= \frac{5}{6} \cdot \left(-\frac{2}{3}\right)$	$= \frac{5}{6} \cdot \left(-\frac{2}{3}\right)$
$= -\frac{5}{9}$	$= \frac{5}{9}$

*11. Jon has 5 marbles. His best friend gives him some more. Then he buys 15 more marbles. The expression $5 + x + 15$ shows the total number of marbles Jon now has. Show two ways to simplify this expression and justify each step.
(12)

12. Multiple Choice What is the value of $\frac{(5x + x)^2(6 - x)}{x}$ when $x = 2$?
(9)

  A 288   **B** 200   **C** 400   **D** 28

*13. Find the value of $(4x^3y^2)^2$ when $x = 2$ and $y = 1$.
(9)

14. Convert 588 ounces to pounds. (Hint: 1 lb = 16 oz)
(8)

© Saxon. All rights reserved.   Saxon Algebra 1

Using the Properties of Real Numbers to Simplify Expressions

LESSON 12

*15. Geometry A wall in a rectangular room is 12 feet by 8 feet. Jose calculated the area using the equations $A = 12 \cdot 8$ and $A = 8 \cdot 12$. Explain why each expression will give him the same answer.

*16. (Interior Decorating) Tim is building a picture frame that is 10 inches long and 6 inches wide. He calculated the perimeter using $P = 2(10 + 6)$. His brother calculated the perimeter for the same frame using $P = 2(6 + 10)$. Will the measurements be the same? Explain.

*17. (Temperature) To convert a temperature from Celsius to Fahrenheit, Marc uses the formula $F = \frac{9}{5}C + 32$. He also uses the formula $F = 32 + \frac{9}{5}C$. Which calculation is correct? Explain.

18. Geometry A rectangle is twice as long as it is wide. If the width of the rectangle measures 2.3 inches, what is the area of the rectangle?

2.3 in.   2.3 in.

19. Multi-Step In each of the first five rounds of a game, Tyra scored 28 points. In each of the next three rounds, she scored −41 points. Then she scored two rounds of −16 points. What is the total number of points that Tyra scored? Explain.

*20. Multiple Choice Order from greatest to least: $\frac{1}{4}, 0.23, -0.24, \frac{1}{3}$.

A $-0.24, 0.23, \frac{1}{4}, \frac{1}{3}$ B $\frac{1}{4}, \frac{1}{3}, 0.23, -0.24$

C $-0.24, 0.23, \frac{1}{3}, \frac{1}{4}$ D $\frac{1}{3}, \frac{1}{4}, 0.23, -0.24$

21. (Sewing) Maria is sewing curtains that require 124 inches of ribbon trim. She can only buy the ribbon in whole yard lengths. How many yards does she need to buy?

22. Error Analysis Student A and Student B simplified the expression $9 - 4 \cdot 2$. Which student is correct? Explain the error.

Student A	Student B
$9 - 4 \cdot 2$	$9 - 4 \cdot 2$
$= 5 \cdot 2$	$= 9 - 8$
$= 10$	$= 1$

23. Write Explain how to use the order of operations to simplify $4(8 - 9 \div 3)^2$.

24. Simplify $x^2kxk^2x^2ykx^2$.

Using the Properties of Real Numbers to Simplify Expressions

LESSON 12

25. Error Analysis Two students evaluate the expression $4t + 5x - \frac{1}{x}$ when $x = 3$. Which student is correct? Explain the error.

Student A	Student B
$4t + 5x - \frac{1}{x}; x = 3$	$4t + 5x - \frac{1}{x}; x = 3$
$= 4t + 5(3) - \frac{1}{3}$	$= 4t + 53 - \frac{1}{3}$
$= 4t + 15 - \frac{1}{3}$	$= 4t + 52\frac{2}{3}$
$= 4t + 14\frac{2}{3}$	

26. (Savings Accounts) The table below shows the transactions Jennifer made to her savings account during one month. Find the balance of her account.

Jennifer's Bank Account

Beginning Balance	$396.25
Withdrawal	$150.50
Deposit	$220.00
Interest (deposit)	$8.00

*27. (Tug of War) In a game of Tug of War, Team A pulls the center of the rope three and a half feet in their direction. Then Team B pulls back five feet before Team A pulls for another eight feet. How far from the starting point is the center of the rope?

28. (Travel) Bill rides a bus for 2.5 hours to visit a friend. If the bus travels at about 60 to 63 miles per hour, how far away does Bill's friend live? (Hint: To find distance, multiply rate by time.)

29. Justify Simplify $2^2 + 24 - (3 - 12)$. Explain your steps.

30. Verify Give an example that illustrates that the sum of a number and its opposite is zero.

Calculating and Comparing Square Roots

LESSON 13

Simplify each expression.

1. $-16 \div -2$
   (11)

2. $\dfrac{4 + 7 - 6}{2 + 7 - 3}$
   (11)

3. $-2 + 11 - 4 + 3 - 8$
   (6)

4. $(-2)(-3) + 11(2) - 3 - 6$
   (6)

Evaluate each expression for the given values.

*5. $3p - 4g - 2x$ for $p = 2$, $g = -3$, and $x = 4$
   (9)

6. $3xy - 2yz$ for $x = 3$, $y = 4$, and $z = 3$
   (9)

*7. $\sqrt{40}$ is between which two whole numbers?
   (13)

*8. **Multiple Choice** Which of the following numbers is a perfect square?
   (13)
   A 200
   B 289
   C 410
   D 150

*9. Solve $b = \sqrt{4}$.
   (13)

10. **Model** Draw a model to compare $\dfrac{5}{12}$ and $\dfrac{1}{3}$.
    (10)

11. Convert 25 feet per hour to yards per hour.
    (8)

12. True or False: The square root of any odd number is an irrational number. If false, provide a counterexample.
    (1)

*13. **Multiple Choice** The area of a square is 392 square meters. The area of a second square is half the area of the first square. What is the side length of the second square?
    (13)
    A 14 meters
    B ≈ 20 meters
    C 196 meters
    D 96 meters

*14. True or False: $xyz = yxz$. Justify your answer using the properties.
    (12)

15. **Verify** Determine whether each statement below is true or false. If false, explain why.
    (4)
    a. $4^2 + 15 \cdot 20$ is equal to 316.
    b. $(4 + 5)^2$ is the same as $4 + 5^2$.

Calculating and Comparing Square Roots

LESSON 13

16. Multi-Step Kristin has several ropes measuring $8\frac{1}{4}$ in., $8\frac{3}{16}$ in., $8\frac{5}{8}$ in., and $8\frac{1}{16}$ in. How should she order them from least to greatest?
  a. Find a common denominator for each measure.
  b. Order the measures from least to greatest.

17. Arrange in order from least to greatest:

$$1.11, \ 1.5, \ 1.09, \ 1.05$$

*18. Are the expressions $(20k^3 \cdot 5v^5)9k^2$ and $900k^3v^5$ equivalent? Explain.

*19. Science The Barringer Meteor Crater in Winslow, Arizona, is very close to a square in shape. The crater covers an area of about 1,690,000 square meters. What is the approximate side length of the crater?

20. Science The time, t, in seconds it takes for an object dropped to travel a distance, d, in feet can be found using the formula $t = \frac{\sqrt{d}}{4}$. Determine the time it takes for an object to drop 169 feet.

*21. Multi-Step The flow rate for a particular fire hose can be found using $f = 120\sqrt{p}$, where f is the flow rate in gallons per minute and p is the nozzle pressure in pounds per square inch. When the nozzle pressure is 169 pounds per square inch, what is the flow rate?

22. World Records The world's largest cherry pie was baked in Michigan. It had a diameter of 210 inches. If the diameter was converted to feet, would it be a rational number? Explain.

23. Find the area of the shaded portion of the circle. The radius of the circle is 4 inches. (Use 3.14 for π.)

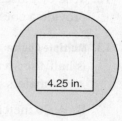
4.25 in.

24. Banking Frank deposited $104.67 into his bank account. Later that day, he spent $113.82 from the same account. Estimate the change in Frank's account balance for that day.

25. Justify Simplify $52 + (1 + 3)^2 \cdot (16 - 14)^3 - 20$. Justify each step.

26. Error Analysis Two students simplify the expression $2 + 3x + 1$. Which student is correct? Explain the error.

Student A	Student B
$2 + 3x + 1$	$2 + 3x + 1$
$= 2 + 1 + 3x$	$= (2 + 3)x + 1$
$= (2 + 1) + 3x$	$= 5x + 1$
$= 3 + 3x$	

Saxon Algebra 1

Calculating and Comparing Square Roots

LESSON 13

*27. Justify Arlene has 30 buttons and buys x packages of buttons. There are 7 buttons in each package. She uses 12 buttons. The number of buttons she now has is represented by the expression $30 + 7x - 12$. Simplify the expression and justify each step using the properties.
(12)

28. Multiple Choice Which of the following expressions will result in a negative number?
(11)

A $(-6)^2$

B $(-6) \div (-6)$

C $-\dfrac{3}{4}(6) \div (-4)$

D $-1 \cdot (-6)^2$

29. (International Banking) A Greek company needs to purchase some products from a U.S. corporation. First, the company must open an account in U.S. dollars. If the account is to hold $1,295,800, how many drachma, the Greek currency, should the company deposit? Use the exchange rate of one Greek drachma for every $0.004 in U.S. currency.
(8)

30. Probability Describe each of the following events as impossible, unlikely, as likely as not, likely, or certain.
(Inv 1)

a. Jim rolls a 10 on a standard number cube.

b. Sarah guesses a number correctly between 1 and 900.

c. Mayra dropped a coin and it landed heads up.

Determining the Theoretical Probability of an Event

LESSON 14

*1. A number cube labeled 1–6 is rolled three times. What is the probability that the next roll will produce a number greater than 4?
(14)

*2. An jar contains 5 green marbles and 9 purple marbles. A marble is drawn and dropped back into the jar. Then a second marble is drawn and dropped back into the jar. Both marbles are green. If another marble is drawn, what is the probability that it will be green?
(14)

3. Convert 20 inches to centimeters (2.54 cm = 1 in.).
(8)

4. Convert 25 feet to centimeters. (Hint: Convert from feet to inches to centimeters.)
(8)

Simplify.

5. $3 - 2 \cdot 4 + 3 \cdot 2$
(4)

6. $-3(-2)(-3) - 2$
(11)

7. $5(9 + 2) - 4(5 + 1)$
(4)

8. $3(6 + 2) + 3(5 - 2)$
(4)

9. Evaluate $\sqrt{31 + z}$ when $z = 5$.
(13)

10. Use <, > or = to compare $\frac{4}{5}$ and $\frac{5}{6}$.
(10)

*11. **Geometry** What is the length of the side of a square that has an area of 49 square centimeters?
(13)

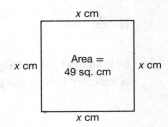

12. **Multiple Choice** Which equation demonstrates the Associative Property of Addition?
(12)
    A $(a + b) + c = a + (b + c)$
    B $ab + c = ba + c$
    C $a(b + c) = ab + ac$
    D $a + (b + c) = a + (c + b)$

*13. **Justify** What must be true of each of the values of x and y if $-xy$ is positive? zero? negative?
(11)

14. Identify the property illustrated in the expression $5 \cdot 6 = 6 \cdot 5$.
(12)

Determining the Theoretical Probability of an Event

LESSON 14

***15. Multiple Choice** A number cube labeled 1–6 is tossed. What is the theoretical probability of rolling an odd number?
(14)

A $\frac{1}{2}$

B $\frac{1}{3}$

C $\frac{1}{4}$

D $\frac{2}{3}$

***16.** A letter is chosen at random from the word probability. What is the probability of randomly choosing the letter b?
(14)

***17. Multiple Choice** A bag contains 4 blue, 6 red, 5 yellow, and 1 orange marble. What is the probability of randomly choosing a blue marble?
(14)

A $\frac{1}{16}$

B $\frac{4}{15}$

C $\frac{1}{4}$

D $\frac{4}{32}$

18. Error Analysis Students were asked to find the square root of 16. Which student is correct? Explain the error.
(13)

Student A	Student B
$\sqrt{16} = 4$	$\sqrt{16} = 8$
$4 \times 4 = 16$	$8 \times 2 = 16$

*19. (Braking Distance) The speed a vehicle was traveling when the brakes were first applied can be estimated using the formula $s = \sqrt{\frac{d}{0.04}}$, where d is the length of the vehicle's skid marks in feet and s is the speed of the vehicle in miles per hour. Determine the speed of a car whose skid marks were 4^2 feet long.
(13)

*20. (Physics) The centripetal force of an object in circular motion can be expressed as $\frac{mv^2}{r}$, where m is mass, v is tangential velocity, and r is the radius of the circular path. What is the centripetal force of a 2-kg object traveling at 50 cm/s in a circular path with a radius of 25 centimeters?
(9)

21. Verify Convert 2.35 pounds to ounces (1 lb = 16 oz). Check to see if your answer is reasonable.
(8)

Determining the Theoretical Probability of an Event — LESSON 14

22. **Write** If a computer program is designed to run until it reaches the end of the number pi (π), will the program ever end? Explain.

23. **Geography** The lowest point in elevation in the United States is Death Valley, California. Death Valley is 86 meters below sea level. Which set of numbers best describes elevations in Death Valley?

24. **Temperature** To convert degrees Celsius to degrees Fahrenheit, use the equation $C = \frac{5}{9}(F - 32)$.
    a. How many terms are in the expression $\frac{5}{9}(F - 32)$?
    b. Identify the constants in the expression.

25. Simplify $-7 + 3 - 2 - 5 + (-6)$.

26. **Error Analysis** Ms. Mahoney, the algebra teacher, has two cakes that weigh 3 pounds and 5 pounds. She cuts the cakes into 16 equal pieces. She asks the students to write an expression that represents the weight of each piece. Which student is correct? Explain the error.

Student A	Student B
uses the expression $3 + 5 \div 16$	uses the expression $(3 + 5) \div 16$

27. **Model** While the Petersen family was waiting for their table, 9 people left the restaurant and 15 people entered. Find the sum of -9 and 15 to determine the change in the number of people in the restaurant. Use algebra tiles to model the situation.

28. **Justify** Simplify $22 - (-11) - 11 - (-22)$. Justify your answer.

29. **Write** Why is the order of operations important when simplifying an expression like $(5 + 7)^2 \div (14 - 2)$?

*30. **Landscaping** Tanisha is building a fence around a square flower bed that has an area of 144 square feet. How many feet of fencing does she need?

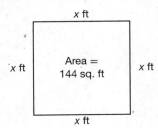

Using the Distributive Property to Simplify Expressions

LESSON 15

Evaluate.

1. $-7(-8 + 3)$
2. $5(-3 - 6)$
3. Evaluate $\sqrt{10{,}000}$.
4. Solve $c = \sqrt{25}$.

*5. **Multi-Step** In a shipment of 800 eggs, the probability of an egg breaking is $\frac{2}{25}$. How many are likely to be broken in the shipment? Justify the answer.

6. The digits 0, 1, 2, 3, 4, 5, 6, 7, 8 and 9 are written on cards that are shuffled and placed face down in a stack. One card is selected at random. What is the probability that the digit is odd and greater than 5?

*7. In a bucket there are 10 balls in a bucket numbered 1, 1, 2, 3, 4, 4, 4, 5, 6, and 6. A single ball is randomly chosen from the bucket. What is the probability of drawing a ball with a number less than 7? Explain.

*8. **Multiple Choice** Simplify the expression $-5(x + 6)$. Which is the correct simplification?
    A $-5 + x - 11$ B $-5x + 1$ C $-5x + 30$ D $-5x - 30$

*9. Find the value of y in the equation $18 - x = y$ if $x = -4$.

10. The water level of the reservoir in Austin, Texas, was 3 feet below normal. After a heavy rain storm, the water level increased to 5 feet above normal. How much did the rain storm change the water level?

11. **Error Analysis** Two students evaluated a numeric expression. Which student is correct? Explain the error.

Student A	Student B
$-8(-5 + 14)$	$-8(-5 + 14)$
$= 40 - 112$	$= -13 + 6$
$= -72$	$= -7$

12. **Write** Evaluate the expression $-8(9 - 15)$ using the Distributive Property. Explain.

*13. (**Surveying**) The county surveyed a piece of property and divided it into equal-sized lots. Use the diagram to write an expression that requires the Distributive Property to evaluate it. Evaluate the expression to find the total number of lots on the property.

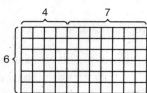

*14. True or False: $m + 0 = m$. Justify your answer using the properties.

Using the Distributive Property to Simplify Expressions
LESSON 15

15. Convert 3.4 yd³ to ft³.
(8)

16. Multi-Step Travis plans to divide his collection of baseball cards among
(15) 8 grandchildren. He will give each child the same number of cards. Each card is worth $14. Write an expression to represent the value of each child's cards. Let c equal the total number of cards in Travis's collection.

17. (Budgeting) Kennedy's teacher asked her to plan the budget for the class party.
(15) Kennedy began by writing the expression $g = b + 7$ to represent that the number of girls equals the number of boys plus seven. Each girl will need $6. Write and simplify an algebraic expression that uses the Distributive Property to show the total cost for girls at the class party.

*18. If a number cube is rolled, what is the probability of it landing on the number 5 or 6?
(14)

*19. Error Analysis Two students are evaluating the expression $\sqrt{36 + z}$ for $z = 13$. Which
(13) student is correct? Explain the error.

Student A	Student B
$\sqrt{36 + z}$	$\sqrt{36 + z}$
$= \sqrt{36 + 13}$	$= \sqrt{36} + z$
$= \sqrt{49}$	$= 6 + 13$
$= 7$	$= 19$

20. Justify The expression 6 · 2 · 4 would be simplified from left to right using the order
(12) of operations. What property would allow this expression to be simplified from right to left?

*21. (Investments) Susan invests the same amount of money in each of 7 stocks. In one
(12) year, her money increased 8 times. The value of her investment is represented by the expression $7x \cdot 8$. Show two methods to simplify the expression and justify each step using the properties.

22. (Age) Rickie is $3\frac{3}{4}$ years older than Raymond. Raymond is $2\frac{1}{2}$ years younger
(10) than Ryan. If Ryan is $14\frac{1}{4}$ years old, how old is Rickie?

23. Write Write the procedure for evaluating the expression $16f^2 g^3 - 4f^8 + 12$ for
(9) $f = 3$ and $g = 5$.

24. Model Use the number line to model $x - 8$ when $x = -6$.
(6)

25. (Astronomy) In astronomy, brightness is given in a value called magnitude.
(3) A -2-magnitude star is 2.512 times brighter than a -1-magnitude star, a -3-magnitude star is 2.512 times brighter than a -2-magnitude star, and so on. If Sirius is magnitude -1.5 and the full moon is magnitude -12.5, how much brighter is the full moon?

Using the Distributive Property to Simplify Expressions

LESSON 15

26. Error Analysis Student A and Student B each solved the absolute-value problem as shown below. Which student is correct? Explain the error.

Student A	Student B
$-\|12 - 15\|$	$-\|12 - 15\|$
$= -\|-3\|$	$= -\|-3\|$
$= -(3)$	$= \|+3\|$
$= -3$	$= 3$

*27. Find the value of y for the given values of x in the equation $x - |x - 2| = y$ if $x = -3$.

28. Verify When simplified, will the expression $3 + \frac{2}{3} + |-5|$ be positive or negative? Explain.

29. Probability Thomas spun a game spinner and recorded the results in the table below.

Outcome	Frequency
Red	3
Blue	5
Yellow	9
Green	8

Use the table to find the experimental probability of each event. Express each probability as a fraction and as a percent.
a. landing on red
b. landing on green
c. not landing on green

30. Geometry What is the perimeter of a square with an area of 121 sq. in.?

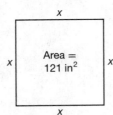

Name _____ Date _____ Class _____

Simplifying and Evaluating Variable Expressions
LESSON 16

Simplify.

1. $2 + 5 - 3 + 7 - (-3) + 5$
   (10)

2. $3(7) + 5 - 3 + 7 - 9 \div 2$
   (4)

3. Represent the following numbers as being members of
   (1) set K: $-2, -1, -4, -1, -3, -1, -5, -3$.

Determine if each statement is true or false. If true, explain why. If false, give a counterexample.

4. The set of whole numbers is closed under multiplication.
   (1)

5. All integers are whole numbers.
   (1)

Simplify by using the Distributive Property.

6. $-4y(d + cx)$
   (15)

7. $(a + bc)2x$
   (15)

Evaluate the expression for the given values.

*8. $pa[-a(-a)]$ when $p = 2$ and $a = -1$
(16)

*9. $x(x - y)$ when $x = \dfrac{1}{5}$ and $y = \dfrac{6}{5}$
(16)

*10. $\left(\dfrac{x-3}{y}\right)^2$ when $x = -5$ and $y = 2$
(16)

11. $4(b + 1)^2 - 6(c - b)^4$ when $b = 2$ and $c = 7$
(9)

 12. **Geometry** The measure of one side of a square is $5x + 1$ meters.
(15) What expression would be used for the perimeter of the square? Explain.

13. Identify the property illustrated in the equation $2 + (1 + 7) = (2 + 1) + 7$.
(12)

*14. **Multiple Choice** A fish tank empties at a rate of $v = 195 - 0.5t$, where v is the
(16) number of liters remaining after t seconds have passed. If the fish tank empties for 20 seconds, how many liters remain?

A 205   B 185

C 175   D 174.5

Saxon Algebra 1

Simplifying and Evaluating Variable Expressions

LESSON 16

*15. **Multi-Step** A solid, plastic machine part is shaped like a cone that is 8 centimeters high and has a radius of 2 centimeters. A machinist has removed some of the plastic by drilling a cylindrical hole into the part's base. The hole is 4 centimeters deep and has a diameter of 1 centimeter.

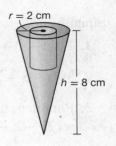

a. Determine the volume of the cone. Use the formula $V = \frac{1}{3}\pi r^2 h$.

b. Determine the volume of the cylindrical hole. Use the formula $V = \pi r^2 h$.

c. Determine the volume of the plastic machine part by subtracting the volume of the cylindrical hole from the volume of the cone.

*16. **(Chemistry)** Boyle's law relates the pressure and volume of a gas held at a constant temperature. This relationship is represented by the equation $P_f = \frac{P_i V_i}{V_f}$. In this equation, P_i and V_i represent the gas's initial pressure and volume. P_f and V_f represent the gas's final pressure and volume. What is the final pressure of the gas if a 3-liter volume of gas at a pressure of 1 atmosphere is expanded to a final volume of 6 liters?

*17. **(Investing)** Jamie wants to determine how much she should invest in a stock. She uses the equation for present value, $V_p = \frac{V_f}{(1 + i)^t}$, in which V_f is the future value, i is the interest rate, and t is the number of years. How much should her present value be if she wants the future value of the stock to be $2000 in 10 years at an interest rate of 0.02? Round the answer to the nearest dollar.

*18. **Multiple Choice** Given the information in the table, which equation best relates x and y?

A $y = x^3 + 5$
B $y = \frac{x^2 + 5}{x}$
C $y = |x^3 + 5|$
D $y = \frac{x^3 + 5}{x}$

x	y
2	13
1	6
-1	4
-2	3

*19. **Measurement** A party planner use the equation $A = Nx^2$ to estimate how much cake is needed for a party with N guests, where x is the width of a square piece of cake. If each piece of cake will be about 3 inches wide, approximate the area of the base of a cake pan for the given number of guests.

a. 50 guests

b. 150 guests

c. 350 guests

Simplifying and Evaluating Variable Expressions

LESSON 16

*20. **Multi-Step** Two teams of students were riding bikes for charity. There were a total of b students on the blue team and they each rode 15 miles. There were a total of r students on the red team and they each rode 3 miles. The students collected $2 for each mile.
(15)
  a. Write an expression for the total number of miles ridden by both teams.
  b. Write and simplify an expression that uses the Distributive Property to show the total amount of money collected.

21. **Error Analysis** A bucket contains 10 balls numbered 1, 1, 2, 3, 4, 4, 4, 5, 6, and 6. A single ball is randomly chosen from the bucket. What is the probability of drawing a ball with a number greater than or equal to 5? Which student is correct? Explain the error.
(14)

Student A	Student B
$P(5 \text{ or } 6) = \dfrac{3}{10}$	$P(5 \text{ or } 6) = \dfrac{2}{10}$ or $\dfrac{1}{5}$

22. **Estimate** $\sqrt{36} + \sqrt{40}$ ◯ $\sqrt{25} + \sqrt{80}$. Verify the answer.
(13)

23. **Multi-Step** John is using square ceramic floor tiles that are each 18 inches long. How many of these tiles will John need to cover a floor with an area of 81 square feet?
(13)

24. **(Oceanography)** *Alvin* (DSV-2), a 16-ton manned research submersible, is used to observe life forms at depths of up to 8000 feet below sea level. After the hull was replaced, *Alvin* was able to dive about 2.6 times the distance as before the hull replacement. About how far was it able to travel after the hull was replaced?
(11)

25. **Analyze** What is the sign of the sum of $-8 + 7$? Explain how the sign is determined.
(10)

26. **Write** Why would you want to convert measures from one unit to another when working with a recipe found in a French cookbook?
(8)

27. **Justify** Evaluate $10(8-6)^3 + 4(|-5 + (-2)| + 2)$. Justify each step.
(7)

28. **Error Analysis** Two students wanted to find out the change in temperature in Calgary, Canada. It was $-1°C$ in the morning and was $-20°C$ by nighttime. Which student is correct? Explain the error.
(6)

Student A	Student B
$-20 - 1$	$-20 - (-1)$
$-20 + (-1)$	$-20 + 1$
-21	-19

Saxon Algebra 1

Simplifying and Evaluating Variable Expressions

LESSON 16

29. (Construction) A father builds a playhouse in the shape of a rectangular prism with a triangular prism on top, as shown in the figure. The volume of the rectangular prism is $(10 \cdot 5.8 \cdot 8)$ ft³, and the volume of the triangular prism is $[\frac{1}{2} \cdot (10 \cdot 5.8)] \cdot 4$ ft³. What is the volume of the whole structure?

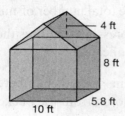

30. (Manufacturing) A manufacturing company produced 500 bowling balls in one day. Of those, 10 were found to be defective. The manufacturer sent a shipment of 250 balls to Zippy Lanes.
  a. What is the experimental probability that a bowling ball with have a defect?
  b. Predict the number of balls in the shipment to Zippy Lanes that will have a defect.

Translating Between Words and Algebraic Expressions

LESSON 17

Expand each algebraic expression by using the Distributive Property.

1. $(4 + 2y)x$

2. $-2(x - 4y)$

3. **Write** What is a term of an algebraic expression?

4. Given the sets $A = \{-3, -2, -1\}$, $B = \{1, 2, 3\}$, and $C = \{-1, 1, -2, 2, -3, 3\}$, are the following statements true or false?
   a. $A \cap C = \{-3, -2, -1\}$
   b. $A \cap B = \{-3, -2, -1, 1, 2, 3\}$
   c. $B \cup C = \{-3, -2, -1, 1, 2, 3\}$
   d. $A \cup B = \{-3, -2, -1\}$

Write the algebraic expressions for each statement.

*5. three times the sum of the opposite of a number and -7

*6. 0.18 of what number is 4.68?

7. Add $4.7 + (-9.2) - 1.9$.

8. Compare $\sqrt{36} + \sqrt{121}$ ◯ $\sqrt{100} + \sqrt{49}$ using <, >, or =.

9. Between which two whole numbers is $\sqrt{15}$?

10. **Justify** True or False: $k = 0 \cdot k$, where k is any real number except for zero. Justify your answer using the properties.

11. Evaluate $(a + 4)^3 + 5x^2$ when $a = -3$ and $x = -1$.

*12. **Justify** True or False: $yx^2m^3 = -4$ when $x = -1$, $y = 2$, and $m = -2$. Justify your answer.

*13. Translate $3(x + 6)$ into word form.

*14. **Multiple Choice** Which expression is the algebraic translation of "4 times the sum of 9 and g"?
   A $4 + 9g$ B $4(9 + g)$ C $4 \cdot 9g$ D $(4 + 9)g$

*15. **Age** Mary is one year younger than twice Paul's age. Write an expression for Mary's age.

*16. **Finance** Miles spent $7 and then received a paycheck that doubled the money he had left. Write an expression to represent how much money he has now.

Saxon Algebra 1

Translating Between Words and Algebraic Expressions

LESSON 17

*17. Multi-Step A produce stand sells apples and bananas. Apples cost $0.20 each and bananas cost $0.10 each.
 a. Choose variables to represent apples and bananas.
 b. Write an expression to represent the total pieces of fruit.
 c. Write an expression to represent how much the fruit costs in dollars.

18. Error Analysis Students are asked to evaluate $\frac{x^2 - 4x}{xy}$ when $x = -2$ and $y = 3$. Which student is correct? Explain the error.

Student A	Student B
$\frac{x^2 - 4x}{xy}$	$\frac{x^2 - 4x}{xy}$
$\frac{(-2)^2 - 4(-2)}{(-2)(3)}$	$\frac{(-2)^2 - 4(-2)}{(-2)(3)}$
$= \frac{4 + 8}{-6}$	$= \frac{-4 - (-8)}{-6}$
$= \frac{12}{-6} = -2$	$= \frac{4}{-6}$
	$= \frac{-2}{3}$

*19. Geometry The figure below has corners that are square and a curved section that is a half circle. The dimensions given are in meters. What is the area of the figure? Use 3.14 for π.

*20. Multi-Step A painter estimates that one gallon of a certain kind of paint will cover 305 square feet of wall. How many gallons of the paint will cover the wall described by the diagram below? (Dimensions given are in feet.)

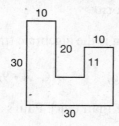

Translating Between Words and Algebraic Expressions

LESSON 17

***21. Error Analysis** Two students simplified an algebraic expression. Which student is correct? Explain the error.
(15)

Student A	Student B
$2r^3t\,(r^5t^2 + 4r^3t^3)$ $= 2r^{15}t^2 + 8r^9t^3$	$2r^3t\,(r^5t^2 + 4r^3t^3)$ $= 2r^8t^3 + 8r^6t^4$

22. Probability There are 400 students in the cafeteria. Of these students, 120 are tenth-graders. What is the probability of randomly selecting a tenth-grader? Express the answer as a percent.
(14)

23. Error Analysis Two students simplify the expression $-7 - x + 7$. Which student is correct? Explain the error.
(12)

Student A	Student B
$-7 - x + 7$ $= x + (-7) + 7$ $= x + (-7 + 7)$ $= x + 0$ $= x$	$-7 - x + 7$ $= -7 + 7 - x$ $= (-7 + 7) - x$ $= 0 - x$ $= -x$

24. Justify Simplify $-\frac{2}{3} \div \left(-\frac{8}{9}\right)$. Show your steps.
(11)

25. Write Must the algebraic expression $x + 7y$ have only one value? Explain.
(9)

26. Measurement Madison used a scale and measured her weight at 85 pounds. How many kilograms does Madison weigh? (Hint: 1 kilogram = 2.2 pounds.)
(8)

27. (Energy Conservation) Wind turbines take the energy from the wind and convert it to electrical energy. Use the formula $P = ad^2v^3\frac{\pi}{4}e$ to find the amount of available energy in the air. Describe the steps you would take to simplify the formula.
(4)

28. (Finance) Tonia deposited a total of $174.52 into her checking account. She also withdrew a total of $186.15. Use addition to find the net change in Tonia's checking account.
(5)

29. Justify What is the first step in simplifying the expression $2 \cdot 4 + 5^2 + (23 - 2)^2$?
(4)

30. (Payroll Accounting) Employees at Wilkinson Glass Company earn x number of dollars per hour. Executives make y number of dollars per hour. Each employee and executive works 40 hours per week. Write and simplify an algebraic expression that uses the Distributive Property to show a weekly payroll for one employee and one executive.
(15)

Name _____ Date _____ Class _____

Combining Like Terms

1. Write an algebraic expression for this statement: The sum of 5 times a number and −8.
(17)

Simplify each expression by adding like terms.

*__2.__ $m + 4 + 3m - 6 - 2m + mc - 4mc$
(18)

*__3.__ $xy - 3xy^2 + 5y^2x - 4xy$
(18)

*__4.__ **Multiple Choice** Simplify $2x^2 + 3x$.
(18)
    A $5x^2$ **B** $5x^3$

    C $6x^3$ **D** cannot be simplified

*__5.__ (Reading) Two classes are keeping track of how many pages they can read. In one
(18) class, the boys read 15 pages per night and the girls read 12 pages per night. In
another class, the boys read 7 pages per night and the girls read 9 pages per night.
Each class has x girls and y boys.
   a. Write expressions representing the number of pages each class read per night.
   b. Write an expression for the number they read altogether.

6. (Justify) After doing three addition problems that included negative numbers, John
(5) found that all three answers were negative. John concluded that any addition problem
involving a negative number must have a negative answer. Is John correct? Explain.
Give a counterexample if necessary.

7. (Geography) The retention pond at Martha's summer home in Florida changed
(6) −3 inches every day for 5 days. After 5 days, the water level was −40 inches.
What was the original water level 5 days ago?

8. (Bowling) A ten-pin bowling ball has a volume of about 5274 cm³. A candlepin
(8) bowling ball has a volume of about 48 in³. About how much greater is the volume
of a ten-pin bowling ball than the volume of a candlepin bowling ball?

Simplify.

9. $\dfrac{-16 + 4}{2(\sqrt{13 - 4})}$ **10.** $-7 - (2^4 \div 8)$
(4, 13)   (4, 13)

*__11.__ $6bac - 7ac + 8acb$ *__12.__ $2x^3y + 4x^3y + 9x^3y$
(18)   (18)

13. $\left|-15 + \sqrt{81}\right|^2$ **14.** $\dfrac{\sqrt{6 - 2}}{2 \cdot \left|-7 + 3\right|}$
(13)   (13)

*__15.__ (Sewing) Susan started with 11 bows. She can tie 4 bows per minute. Analise ties twice
(18) as many per minute.
   a. Write expressions representing the number of bows each girl will have after
       x minutes.
   b. Write an expression for the number they will have altogether after x minutes.

Combining Like Terms

LESSON 18

16. Multi-Step Marshall, Hank, and Jean are all cousins. Marshall is 3 years older than Hank. Hank is twice the age of Jean.
   a. Write expressions to represent the ages of the cousins. Assign the variable j to represent Jean.
   b. If Jean is 12 years old, how old are the other cousins?
   c. If Hank was 14, how old would Jean be?

*17. Justify Simplify $8x + x(2x + 5)$ and explain each step.

18. Evaluate $\dfrac{8ak}{4k(2a - 2c + 8)}$ when $a = \dfrac{1}{2}$, $c = 3$, and $k = -2$.

*19. True or False: $pm^2 - z^3 = 27$ when $p = -5$, $m = 0$, and $z = -3$. Justify your answer.

20. Error Analysis Two students are asked to evaluate $x^2 y - |4x|^2 z$ when $x = -2$, $y = \dfrac{1}{2}$, and $z = -1$. Which student is correct? Explain the error.

Student A	Student B
$x^2 y - \|4x\|^2 z$	$x^2 y - \|4x\|^2 z$
$(-2)^2 \left(\dfrac{1}{2}\right) - \|4(-2)\|^2 (-1)$	$(-2)^2 (-1) - \|4(-2)\|^2 \left(\dfrac{1}{2}\right)$
$= 4\left(\dfrac{1}{2}\right) - \|-8\|^2 (-1)$	$= 4(-1) - \|-8\|^2 \left(\dfrac{1}{2}\right)$
$= 2 - (8)^2 (-1) = 66$	$= -4 - (8)^2 \left(\dfrac{1}{2}\right) = -36$

21. Predict How can finding a common denominator tell you that $\dfrac{1}{9} - \dfrac{2}{20}$ will result in a positive number?

*22. Multi-Step Tamatha picks 10 peaches a minute for x minutes. Her grandmother picks 12 peaches a minute for 3 fewer minutes.
   a. Write an expression to represent the number of minutes the grandmother picks peaches.
   b. Write an expression for the number of peaches they pick together and then simplify.

23. Geometry Translate the Pythagorean Theorem into symbols. In a right triangle, the sum of the squares of the legs of the triangle is equal to the square of the hypotenuse. Let a and b be the legs of the triangle and c be the hypotenuse.

Combining Like Terms

LESSON 18

24. Measurement A railing is being built around a rectangular deck.

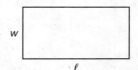

  a. Write an expression to represent the number of feet of railing needed.

  b. The width is doubled. The length is tripled. Write an expression to represent the number of feet of railing needed.

25. Multiple Choice Simplify the expression $7(10 - y)$. Which expression is correct?

  A $70 - y$ **B** $70 - 7y$

  C $70 - 7 + y$ **D** $70y - 7y$

26. Justify Simplify the expression $-m(mn^2 - m^2n)$ and explain your method for simplifying.

 27. Probability A number is chosen at random from the numbers 1 through 5. What is the probability that an odd number will be chosen?

28. Carpentry A new company buys 140 square feet of carpet to cover the floor in one of its square offices. The carpet is 4 square feet too small. What is the length of the office floor?

29. Verify The Commutative Property states that $6 \cdot 4 = 4 \cdot 6$. Show that the Commutative Property does not apply to division.

30. Error Analysis Two students translate the phrase "the sum of the squares of 8 and p" into an algebraic expression. Which student is correct? Explain the error.

Student A	Student B
$(p + 8)^2$	$p^2 + 8^2$

Solving One-Step Equations by Adding or Subtracting

LESSON 19

1. Simplify $-3x^2ym + 7x - 5ymx^2 + 16x$.
(18)

Solve each equation.

*2. $x + 5 = 7$
(19)

*3. $x + 5 = -8$
(19)

*4. $x - 6 = 4$
(19)

5. Write the algebraic expression for the phrase "seven times the sum of a number and -5."
(17)

6. Expand the expression $-3(-x - 4)$ by using the Distributive Property.
(15)

7. Simplify $xm^2xm^3x^3m$.
(3)

8. Identify the property illustrated by $3 + 8 = 8 + 3$.
(12)

9. **Write** True or False: $-5^4 = (-5)^4$. Explain.
(11)

10. Sandra lost 8 points for incorrect answers on her quiz, but gained 5 points for a bonus question. What is the sum of points Sandra lost and gained?
(10)

*11. **Error Analysis** A teacher asked two students to solve the following equation for x: $x + \frac{1}{3} = \frac{4}{9}$. Which student is correct? Explain the error.
(19)

Student A	Student B
$x + \frac{1}{3} + \frac{1}{3} = \frac{4}{9} + \frac{1}{3}$	$x + \frac{1}{3} - \frac{1}{3} = \frac{4}{9} - \frac{1}{3}$
$x = \frac{4}{9} + \frac{3}{9}$	$x = \frac{4}{9} - \frac{3}{9}$
$x = \frac{7}{9}$	$x = \frac{1}{9}$

*12. **Multiple Choice** A swimming pool is $\frac{4}{5}$ full. A maintenance man removes some of the water so that the pool is $\frac{1}{3}$ full. As a fraction of the pool's total capacity, how much water did the maintenance man remove?
(19)

    A $\frac{1}{5}$ **B** $\frac{3}{2}$ **C** $\frac{7}{15}$ **D** $\frac{17}{15}$

*13. **Chemistry** Many chemists use kelvins to describe temperatures. To convert from a temperature in degrees Celsius to kelvins, a chemist will use the equation $T_{Celsius} + 273.15 = T_{kelvin}$. If a gas cools to a temperature of 325.20K, what is its temperature in degrees Celsius?
(19)

Solving One-Step Equations by Adding or Subtracting

LESSON 19

*14. **Business** A movie theater needs to sell 3500 tickets over a single weekend to cover its operating expenses before it starts making a profit. If it sells 1278 tickets on Friday, what is the minimum number of tickets it needs to sell over the rest of the weekend in order to make a profit? Write an equation and then solve it.

*15. **Write** Jeremy is solving the equation $x - 2.5 = 7.0$. What must he do to both side of the equation in order to isolate x?

*16. **Multiple Choice** Given the information in the table, which equation best relates a and b?

a	b
−5	0
0	−5
5	−10
−10	5

A $a - b = 5$ **B** $a - b = -5$ **C** $-5 - b = a$ **D** $a - 5 = b$

17. **Error Analysis** Two students simplify the expression $6x + 8 - 4x - 2$. Which student is correct? Explain the error.

Student A	Student B
$2x + 6$ $= 8x$	$2x + 6$

18. **Geometry** Write an expression to represent the sum of the degrees in the triangle.

19. **Multi-Step** At a family camp, the big race is on the final day. Each family member runs for t hours and the family that runs the farthest wins. The rate each person ran is shown in the chart. To find how far they ran, multiply their rate by the amount of time they ran.

Family Member	Rate (mph)
Julio	4
Jorge	5
Sam	3

  a. Write an expression to represent how far each person ran.
  b. Write an expression to represent how far the family ran.
  c. How far did they run if each person ran $\frac{1}{6}$ hour?

Solving One-Step Equations by Adding or Subtracting

LESSON 19

20. Error Analysis Two students translate the phrase "five more than the product of a number and three" into an algebraic expression. Which student is correct? Explain the error.
(17)

Student A	Student B
$\frac{x}{3} + 5$	$3x + 5$

*21. When $x = 1.5$ and $y = -2$, what is the value of $\left| x^2 + y^3 \right|$?
(16)

22. Multiple Choice A store owner makes a pyramid-shaped display using a stack of soup cans. She arranges the cans so that the highest level of the stack has one can. The second-highest level has four cans arranged in a square supporting the top can. The third-highest level has nine cans arranged in a square that supports the second-highest level. Which of the following expressions best represents the number of cans in the lowest level of a display that is l levels high?
(16)
   A $3l$ **B** $4l$ **C** l^2 **D** l^3

23. Write **a.** What is the sign of the result when a negative value is cubed?
(5, 16)
       b. What is the sign of the result when the absolute value of an expression is taken?

24. (**Sites**) Use the table to answer the questions.
(14)

   a. If a building is randomly chosen, what is the probability that it is exactly 1250 feet tall?

   b. What is the probability that a building exactly 1046 feet tall is chosen?

Building	Height (ft)	Year Built
Sears Tower	1451	1974
Empire State Building	1250	1931
Aon Center	1136	1973
John Hancock Center	1127	1969
Chrysler Building	1046	1930
New York Times Building	1046	2007
Bank of America Plaza	1023	1992

   c. What is the probability that a building that was built between 1960 and 1980 is chosen?

25. Write What is a perfect square?
(13)

26. (**Personal Finance**) The expression $P(1 + i)^2$ can be used to find the value of an investment P after 2 years at an interest rate of i. What is the value of an investment of $500 deposited in an account with a 3% interest rate after 2 years? (Hint: Remember to convert the percent to a decimal before calculating.)
(9)

Solving One-Step Equations by Adding or Subtracting

LESSON 19

27. **(Racing)** The formula for the cylindrical volume of an engine on a dragster is
    $\left(\frac{\pi}{4}\right)b^2 s$, where b is the inside diameter (the bore) and s is the distance that the piston moves from its highest position to its lowest position (the stroke). Following the order of operations, describe the steps you would take to simplify the formula.

28. **Justify** What is the additive inverse of 12? Justify your answer.

29. **Multi-Step** Theater tickets cost $14 dollars for adults, a, and $8 for children, c. Additionally, each person who went to the theater on Thursday made a $5 contribution to charity. Write an expression using the Distributive Property to show the amount of money that the theater received on Thursday. Simplify the expression.

30. **Probability** What is the probability that an ace will be chosen from a full deck of 52 playing cards? What is the probability of another ace being drawn if the first ace drawn is not returned to the deck?

Name _____ Date _____ Class _____

Graphing on a Coordinate Plane — LESSON 20

Simplify.

1. $(+3) + (-14)$
   (5)

2. $4xyz - 3yz + zxy$
   (18)

3. $3xyz - 3xyz + zxy$
   (18)

Solve.

4. $x - 4 = 10$
   (19)

5. $x + \dfrac{1}{5} = -\dfrac{1}{10}$
   (19)

*6. Graph the ordered pair $(3, -4)$ on a coordinate plane.
(20)

*7. Graph the ordered pair $(0, 5)$ on a coordinate plane.
(20)

*8. **Multiple Choice** Which ordered pair is associated with point Z?
(20)
   A $(3, 0)$ B $(0, 3)$
   C $(-3, 0)$ D $(0, -3)$

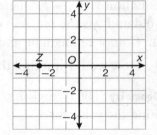

*9. (**Babysitting**) Ellen charges $3 plus $1 per child for an hour of
(20) babysitting. To determine her hourly rate, she uses the formula $r = 3 + c$, where r is the rate and c is the number of children Complete the table and graph the solutions.

c	r
1	
2	
3	
4	

*10. (**Reading**) Thomas read 10 pages in a book before starting
(20) his speed-reading lessons. After his lessons, he could read 3 pages per minute. The equation $y = 3x + 10$ calculates the total number of pages read after x minutes. Complete the table.

x	y
15	
20	
30	
50	

Saxon Algebra 1

Graphing on a Coordinate Plane

LESSON 20

*11. Error Analysis Two students completed an *x/y* chart for the equation $y = 3 + 2x$ to find a solution to the equation. Which student is correct? Explain the error.

Student A
x | y
2 | 10
$y = 3 + 2(2)$
$y = 5(2)$
$y = 10$
(2, 10) is a solution.

Student B
x | y
2 | 7
$y = 3 + 2(2)$
$y = 3 + 4$
$y = 7$
(2, 7) is a solution.

*12. Multi-Step For a lemonade stand, profit depends on the number of cups sold. Profit is represented by the equation $y = x - 5$, where x is the number of cups sold and y is the profit in dollars.

x	5	10	20	50
y				

  a. Complete the table and graph the solutions.

  b. How would you find the profit if 30 cups were sold?

13. Geometry The triangle has a perimeter of 24 centimeters. Find the value for x.

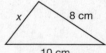

*14. Multi-Step To climb to the highest observation deck in the Empire State Building, you have to walk up 1860 steps. Starting at the lowest step, a security guard walks up $\frac{1}{4}$ the total number of steps during his morning rounds. At the end of his afternoon rounds, he stands on the 310th step. How many steps did he walk down during the afternoon rounds?

*15. Error Analysis Two students simplify the expression $5x^2 + 7x^2$. Which student is correct? Explain the error.

Student A	Student B
$12x^2$	$12x^4$

16. Write Why do mathematicians use symbols rather than words?

17. Multiple Choice Which equation demonstrates the Associative Property of Addition?
  A $6 - 3c = 3c - 6$
  B $c^3 - 6 = c^3 + 6$
  C $(6 - c)^3 = (c - 6)^3$
  D $(6 + c^3) - 4 = 6 + (c^3 - 4)$

18. Evaluate each of the following expressions when $a = 2$.
  a. a^2 **b.** $-a^2$ **c.** $-a^3$ **d.** $(-a)^3$ **e.** $|(-a)^2|$

Graphing on a Coordinate Plane

LESSON 20

19. Analyze A school is holding a blood drive. If 3 students out of the 50 who give blood are Type A, what is the probability that a randomly selected student is Type A? Write the probability as a decimal.
(14)

20. Evaluate $\sqrt{441} + \sqrt{1089}$.
(13)

21. Analyze Use an example to show that the Associative Property holds true for multiplication.
(12)

22. Statistics Use the data in the table to find the average yearly change in the deer population during a five-year period.
(11)

Deer Population

Year	Decrease in Number of Deer
2000	10
2001	7
2002	9
2003	10
2004	12

23. (Stocks) A Stock Market Report shows the value of stocks in points. The value of the stock is determined by the number of points. If a stock is at $79\frac{5}{7}$ points and it drops 3 points, what is the value of stock?
(10)

24. (Nutrition) One hundred grams of honey contains about 0.3 grams of protein. How many milligrams is 0.3 grams of protein?
(8)

25. (Tiling) A contractor lays patterned tile floors. He often begins with a polygon and makes diagonal lines that pass from corner to corner of the polygon. He uses the following expression $\frac{n^2 - 3n}{2}$, where n equals the number of sides of the polygon to find the number of diagonal lines for any given polygon.
(7)
a. Find the number of diagonals for a hexagon.
b. **Model** Check your work by drawing the diagonals in a hexagon.

26. Write and simplify a mathematical expression that shows "fourteen minus the quotient of three squared and the sum of three plus six."
(4)

27. Verify Indicate whether each statement is true or false. If the statement is false, explain why.
(2)
a. The coefficient of x in the expression $x + 3$ is 3.
b. The factors of the expression $\frac{2mn}{5}$ are $\frac{2}{5}$, m, and n.

28. Name the coefficient(s), variables, and number of terms in the expression $b^2 - 4ac$.
(2)

Graphing on a Coordinate Plane

29. Multi-Step Pencils cost ten cents and erasers cost five cents.
(17)
  a. Write an expression to represent the total number of school supplies purchased.

  b. Write an expression to represent how much the supplies cost in cents.

*30. Analyze A person runs 5 miles per hour. The equation $d = 5t$ tells how far the
(20) person has run in t hours. Make a table when $t = 0, 1, 2,$ and 4 hours. Graph the ordered pairs in a graph and connect all the points. What do you notice?

Name _____ Date _____ Class _____

Solving One-Step Equations by Multiplying or Dividing
LESSON 21

*1. (Safety) For every 4 feet a ladder rises, the base of a ladder should be placed 1 foot away from the bottom of a building. If the base of a ladder is 7 feet from the bottom of a building, find the height the ladder rises up the building?
(21)

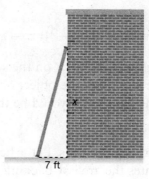

2. What is a term of an algebraic expression?
(2)

*3. Write Explain how to use inverse operations to solve $\frac{2}{3}x = 8$.
(21)

*4. (Physics) In physics equations, a change in a quantity is represented by the delta symbol, Δ. A change in velocity, Δv, is calculated using the equation $\Delta v = v_f - v_i$, where v_i is the initial velocity and v_f is the final velocity. If a cart has an initial velocity of 5 miles per second and experiences a change in velocity of 2 miles per second, what is the final velocity of the cart?
(19)

5. Graph the ordered pair $(-2, 6)$ on a coordinate plane.
(20)

6. Geometry The length of a rectangular picture frame is 3 times the width.
(3)
   a. Draw a picture of the picture frame and label the dimensions.
   b. Write an expression for the area of the frame.

7. Complete the table for $y = 2x + 7$.
(20)

x	−5	1	4
y			

*8. Solve $\frac{x}{3} = 5$.
(21)

*9. Multiple-Choice Which step can you use first to solve $-\frac{x}{9} = -52$?
(21)
   A Multiply both sides by $\frac{1}{9}$.
   B Multiply both sides by -9.
   C Divide both sides by -52.
   D Divide both sides by 52.

Saxon Algebra 1

Solving One-Step Equations by Multiplying or Dividing

LESSON 21

*10. **Estimate** Alan makes $1 for each snow cone he sells. Alan calculates his profit by
(19) subtracting the daily cost of $195 to run the stand from the total number of snow cones he sells each day. How many snow cones does Alan need to sell to make a profit of $200 a day?

11. **Write** Explain which terms in $3z^2y + 2yz - 4y^2z - z^2y + 8yz$ can be combined.
(18)

12. (**Astronomy**) The relative weight of an object on the surface of Jupiter can be found
(9) using $2.364w$, where w is the weight of the object on Earth. The space shuttle weighs about 4,500,000 pounds on earth. What would be the weight of the space shuttle on Jupiter?

13. **Geometry** The length of a frame is 8 inches. Let w be the width of the frame.
(20) The formula $A = 8w$ calculates the area of backing needed for a framed picture. Complete the table and graph the solutions.

w	A
2	
4	
6	
8	

14. Simplify $-4 - 3 + 2 - 4 - 3 - 8$.
(6)

15. **Multiple Choice** Simplify $5p + 7 - 8p + 2$.
(18)
   A $9 - 3p$ **B** $3p - 9$ **C** $13p + 5$ **D** $13p - 9$

16. **Verify** Evaluate $\frac{2}{3}\left(4 + \frac{3}{4}\right)$ using two different methods. Verify the solution of each
(15) method.

17. **Probability** A letter of the alphabet is randomly chosen. What is the probability that
(14) the letter is a vowel?

18. **Justify** The area of a square is 100 square feet, what is the length of each side?
(13) Explain.

19. **Measurement** A picture framer calculates the amount of materials needed
(12) using $2l + 2w$. If the framer used $2w + 2l$, would the results be the same? Explain.

Solving One-Step Equations by Multiplying or Dividing

LESSON 21

*20. **Multi-Step** Alda's school is 1200 yards from her house. She walks 150 yards per minute. The equation $y = 1200 - 150x$ represents how far she will be from the school after x minutes.
(20)

   a. Complete the table and graph the solutions.

x	y
1	
4	
6	
8	

   b. What does it mean to say that after 8 minutes, she is 0 yards from the school?

*21. **Analyze** A student says that to solve $-\frac{3}{4}x = 12$ you should divide each side by $-\frac{3}{4}$. Another student says that to solve the equation you should multiply each side by $-\frac{4}{3}$. Will both methods result in the correct solution? Explain.
(21)

22. **Multi-Step** Determine whether $4^3 \cdot \left(\frac{1}{4}\right)^3 = 1$.
(3)

   a. Simplify the expressions 4^3 and $\left(\frac{1}{4}\right)^3$.

   b. Write an expression for the multiplication of 4^3 and $\left(\frac{1}{4}\right)^3$ without using exponents. Then check to see if the product of the expressions is 1.

23. **Write** Will dividing two integers ever produce an irrational number? Explain.
(1)

24. **Analyze** If a student is converting from 225 square units to 22,500 square units, what units of measure is he or she most likely converting?
(8)

25. **(Golf)** A round of golf takes 4.5 hours and each hole takes 0.25 hours. In the equation $y = 4.5 - 0.25x$, x is the number of holes played and y is the remaining time to finish the round. Make a table for 4, 8, 12, and 16 holes and then graph the ordered pairs in your table.
(20)

26. Simplify $-|15 - 5|$.
(5)

*27. **(Pricing)** One fruit stand has s strawberries and k kiwis to sell. Another stand has twice as many strawberries and four times as many kiwis to sell.
(18)

   a. Write expressions representing the number of strawberries and kiwis each stand has to sell.

   b. Write an expression for the total number of pieces of fruit.

Solving One-Step Equations by Multiplying or Dividing

LESSON 21

***28.** Sketch a graph to represent the following situation: A tomato plant grows at a slow
(Inv 2) rate, and then grows rapidly with more sun and water.

29. Write an algebraic expression for "0.21 of what number is 7.98?"
(17)

30. Error Analysis Two students solve $x - 5 = 11$. Which student is correct? Explain
(19) the error.

Student A	Student B
$x - 5 = 11$	$x - 5 = 11$
$\underline{+5 = +5}$	$\underline{-5 = -5}$
$x = 16$	$x = 6$

Name _____ Date _____ Class _____

Analyzing and Comparing Statistical Graphs

LESSON 22

***1.** True or False: A stem-and-leaf plot can help analyze change over time. If false, explain why.
(22)

2. Complete the table for $y = -3x - 9$.
(20)

x	−1	0	1
y			

3. Simplify $2p(xy - 3k)$.
(15)

4. Solve $y - 3 = 2$.
(19)

*5. Solve $x - \dfrac{1}{4} = \dfrac{7}{8}$.
(19)

6. Solve $4x = 2\dfrac{2}{3}$.
(21)

7. Solve $7x = 49$.
(21)

*8. Choose an appropriate graph to display the change in profit of a company over several years. Explain your choice.
(22)

9. Verify Determine whether each statement below is true or false. If false, provide a counterexample.
(1)
   a. The set of integers is closed under division.
   b. The set of irrational numbers is closed under division.
   c. The set of integers is closed under addition.

*10. (**Racing**) The table shows the Indianapolis 500 fastest lap times to the nearest second, every 5 years since 1960. Make an appropriate graph to display the data. Then make a conclusion about the data.
(22)

Fastest Lap Times in the Indianapolis 500

Year	1960	1965	1970	1975	1980	1985	1990	1995	2000	2005
Time (seconds)	62	57	54	48	47	44	40	40	41	39

11. Graph the ordered pair $(-4, -1)$ on a coordinate plane.
(20)

© Saxon. All rights reserved.   71   Saxon Algebra 1

Analyzing and Comparing Statistical Graphs

LESSON 22

12. Error Analysis Two students plotted the point (−4, 3). Which student is correct? Explain the error.

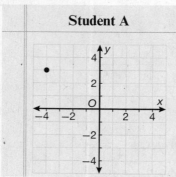
Student A

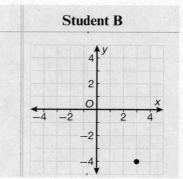
Student B

13. Multiple Choice A baker needs 25 eggs for all the cakes she plans to bake. She only has 12 eggs. If x is the number of eggs she will buy to complete her ingredients list, which of the following equations best represents how she can find x?

   A $12 - x = 25$ **B** $12 + x = 25$ **C** $25 + x = 12$ **D** $x - 12 = 25$

*14. Analyze For the equation $x = 14 - y$, what must be true of each value of x if y is

   a. greater than 14?

   b. equal to 14?

   c. less than 14?

*15. Multiple Choice Which graph would best compare the ages of people living in two different cities?

   A circle graph   **B** stem-and-leaf plot

   C double-line graph   **D** double-bar graph

16. Travel A man travels 25 miles to work. On his way home, he stops to fill up with gas after going d miles. Write an expression to represent his distance from home.

17. Verify Show that each equation is true for the given values of x and y.

   a. $x\left(\dfrac{y}{y-x}\right)^2 = -\dfrac{4}{9}$; $x = -4$ and $y = 2$ **b.** $|(x-y)^3| = 27$; $x = -1$ and $y = 2$

18. Write What is a sample space?

*19. Endangered Animals The table shows the number of threatened or endangered animal species as of July 22, 2007. Make an appropriate graph to display the data. Then make a conclusion about the data.

Number of Threatened or Endangered Species in the U.S. and Foreign Countries

	Mammals	Birds	Reptiles	Amphibians	Fish	Clams	Snails	Insects	Arachnids	Crustaceans
U.S.	81	89	37	23	139	70	76	57	12	22
Foreign	276	182	81	9	12	2	1	4	0	0

Analyzing and Comparing Statistical Graphs

LESSON 22

20. Generalize Use the pairs of equations. What can be concluded about the Commutative Property?
(12)

$9 - 5 = 4$ and $5 - 9 = -4$

$12 - 6 = 6$ and $6 - 12 = -6$

$7 - 3 = 4$ and $3 - 7 = -4$

21. (**Landscape Design**) Wanchen is planting a garden the shape of a trapezoid in her yard. Use the diagram to find the area of her garden.
(4)

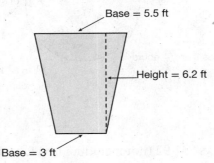

Base = 5.5 ft
Height = 6.2 ft
Base = 3 ft

22. Simplify $-7 + (-3) + 4 - 3 + (-2)$.
(6)

*23. Geometry An arc of a circle is a segment of the circumference of a circle. If an arc measures 16 inches and is $\frac{4}{9}$ the circumference of a circle, what is the circumference of the circle?
(21)

24. Multi-Step A house has an area of 1200 square feet. The owners add on a new room that is 15 feet long and 20 feet wide. What is the area of the house now?
(4)

a. Write an expression to represent the area of the new room.

b. Write an expression to represent the total area of the house now.

c. Find the area of the house now.

25. Write Describe a situation that could be represented by the expression $2d - w$.
(9)

26. Simplify $3ab^2 - 2ab + 5b^2a - ba$.
(18)

73

Saxon Algebra 1

Analyzing and Comparing Statistical Graphs

LESSON 22

***27.** The circle graph shows the result of a poll on the sleeping habits of children
(10) ages 9–12. What portion of the children said they slept the recommended $9\frac{1}{2}$ to
$10\frac{1}{2}$ hours for their age group? Express the answer as a decimal rounded to
the nearest hundredth.

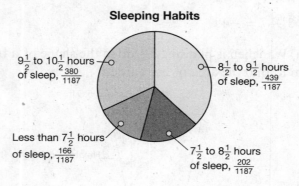

28. Simplify $x^2yy\dot{y}x^3yx$.
(3)

***29. Multi-Step** Enrique pays $31.92 (not including tax) for 6 books that are on sale. Each
(21) book costs the same amount. Enrique pays $\frac{4}{5}$ of the original cost of the books.
   a. What is the sale price of each book?
   b. What was the original cost of each book?

30. Probability In a standard deck of cards, there are
(Inv 1) 13 cards in each of four suits: hearts, diamonds,
clubs, and spades. Jose randomly draws a card
from a deck and replaces it after each draw. His
results are recorded in the table. Find the experimental
probability of each event.
   a. drawing a heart
   b. not drawing a club

Outcome	Frequency
Hearts	8
Diamonds	8
Clubs	6
Spades	4

Name _____ Date _____ Class _____

Solving Two-Step Equations LESSON 23

1. Evaluate $(x - y) - (x - y)$ for $x = 3.5$ and $y = 2.5$.
   (9)

2. **Write** Explain how to graph the point $(-2, 4)$.
   (20)

*3. **Multiple Choice** What is the value of x in the equation $3x + 5 = 32$?
   (23)
      A 24   **B** 9   **C** 81   **D** $12\frac{1}{3}$

4. **Error Analysis** Two students solve $-12x = -72$. Which student is correct? Explain the
   (21) error.

Student A	Student B
$-12x = -72$	$-12x = -72$
$-\dfrac{12x}{12} = -\dfrac{72}{12}$	$\dfrac{-12x}{-12} = \dfrac{-72}{-12}$
$x = -6$	$x = 6$

*5. (**Altitude**) A plane increases altitude by 350 meters every minute. If the plane
   (17) started at an altitude of 750 meters above sea level, what is the plane's altitude
   after 6 minutes?

6. **Verify** Is $x = 9$ the solution for $3x - 8 = 22$? Explain. If false, provide a correct
   (23) solution and check.

7. **Justify** Find a counterexample to the following statement: A rational number that is
   (1) not an integer, such as $\frac{3}{5}$, multiplied by any integer will produce a rational number
   that is not an integer.

*8. **Multi-Step** Three hundred people were surveyed as they left a movie
   (22) theater. They were asked which type of movie they like best.
   The circle graph shows the survey results.
   a. Which type of movie was most popular?
   b. How many people liked horror movies the best?
   c. How many more people liked action movies than dramas?

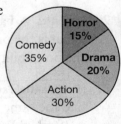

*9. Choose an appropriate graph to display the number of different types of DVDs sold
   (22) at two video stores. Explain your answer.

*10. A class of 20 students answered a survey about their favorite places to go on
   (22) vacation. Use the data in the table to make a bar graph.

Beach	Amusement Park	Mountains	Museums
5	8	3	4

*11. **Geometry** A circle has a circumference of $\frac{8}{9}\pi$ meters. What is the radius of the circle?
   (21)

75   Saxon Algebra 1

Solving Two-Step Equations
LESSON 23

12. Multiple Choice Which equation has solutions that are represented by the graphed points?
 A $y = 2x + 1$
 B $y = 2x + 3$
 C $y = -2x$
 D $y = -2x - 1$

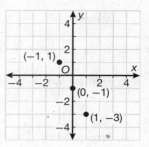

13. (Coins) Jenny and Sam took the coins out of their pockets. Jenny has x quarters and y dimes. Sam has h half dollars and z nickels.
 a. Write expressions representing the value of the coins, in cents, in each person's pocket.
 b. Write an expression for the total number of cents they have.

14. Verify "12 more than the product of x and 3" can be written as $3x + 12$ or $12 + 3x$. Substitute 2 for x and show that the expressions are equivalent.

15. Use $>$, $<$, or $=$ to compare the expressions.
$$24 + \frac{16}{4} - (4 + 3^2) \cdot 2 \ \bigcirc \ 24 + \left(\frac{16}{4} - 4\right) + 3^2 \cdot 2$$

Solve.

16. $y - \frac{1}{2} = -2\frac{1}{2}$

*17. $2x + 3 = 11$

*18. $3x - 4 = 10$

*19. Solve $2.2x + 2 = 8.6$

*20. Statistics A basketball player attempted 1789 free throws. He made 801 of them. What is the probability that the player will make the next shot he attempts? Write the probability as a decimal rounded to the nearest hundredth.

21. (Presidential Facts) Many of our first 43 U.S. Presidents had the same first name. Use the table.
 a. If a U.S President is chosen at random, what is the theoretical probability of choosing one whose name is George?
 b. What is the probability of choosing one whose name is William or John?
 c. What is the probability of choosing a president whose name is not shown in the table?

Names	Number of Presidents
James	6
John	4
William	4
George	3

22. (Carpentry) Alice wants to add a square porch to the back of her house. The area of the porch is 361 square feet. What is the length of each side of the porch?

Solving Two-Step Equations LESSON 23

23. (**Meteorology**) The highest temperature ever recorded at the South Pole is −13.6°C.
(11) The lowest temperature is about 6 times lower than the highest temperature recorded. Approximately what is the lowest temperature recorded?

24. Model On Monday the low temperature was −4°F. The temperature rose 21°F to the
(5) high temperature for that day. What was the high temperature on Monday? Use a number line or thermometer to model the addition.

25. Multi-Step Paula's bank statement showed the following transactions for last month. The
(5) beginning balance was $138.24. There was a withdrawal of $46.59, then a deposit of $29.83, plus $1.87 in interest added. What was the balance after these transactions?

26. Simplify $4 \div 2 + 6^2 - 22$.
(4)

27. Write Explain how to simplify $a^3b^2ac^5a^4b$.
(3)

28. Convert 332 meters per second to centimeters per second.
(8)

29. Multiple Choice In 2000, the U.S. economy gained $111,349 million from the sale of
(10) goods exported to Mexico. However, the U.S. economy lost $135,926 million from the sale of goods imported from Mexico. What was the U.S. balance of trade with Mexico in the year 2000?

  A $247,275 million

  B $24,926 million

  C $1.2 million

  D −$24,577 million

30. Probability Describe each of the following events as impossible, unlikely, as likely
(Inv 1) as not, likely, or certain.

  a. Tanisha buys a new pair of shoes and the first shoe she pulls out of the box is for the left foot.

  b. Ralph rolls a number less than 7 on a standard number cube.

  c. November will have 31 days.

Name _____ Date _____ Class _____

Solving Decimal Equations

LESSON 24

1. Multiple Choice What is the solution of $\frac{4}{5}x = -24$ for x?
(21)
    A -30 **B** $-\frac{96}{5}$ **C** $\frac{96}{5}$ **D** 30

2. Simplify $3(2x + 5x)$ using the two different methods shown below.
(18)
    a. Combine like terms, and then multiply.
    b. Distribute, and then combine like terms.

*3. Solve $0.45x - 0.002 = 8.098$.
(24)

*4. Justify If you multiply both sides of an equation by a constant c, what happens
(21) to the solution? Explain your answer.

*5. (**Stock Market**) An investor buys some stock at $6.57 a share. She spends $846.25 which
(24) includes a transaction fee of $25. How many shares of stock did she buy?

*6. Multiple Choice 0.8 is 0.32 of what number?
(24)
    A 2.5   **B** 0.25   **C** 0.4   **D** 4

*7. Verify Solve $0.45x + 0.9 = 1.008$. Will both methods shown below result in the same
(24) solution? Verify by using both methods to solve.

Method I: Multiply both sides of the equation by 1000 first.

Method II: Subtract 0.9 from both sides first.

8. Identify the coefficient, the variable(s), and the number of terms in the
(2) expression $\frac{9}{5}C + 32$.

*9. Verify Solve $0.25x + \frac{1}{2} = 0.075$. Will both methods shown
(24) below result in the same solution? Explain.

Method I: First write the fraction as a decimal.

Method II: First write the decimals as fractions.

*10. Error Analysis Two students use the circle graph to find the
(22) total percent of students who have fewer than two siblings.
Which student is correct? Explain the error.

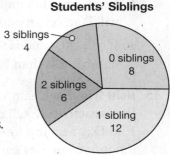

Students' Siblings

3 siblings 4
0 siblings 8
2 siblings 6
1 sibling 12

Student A	Student B
0 siblings or 1 sibling = 20 20% of the students	Total students: $8 + 12 + 6 + 4 = 30$ 0 siblings or 1 sibling = 20 $\frac{20}{30} \approx 67\%$ of the students

© Saxon. All rights reserved.   Saxon Algebra 1

Solving Decimal Equations

LESSON 24

*11. Measurement The graph shows an estimation of the changes in the diameter of a tree, in inches, every 20 years. What was the approximate circumference of the tree when the tree was 100 years old? Use 3.14 for π. Round the answer to the nearest tenth.

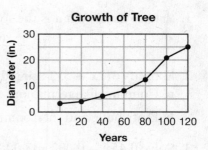

*12. The circle graph shows the amount of money Will spent on different snacks at a store. If Will spent $12, how much money did he spend on each item?

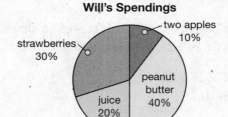

13. Graph the ordered pair on a coordinate plane (2, 1).

14. Probability A spinner is divided in equal sections and labeled as shown in the diagram.
   a. If x is an even number, what is the probability the spinner will land on an even number?
   b. If x is an odd number, what is the probability the spinner will land on an odd number?
   c. If $x = 20$, what is the probability the spinner will land on a number less than 20?

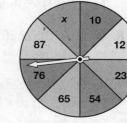

15. Multi-Step Each camp counselor at Camp Wallaby walked 6 miles for a health and fitness activity. Each camper walked 2 miles. The camp leader paid $0.50 into a Fun Day account for every mile walked. Write an expression to represent the total amount of money earned from walking by counselors and campers.

16. Verify Are the expressions below equivalent? Explain.
$(11w^4 \cdot 3z^9)(2w^7z^2) \stackrel{?}{=} 66w^{11}z^{11}$

17. (Contests) Miguel entered a contest offering prizes to the top 3 finishers. The probability of winning 1st is 12%, the probability of winning 2nd is 18%, and the probability of winning 3rd is 20%. What is the probability that Miguel will not win any prize?

Solving Decimal Equations

LESSON 24

18. (Retailing) Use the circle graph.
(14)
  a. What is the probability that a randomly chosen person who purchased a shirt paid more than $40.00?
  b. What is the probability that a randomly chosen person who purchased a shirt paid $30 or less?

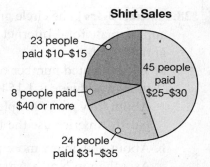

Shirt Sales
- 23 people paid $10–$15
- 45 people paid $25–$30
- 8 people paid $40 or more
- 24 people paid $31–$35

19. (Construction) To calculate the amount of fencing for a rectangular area, Kelvin
(12) uses the formula $P = 2(l + w)$. Bonnie uses the formula $P = 2(w + l)$. Will their calculations of the perimeter be the same? Explain.

20. Write Explain how to simplify $\frac{4}{7} \div \left[\left(-\frac{3}{8}\right) \cdot \left(-\frac{8}{3}\right)\right]$.
(11)

21. Simplify $\frac{2}{5} \div \left(-\frac{7}{2}\right) \cdot \left(-\frac{5}{2}\right)$.
(4)

22. Write Explain why $k^2 \cdot m \cdot b^4 \cdot c^3$ cannot be simplified using the Product Rule
(3) of Exponents.

23. Justify Write the expression so there are no parentheses. Justify your change
(15) with a property. $6(ab + ef)$

24. Generalize Some real numbers can contain patterns within them, such as
(1) .21.12122122212222…
  a. Find a pattern in the number above. Is the pattern you found a repeating pattern?
  b. Is this number a rational number or an irrational number? Explain.

25. Convert 630 cubic centimeters to cubic inches. (Hint: 1 in. = 2.54 cm)
(8)

26. Multi-Step The temperature at 6 a.m. was 30°C. If the temperature increases by
(10) 2 degrees every half hour, what will the temperature be by 9 a.m.? What time will it be when the temperature is 50°C?

27. Simplify $-|10 - 7|$.
(5)

Solving Decimal Equations — LESSON 24

***28.** (Internet Usage) The circle graph shows approximate total Internet usage in the world.

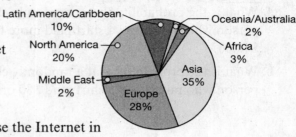

Internet Usage per Region

a. The estimated number of Internet users worldwide is 1,154,358,778. About how many people in North America use the Internet?

b. About how many more people use the Internet in Asia than in North America?

29. Geometry A small square park is 784 square yards. A row of trees was planted on one side of the park. One tree was planted at each corner. Then one tree was planted every seven yards between the corner trees. How many trees were planted in the row?

30. (Quality Control) Elite Style inspects 500 hair dryers manufactured and finds 495 to have no defects. There are 20,000 hair dryers in their warehouse.

a. What is the experimental probability that a hair dryer will have no defects?

b. Predict the number of hair dryers that will have no defects in the warehouse.

Name _____ Date _____ Class _____

Differentiating Between Relations and Functions

LESSON 25

1. Solve $0.3 + 0.05y = 0.65$.
(24)

2. Verify Verify that the following solutions are correct for each equation given.
(19)
 a. $103 + x = 99$ when $x = -4$

 b. $\frac{1}{2} - x = \frac{3}{4}$ when $x = -\frac{1}{4}$

*3. Make a table to determine whether $y = x + 2$ represents a function.
(25)

*4. (Hiking) A hiker can average 15 minutes per mile. Write a rule in function notation to describe the time it takes the hiker to walk m miles.
(25)

5. Subtract $3.16 - 1.01 - 0.11$.
(10)

*6. Multiple Choice Which set of ordered pairs represents a function?
(25)
 A $\{(1, 1); (2, 2); (3, 3); (4, 4)\}$

 B $\{(1, 0); (2, 1); (1, 3); (2, 4)\}$

 C $\{(1, 1); (1, 2); (1, 3); (1, 4)\}$

 D $\{(10, 1); (10, 2); (12, 3); (12, 4)\}$

*7. A square has a side length of s. Write a rule in function notation to represent the perimeter.
(25)

*8. Generalize If a set of ordered pairs is a function, are the ordered pairs also a relation? Explain.
(25)

*9. Analyze A student draws a circle on a coordinate plane. The center of the circle is at the origin. Is this circle a function or a relation? Explain.
(25)

*10. (Photography) A student is making a pinhole camera. What is the circumference of the pinhole in the box? Use 3.14 for π and round to the nearest hundredth.
(24)

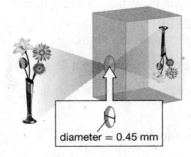
diameter = 0.45 mm

*11. (Astronomical Unit) An astronomical unit is the average distance from the Sun to the Earth. 1 AU (astronomical unit) is approximately equal to 93 million miles. If Jupiter is about 5.2 AU from the Sun, about how many miles is it from the Sun?
(8)

Saxon Algebra 1

Differentiating Between Relations and Functions

LESSON 25

12. Write Describe a possible situation for the discrete graph.
(Inv 2)

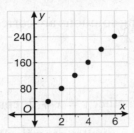

*13. (**Movie Club**) Stephen belongs to a movie club in which he pays an annual fee of $39.95
(23) and then rents DVDs for $0.99 each. In one year, Stephen spent $55.79. Write and solve an equation to find how many DVDs he rented.

14. Error Analysis Two students solve $5a + 4 = 34$. Which student is correct? Explain
(23) the error.

Student A	Student B
$5a + 4 = 34$	$5a + 4 = 34$
$\dfrac{5a + 4}{5} = \dfrac{34}{5}$	$ +4 +4$
$a + \dfrac{4}{5} = \dfrac{34}{5}$	$5a = 38$
$a = \dfrac{34}{5} - \dfrac{4}{5}$	$a = \dfrac{38}{5}$
$a = \dfrac{30}{5}$	
$a = 6$	

15. Multiple Choice The graph shows the points scored
(22) by Michaela and Jessie during the first five basketball games of the season. What conclusion can you make from the graph?

 A Michaela is the best player on the team.

 B Michaela usually scores more points than Jessie.

 C Neither player will score more than 18 points in the next game.

 D Jessie does not play as much as Michaela.

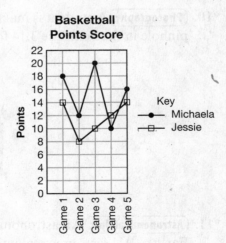

*16. Write Two sets of data represent the number of
(22) bottles of water and the number of bottles of juice a store sells each month. Give reasons why the following types of graphs would be appropriate to represent the data: a double-bar graph, a double-line graph, and two stem-and-leaf plots.

Differentiating Between Relations and Functions

LESSON 25

17. Choose an appropriate graph to display the portion of students in a class who have birthdays in each month. Explain your choice.
(22)

18. Find the value of $3z - 2(z - 1)^2 + 2$ for $z = 4$.
(9)

19. (**Cooking**) A recipe calls for 2.5 cups of orange juice for a batch of fruit drink. In the equation, $y = 2.5x$, y represents the number of cups of orange juice and x represents number of batches of fruit drink. Make a table when $x = 1, 2, 3$, and 4 batches of fruit drink. Then graph the ordered pair in your table.
(20)

20. Analyze "Three more than x" can be written as $x + 3$ or $3 + x$. Can "three less than x" be written as $x - 3$ or $3 - x$? Explain.
(17)

21. Multi-Step Population growth for a certain type of animal is determined by the formula $N_n = N_i 2^n$, where N_i is the initial population size and N_n is the population size after n generations. If the initial population is 45, what is the difference between the population size after the fourth generation and the population size after the sixth generation?
(16)

22. (**Architecture**) An architect is designing a very large square mall. Estimate the total area of the mall, if each side length is approximately 4890 feet.
(13)

23. Write True or False. The expression $12 - 8 - 2$ could be simplified using the Associative Property. Explain.
(12)

24. Solve $1\frac{1}{2}y = 6\frac{3}{4}$.
(21)

25. Solve $\frac{1}{8}m - \frac{1}{4} = \frac{3}{4}$.
(23)

26. Verify Determine if each statement below is true or false. If false, explain why.
(9)
a. $\dfrac{(5 - x)^3 + 12}{(4x)} \stackrel{?}{=} \dfrac{41 - x}{x^3}$ for $x = 2$.

b. $\dfrac{(5 - x)^3 + 12}{(4x)} = \dfrac{41 - x}{x^3}$ for $x = 3$.

27. Measurement The distance between City A and City C is 312.78 miles. City B lies on a point on a direct line between Cities A and C. If the distance between City C and City B is 191.9 miles, what is the distance between Cities A and B?
(6)

28. (**Computer Engineering**) Eight bits, or 2^3 bits, equal one byte. How many bits are in 64, or 2^6, bytes?
(3)

29. Geometry The measure of the length of a rectangle is $4x - y$ feet and the width is xy. What expression would show the area of the rectangle? Explain.
(15)

30. Simplify $|-2 - 3| - 4 + (-8)$.
(6)

Solving Multi-Step Equations

LESSON 26

*1. Solve for x in the equation $\frac{3}{4} + \frac{1}{2}x + 2 = 0$.

*2. **Multiple Choice** A vending machine will only accept quarters in change. What are the independent and dependent variables that describe the amount of money in change held by the vending machine?

A Independent variable: value of 1 quarter; dependent variable: number of quarters

B Independent variable: value of 1 quarter; dependent variable: value of the quarters

C Independent variable: value of the quarters; dependent variable: number of quarters

D Independent variable: number of quarters; dependent variable: value of the quarters

*3. **Multiple Choice** Which one of the expressions below can be simplified by combining like terms?
A $6(5x + 1)$
B $2x(3 + 8)$
C $7x + 5$
D $9x - 6y + 4$

4. A table shows temperature changes over a period of a week.
  a. Why would a circle graph inaccurately display the data?
  b. Which type of graph would best display the data?

*5. (**Digital Technology**) The average size for the memory storage of an mp3 player is 2 gigabytes (GB). The average size of an mp3 song is 5.5 megabytes (MB). About how many songs can you store on a 2-gigabyte player if the player requires 16 megabytes for its own use? (Hint: 1 gigabyte = 1024 megabytes)

*6. **Write** Describe two different methods for solving $12(x + 7) = 96$.

*7. **Justify** Solve $-5(3x - 7) + 11 = 1$. Justify each step with an operation or property.

*8. **Verify** Draw the graph of a function. Check to see if your graph is truly a function.

*9. Use the graph. Determine whether the relation is a function.

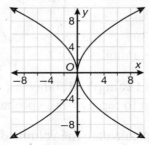

10. Solve $0.4m + 2.05 = 10.45$.

Solving Multi-Step Equations

LESSON 26

11. Error Analysis Two students solved $0.4x - 0.08 = 6.32$. Which student is correct? Explain the error.
(24)

Student A	Student B
$0.4x - 0.08 = 6.32$	$0.4x - 0.08 = 6.32$
$10(0.4x) - 100(0.08) = 100(6.32)$	$100(0.4x) - 100(0.08) = 100(6.32)$
$4x - 8 = 632$	$40x - 8 = 632$
$+8 +8$	$+8 +8$
$4x = 640$	$40x = 640$
$\dfrac{4x}{4} = \dfrac{640}{4}$	$\dfrac{40x}{40} = \dfrac{640}{40}$
$x = 160$	$x = 16$

12. Verify Is $x = 8$ a solution for $7x - 12 = 44$? Explain. If false, provide a correct solution and check.
(23)

*13. Multi-Step Emil cooks 64 hot dogs. He uses 5 packages of hot dogs plus 4 hot dogs left over from a meal earlier in the week. How many hot dogs are in each package?
(23)
  a. Write an equation to find the number of hot dogs in a package.
  b. Solve the equation, and then check the solution.

*14. (Kangaroos) A large kangaroo can travel 15 feet in each hop. Write and solve an equation to find how many hops it takes for the kangaroo to travel one mile. (Hint: 5,280 feet = 1 mile)
(21)

15. Verify Show that the graphed point is a solution to the equation $y = 2x + 9$.
(12)

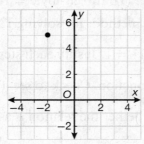

16. Analyze Determine whether $3p^2qd^3$ and $(2qdp \cdot -5d^2p)$ are terms that can be combined. Explain your reasoning.
(18)

17. Probability The probability of rain on Monday is a. It is twice as likely to rain on Tuesday. Write an expression to represent the probability of rain on Tuesday.
(17)

18. (Biology) A biologist wants to calculate the volume of a spherical cell. She uses the equation for the volume of a sphere, which is $V = \frac{4}{3}\pi r^3$. If the cell has a radius of 2 micrometers, what is its volume? Use 3.14 for π and round to the nearest tenth.
(16)

Solving Multi-Step Equations

LESSON 26

19. **(Employment)** Jim manages a restaurant that is currently hiring employees. On Tuesday, he interviewed 2 waiters, 2 line cooks, 3 dishwashers, and 1 chef. On Thursday, he interviewed 2 waiters, 1 line cook, 2 dishwashers, and 3 chefs. What is the probability that a randomly selected person interviewed applied to be a waiter?

20. **Verify** Compare the following expression using $<, >, =$. Verify your answer.

$\sqrt{324} - \sqrt{144} \bigcirc \sqrt{400} - \sqrt{289}$

21. Evaluate.

$\dfrac{6}{2}[5(3+4)]$

22. **Model** Use a number line to model $-8 - (-4) - (-6)$. Then simplify the expression.

23. Simplify.

$2 \cdot (3+4)^2 + 15$

24. Subtract $\dfrac{1}{4} - \dfrac{1}{3}$.

25. **Probability** The probability of rolling a 4 on a six-sided number cube is $\dfrac{1}{6}$. To find the probability of rolling a 6-sided number cube and getting a 4 five times in a row, multiply the probability $\dfrac{1}{6}$ by itself five times. Write the answer using an exponent.

26. **(Investing)** To find the amount of money earned on a bank deposit that earns quarterly compounded interest, the formula $A = P\left(1 + \dfrac{r}{4}\right)^{4t}$ is used.

P = principal, (the amount originally deposited)

r = the interest rate

t = time in years.

a. How many terms are in $P\left(1 + \dfrac{r}{4}\right)^{4t}$?

b. How many variables are in $P\left(1 + \dfrac{r}{4}\right)^{4t}$?

c. What is the coefficient of t?

27. Identify the coefficient, the variable(s), and the number of terms in $\dfrac{1}{3}Bh$.

28. **Multi-Step** The Noatak National Preserve in Alaska covers 6,574,481 acres. One acre is equal to 4840 square yards. What is the area of the preserve in square miles?
a. Find the area of the preserve in square yards.
b. Convert square yards to square miles. (Hint: 1 mile = 1760 yards)

29. **Geometry** To find the volume of a rectangular-prism shaped sunscreen bottle, Jagdeesh uses the formula $V = lwh$. Betty uses the formula $V = wlh$. Will the volume of the bottle be the same? Explain.

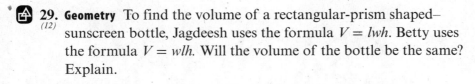

Solving Multi-Step Equations LESSON 26

30. The spinner in a board game is divided into four equal sections colored blue, red,
(Inv 1) green, and yellow. Conduct a simulation using random numbers to determine the
number of times the spinner lands on blue in 30 spins. Use the random number
generator in a graphing calculator to simulate the spins.

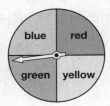

Identifying Misleading Representations of Data

LESSON 27

1. Simplify $(-2 + 3) \div (4 - 5 + 3)$.
   (11)

Solve.

2. $0.5x - 0.2 = 0.15$
   (23)

3. $\frac{1}{4} + \frac{2}{5}x + 1 = 2\frac{1}{4}$
   (26)

4. **Multiple Choice** Which is the solution to the equation below?
   (24)
   $-0.4n + 0.305 = 0.295$
   A 0.025 **B** -0.025 **C** 0.0004 **D** -0.7375

5. **Analyze** On a coordinate plane, a student draws a graph of two parallel lines
   (25) perpendicular to the y-axis. Does the graph represent a function?

6. Identify the property illustrated by $3 \cdot (9 \cdot 5) = (3 \cdot 9) \cdot 5$.
   (12)

*7. **Automotive Safety** The stopping distance d required by a moving vehicle
   (25) is dependent on the square of its speed s. Write a rule in function notation
   to represent this information.

*8. A petting zoo contains 10 species of animals. The graph shows
   (27) percentages of the 5 most numerous types of animals at the zoo.
   Give reasons why the circle graph may be misleading.

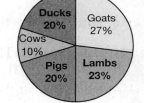

*9. **Justify** Is 4 a solution to the equation $5x + 8 - 3x + 4 = 20$?
   (26) Justify your answer.

*10. The bar graph shows results of a taste test
    (27) of four different brands of yogurt. True or
    False. Twice as many people preferred
    Brand A over Brand D.

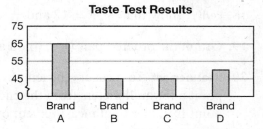

*11. **Analyze** Using the set of data values 125,000, 105,000, 162,000, 112,000, and 148,000
    (27) without using a broken axis or very large intervals, how could a student make a
    reasonably sized graph of the data?

Identifying Misleading Representations of Data

LESSON 27

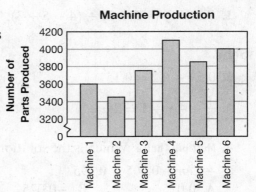

Machine Production

*12. **(Production)** A company has 6 machines to produce parts for its product. A manager uses the bar graph showing the number of parts produced by each machine each day. What incorrect conclusions might the manager make about the efficiency of the machines?
(27)

*13. True or False: Large intervals on a scale can make changes in data appear less than they actually are. If false, explain why.
(27)

*14. **Multi-Step** Three friends rented a kayak. It cost $4 per hour per person to rent the kayak, plus $2 for each life jacket, and $3 to park the car. It cost $57 in all. How many hours did they spend kayaking?
(26)

*15. **Geometry** The formula for the surface area of a square pyramid is $S = \left(4 \cdot \frac{1}{2}bh\right) + b^2$. If the measure of b is 5 m, what is the largest slant height possible for the total surface area to be no more than 150 m²?
(26)

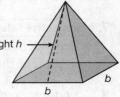

slant height h

16. **Justify** What is the first step in solving $0.35 + 0.22x = 1.67$?
(24)

*17. **(Phone Charges)** The length of the first ten calls Tyrese made one month were 13, 28, 6, 10, 13, 22, 31, 12, 2, and 9 minutes. In a stem-and-leaf plot of the data which digit would appear the most in the leaves column?
(22)

18. **Verify** Show that $-\frac{3}{4}x = 12$ and $\frac{5}{32}x = -2\frac{1}{2}$ have the same solution.
(21)

19. Graph the ordered pair $(-1, 0)$ on a coordinate plane.
(20)

20. **Measurement** To measure the length of a steel rod, an engineer uses a reference point a few millimeters from the end of the ruler. She then subtracts this reference point from her final measurement of 325 mm. If the rod's length is 318 mm, what reference point did she use?
(19)

21. Convert 37 American dollars to Indian rupees. (Hint: 1 rupee = $0.025)
(8)

22. **Statistics** Absolute deviation is the absolute value of the difference between a value in a data set and the mean of the data set. For the data set {8, 9, 11, 12, 15}, the mean is 11, so the absolute deviation for the value 15 is $|15 - 11| = |4| = 4$. What is the absolute deviation for each of the other numbers in the above data set?
(5)

23. **(Fundraising)** The cheerleaders made $3 profit on each item sold in a fundraiser. They sold x calendars and y candles in total. Write and simplify an algebraic expression to find the total profit.
(15)

Identifying Misleading Representations of Data

LESSON 27

24. **Write** A coin is tossed 8 times. What is the probability that the next time the coin is tossed the result will be heads? Explain.
(14)

25. Simplify $11 \cdot 3 + 7$.
(4)

26. **Multi-Step** A vending machine has q quarters and d dimes.
(9)
   a. Write an expression with variables to represent the value of the money.

   b. Find the value of the change in the machine if there are 21 quarters and 13 dimes.

27. Use $<$, $>$, or $=$ to compare the expressions. $\frac{1}{3} + \frac{1}{5} \cdot \frac{2}{15} \bigcirc \left(\frac{1}{3} + \frac{1}{5}\right) \cdot \frac{2}{15}$
(7)

28. **Write** A man runs up and down stairs. If the number of stairs he runs up plus the number of stairs he runs down is the total number of stairs, describe his position at the end of his run.
(6)

29. **Write** Show the steps for simplifying $10 \cdot 4^2 + 72 \div 2^3$.
(4)

30. **(Accounting)** Accountants prepare financial reports for businesses. Identify the set of numbers that best describes the numbers in a financial report. Explain your choice.
(1)

Name _____ Date _____ Class _____

Solving Equations with Variables on Both Sides
LESSON 28

1. Solve for y: $\frac{3}{4}y = 4\frac{7}{8}$.

*2. Solve for p: $3p - 4 - 6 = 2(p - 5)$.

*3. Formulate You have $3 in bills and a certain number of nickels in one pocket. In the other pocket you have $2 in bills and a certain number of dimes. You have the same number of dimes as nickels and the same amount of money in each pocket. Write an equation to find the number of dimes and nickels you have.

4. Error Analysis Two students used the Distributive Property to solve the same multi-step equation. Which student is correct? Explain the error.

Student A	Student B
$4x - 2(12 - x) = 18$	$4x - 2(12 - x) = 18$
$4x - 24 - x = 18$	$4x - 24 + 2x = 18$
$3x - 24 = 18$	$6x - 24 = 18$
$3x = 42$	$6x = 42$
$x = 14$	$x = 7$

*5. Wages A worker at one farm is paid $486 for the week, plus $0.03 for every pound of apples she picks. At another farm, a worker is paid $490 for the week, plus $0.02 for every pound of apples. For how many pounds of apples are the workers paid the same amount?

*6. Multiple Choice What is the value of x when $(x + 15)\frac{1}{3} = 2x - 1$?

A $\frac{18}{5}$ **B** $\frac{5}{18}$ **C** $\frac{18}{7}$ **D** $\frac{7}{18}$

*7. Error Analysis Two students solved the same multi-step equation. Which student is correct? Explain the error.

Student A	Student B
$3x - 4 = 2x - (4 + x)$	$3x - 4 = 2x - (4 + x)$
$3x - 4 = 2x - 4 + x$	$3x - 4 = 2x - 4 - x$
$3x - 4 = 3x - 4$	$3x - 4 = x - 4$
$0 = 0$	$2x = 0$
All real numbers.	$x = 0$

*8. Generalize If the equation $yx = zx$ is true, when yx is positive, $x \neq 0$, and z is a negative integer. Will x be positive or negative?

Solving Equations with Variables on Both Sides

LESSON 28

9. Geometry The graph shows areas of several square sheets of paper.

a. About how many times greater does the area of Sheet 4 appear to be than that of Sheet 3?

b. The squares have side lengths of 9, 10, 8, and 11 inches. About how many times greater is the area of Sheet 4 than the area of Sheet 3?

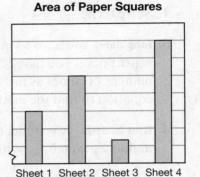

Area of Paper Squares

Sheet 1   Sheet 2   Sheet 3   Sheet 4

10. True or False: A broken scale can make changes in data appear less than they actually are. If false, explain why.

11. Multi-Step Average home prices in several cities are shown in the table.

City	Woodside	Reefville	Boynton	Dunston	York
Average Home Price (in thousands)	$265	$210	$320	$375	$350

a. Make a bar graph with a scale from 200 to 400. Use intervals of 40.

b. Without looking at the scale, what conclusions might the graph lead to?

c. Why might a real estate agent who sells houses only in Reefville want to show potential clients a graph like this?

*12. Multi-Step Use the graph.

a. Give the domain and range of the relation.

b. Determine whether the relation is a function. Explain.

*13. Multiple Choice Use the graph shown. A conservation group has been working to increase the population of a herd of Asian elephants. The graph shows the results of their efforts. Which relations represent the data in the graph?

A {(1, 4.5), (2, 6), (3, 10), (4, 14.5)}

B {(1, 5), (2, 6), (3, 10), (4, 15)}

C {(4.5, 1), (6, 2), (10, 3), (14.5, 4)}

D {(5, 1), (6, 2), (10, 3), (15, 4)}

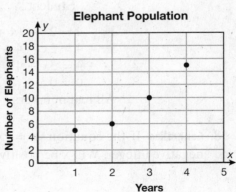

Elephant Population

14. 0.28 of what number is 18.2?

Solving Equations with Variables on Both Sides

LESSON 28

15. **(Internet Access)** At a local diner, customers can enjoy wireless Internet access.

    a. Write an equation that can be used to find the cost of being online for m minutes.

    b. **Estimate** You know it will require $1\frac{1}{2}$ to 2 hours to get your research done online. About how much will it cost to do your work at the diner?

16. Determine whether the statement is true or false. If false, explain why.
    A line graph can help analyze change over time.

17. Use the coordinate grid.
    Find the coordinates of point K.

    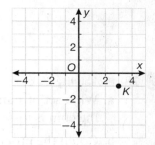

*18. Silvia had $247 in her savings account. She made a deposit into her savings account. Now, her account has $472. Write an equation that shows the amount of money that is currently in her account. Determine how much money was deposited.

19. **(Cooking)** George has already sliced 1 carrot and continues to slice 6 carrots per minute. Frank has already sliced 16 carrots and continues to slice 4 carrots per minute.
    a. Write expressions representing the number of carrots sliced by each person in m minutes.
    b. Write an expression for the total number of carrots sliced.

20. **(Grades)** A student raised her grade by 13 points. Write an expression to represent her new grade.

*21. **(Astronomy)** The gravitational force between two objects can be approximated by using $F = \frac{m_1 m_2}{d^2}$, where F is the gravitational force in newtons, m_1 is the mass in kilograms of the first object, m_2 is the mass in kilograms of the second object, and d is the distance between them expressed in meters. If the mass of a satellite is 500 kilograms, the mass of a small asteroid is 1500 kilograms, and the distance between them is 1000 meters, what is the gravitational force between the satellite and the asteroid?

22. **Verify** Use two different methods to evaluate $8(10 - 4)$. Verify that each method gives the same result.

23. Evaluate $\sqrt{49} + 4^2$.

Solving Equations with Variables on Both Sides

LESSON 28

24. True or False: $(b + c) + d = b + (c + d)$. Justify your answer.
(12)

25. Evaluate the expression $3n^2p^5 + 4(n - 8)^2$ for the given values $n = -3$ and $p = 1$.
(9)

26. Write Give a counterexample for the following statement: The set of irrational numbers is closed under subtraction.
(1)

27. Measurement Find the perimeter of the polygon.
(18)

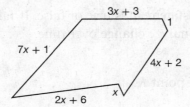

28. Probability A coin is tossed 3 times.
(1)
  a. What set of numbers could be used to express the different probabilities of how the coin is tossed?
  b. Could the probability ever be a whole number? Explain.

29. Write When a student converts from miles to feet, will the student multiply or divide? Explain.
(8)

*30. Write Describe a situation that could be represented by a continuous graph.
(Inv 2)

Name _____ Date _____ Class _____

Solving Literal Equations

LESSON 29

*1. Solve $3x + 2y = 5 - y$ for y.
(29)

*2. Solve $-2y + 6y - x - 4 = 0$ for y.
(29)

3. **Average Cost** Boris uses a coupon for $35 off any framing order of $50 or more. He wants to frame 5 photographs. With the coupon it will cost $107.50 not including tax. What is the average cost to frame a photograph without the coupon?
(24)

*4. **Multiple Choice** The floor area of recreation center will be a rectangle with a length of 130 feet and a width of 110 feet. Which formula can be used to find the area of the recreation center?
(29)
  A $A = s^2$
  B $A \cdot l = w$
  C $A = lw$
  D $A = \frac{1}{2}ab$

5. Give the domain and range of the relation.
(25)
{(9, 3); (8, 1); (8, 2); (8, 3); (7, 0); (7, 4); (6, 2)}

*6. **Multiple Choice** Which operation should be performed first when solving the multi-step equation $12x + 6(2x - 1) + 7 = 37$?
(26)
  A Divide both sides of the equation by 12.
  B Multiply $(2x - 1)$ by 6.
  C Add 1 to both sides of the equation.
  D Subtract 37 from both sides of the equation.

*7. **Write** Explain how to solve $5x + 4z = 10z - 2x$ for x.
(29)

*8. **Basketball** Lee's basketball team played 22 games this season. Lee scored an average of t three-pointers and s two-pointers per game during the season. Write an expression to show how many total points he scored during the season.
(17)

9. **Analyze** The measures of the angles in the two triangles can be found using the equations $x + x + 90 = 180$ and $3y + y + 90 = 180$. Which triangle will contain the smallest angle?
(26)

*10. **Error Analysis** Two students are making a bar graph of test scores ranging from 85 through 100. They want to emphasize the difference in the range of scores. Which student is correct? Explain the error.
(27)

Student A	Student B
Use a vertical scale from 75 to 100 in increments of 5.	Use a vertical scale from 0 through 100 in increments of 10.

11. **Analyze** The expression xy^3 has a positive value. What must be true of the value of x if y is negative?
(16)

Solving Literal Equations
LESSON 29

*12. What should you watch for when analyzing a circle graph?
(27)

13. Round $\sqrt{26}$ to the nearest integer.
(13)

*14. Geometry A triangle and a rectangle have the same area. Use the diagrams to find the area.
(28)

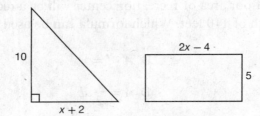

*15. Multi-Step A gym charges for attending exercise classes. Raquel pays $10 to attend each class because she is a non-member. Since Viola is a member, she pays $15 per month for a membership fee, but only $5 for each class.
(28)
  a. Write an expression to show how much it cost Raquel per month for c exercise classes.
  b. Write an expression to show how much it cost Viola per month for membership and c exercise classes.
  c. Set the two expressions equal to each other. Solve this equation to determine how many classes they would each have to attend to have the same cost.

16. Verify Is $x = 11$ a solution for $6x + 8 = 74$? Explain. If false, provide a correct solution and check.
(23)

17. Choose an appropriate graph to display the average grade received by students on a science test. Explain your choice.
(22)

18. Measurement To convert between feet and inches, use the equation $i = 12f$ where i is the number of inches and f is the number of feet. Complete the table and graph the solutions.
(20)

f	i
3	
5	
8	
10	

19. Verify True or False. A repeating decimal multiplied by a variable is an irrational number. If the statement is false, give a counterexample.
(2)

Solving Literal Equations — LESSON 29

20. (Astronomy) The temperature on the surface of Mars varies by 148°F. The highest temperature is about 23°F. What is that lowest temperature on the surface of Mars?
₍₆₎

21. (Consumer Economics) A strawberry container costs $1 and the strawberries cost $2 per pound. Write an expression to represent the total cost for a container with s pounds of strawberries.
₍₁₇₎

22. Simplify $x^2 - 3yx + 2yx^2 - 2xy + yx$.
₍₁₈₎

23. Solve $-3y + \frac{1}{2} = \frac{5}{7}$.
₍₂₃₎

24. Solve $k + 4 - 5(k + 2) = 3k - 2$.
₍₂₈₎

25. Generalize The value of $z + 2$ is an odd integer. What generalizations can be made about z using this information?
₍₉₎

26. Which expression is greater: $\frac{1}{3} - 1$ or $\frac{1}{2} - 1$.
₍₁₀₎

27. Justify Simplify $(3 + 5) - 2^3$. Justify each step using the order of operations or mathematical properties.
₍₇₎

28. Analyze Find the value of y when $x = 2$.
₍₆₎
$$-x - (-2) = y$$

29. (Population Growth) The expression $303{,}000{,}000 \times (1.015)^t$, where t stands for years, represents the population growth of the U.S.A. Based on this expression, about how many people will live in the U.S.A. eight years from now?
₍₃₎

30. Probability On the first street in Hidden Oaks subdivision, 5 out of 20 families own trucks.
_(Inv 1)
   a. What is the probability that a randomly selected family in the subdivision owns a truck?

   b. Predict the number of truck-owning families you can expect among the 140 families living in the subdivision.

Graphing Functions — LESSON 30

Solve.

1. (19) $x + \dfrac{1}{2} = 2\dfrac{1}{5}$

2. (23) $0.4x - 0.3 = -0.14$

3. (26) $\dfrac{1}{3} + \dfrac{5}{12}x - 2 = 6\dfrac{2}{3}$

4. (26) $\dfrac{2}{3} - \dfrac{4}{9}x + 1 = 2\dfrac{7}{9}$

*5. (28) Solve and check $x - 4(x - 3) + 7 = 6 - (x - 4)$.

*6. (25) **Verify** Is the statement below true or false? Explain.

The graph of a circle shows that the equation of the circle, $x^2 + y^2 = 1$, is a function.

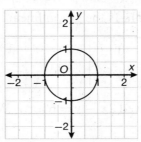

*7. (30) **(Savings)** For every dollar Mirand deposits into her checking account, she deposits 1.5 times as much into her savings account, which started with $50. So, $s = 1.5c + 50$, where s is the amount in savings and c is the amount deposited in checking. Which graph represents this equation?

Graph 1
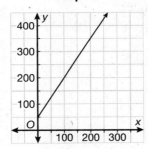

Graph 2
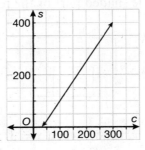

*8. (30) **Multiple Choice** Which equation represents the line on the graph?

A $y = x + 10$

B $y - x = 10$

C $-x = 10 + y$

D $y = -x + 10$

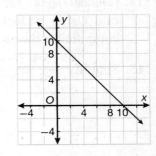

Graphing Functions
LESSON 30

***9. Multi-Step** The table shows the total number of shrubs a gardener plants after each half hour.
(30)

Time (hours)	0.5	1	1.5	2
Number of Shrubs	1	3	7	8

  a. Plot these data on a coordinate grid.

  b. Is the graph a function? Explain.

  c. Predict Can you predict the number of shrubs the gardener will plant in 3 hours? Why or why not?

*10. Use the table to make a graph. Is the graph linear? Explain.
(30)

x	0	1	2	3
y	4	7	10	13

11. Error Analysis Two students solved $2x - y = 6$ for y. Which student is correct? Explain the error.
(29)

Student A	Student B
$2x - y = 6$	$2x - y = 6$
$2x = y + 6$	$2x = y + 6$
$2x - 6 = y$	$2x + 6 = y$

12. Geometry What is the area of the shaded part of the rectangle?
(29)

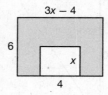

*13. Multi-Step Solve $\frac{x}{2} + \frac{y}{3} = 2$ for y. Find y when $x = 3$.
(29)

*14. Consumer Math Joel deposited money in an account that has a certain annual interest rate. Using the formula $i = prt$, or interest = principal · rate · time, how could he compute for the rate if the numeric value of the other items was given?
(29)

15. Simplify $\frac{3 + 7(-3)}{-7 - 2(-3)}$.
(11)

Graphing Functions

LESSON 30

16. Error Analysis Two students solved the same multi-step equation. Which student is correct? Explain the error.
(28)

Student A	Student B
$2x - 4(3x + 6) = -6(2x + 1) - 4$	$2x - 4(3x + 6) = -6(2x + 1) - 4$
$2x - 12x + 6 = -12x + 1 - 4$	$2x - 12x - 24 = -12x - 6 - 4$
$-10x + 6 = -12x - 3$	$-10x - 24 = -12x - 10$
$2x = -9$	$2x = 14$
$x = -4\frac{1}{2}$	$x = 7$

17. Measurement On a map, 1 centimeter represents 50 kilometers. The actual distance between two cities is 675 kilometers. Find the distance between the two cities on the map.
(21)

*18. Multiple Choice What would make the graph of basketball scores less misleading?
(27)
A Using a broken scale on the horizontal axis
B Using a broken scale on the vertical axis
C Using larger intervals
D Using smaller intervals

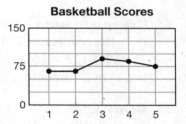

19. Generalize What effect do large intervals have on the appearance of a graph?
(27)

*20. Sales The circle graph shows the amounts of orange juice and fruit punch sold each month. Explain why this graph is misleading and determine what may be a more appropriate graph to compare the sales of the two beverages.
(27)

21. Model Make a stem-and-leaf plot of the following temperatures in Woodmont:
(22)
72°F, 74°F, 63°F, 62°F, 63°F, 78°F, 65°F, 51°F, 53°F, 53°F, 61°F, 80°F.

22. Hair Growth Hair grows approximately half an inch each month. John's hair is 2 inches long. Let m be the number of months. The formula $h = 2 + 0.5m$ calculates the length John's hair will be in m months if he does not cut it. Complete the table and graph the solutions.
(12)

m	h
4	
6	
10	
20	

Graphing Functions

LESSON 30

23. A television remote has a key for each of the channels 0 through 9. If one key is chosen at random, what is the chance that channel 5 is chosen? Write your answer as a percent.
(14)

24. Multi-Step Todd has 18 boxes of cards with x cards in each box. He divides them equally with 5 friends. The expression $18x \div 6$ represents the number of cards each person has. Simplify the expression and justify each step.
(12)

25. Justify Evaluate $5.2 - 1.6 + 4.08 + 8$. Justify each step.
(10)

26. Generalize In unit analysis, you often need to apply unit ratios multiple times to convert to the desired units. For example, $4518 \text{ cm}^2 \cdot \frac{1 \text{ m}}{100 \text{ cm}} \cdot \frac{1 \text{ m}}{100 \text{ cm}} = 0.4518 \text{ m}^2$ converts from square centimeters to square meters. State a general rule for applying a unit ratio the correct number of times to perform a unit conversion.
(8)

27. Justify Evaluate $3 + \left(\frac{5-2}{4} + 2^2\right)$. Justify each step.
(7)

28. (**Consumer Economics**) A gym charges \$2 a visit for the first 15 visits in a month. After that, the cost is reduced to $\frac{1}{4}$ of the price per visit. Use the expression $15 \cdot \$2 + (23 - 15) \cdot \frac{1}{4} \cdot \$2$ to show how much someone will pay if they go to this gym 23 times in a month.
(4)

29. Analyze Given the equations $a = (1.01)^x$ and $b = (0.99)^x$, which value, a or b, grows smaller as the exponent x grows larger?
(3)

30. Write Write a possible situation that could be represented by the graph at the right.
(Inv 2)

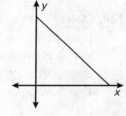

Name _____ Date _____ Class _____

Using Rates, Ratios, and Proportions
LESSON 31

Simplify.

1. $7 - 4 - 5 + 12 - 2 - |-2|$
(10)

2. $-6 \cdot 3 + |-3(-4 + 2^3)|$
(11)

Solve.

3. $-0.05n + 1.8 = 1.74$
(24)

4. $-y - 8 + 6y = -9 + 5y + 2$
(28)

*5. **Multiple Choice** What is the value of x when $2x - 4.5 = \frac{1}{2}(x + 3)$?
(28)

    A 9   **B** 2.4   **C** 2   **D** 4

6. Solve for y: $4 + 2x + 2y - 3 = 5$.
(29)

7. Simplify $4k(2c - a + 3m)$.
(15)

Evaluate.

8. $3x^2 + 2y$ when $x = -2$ and $y = 5$
(16)

9. $2(a^2 - b)^2 + 3a^3b$ when $a = -3$ and $b = 2$
(16)

*10. If 10 boxes of cereal sell for $42.50, what is the unit price?
(31)

*11. **Geometry** In the diagram, $\triangle ABC$ and $\triangle XYZ$ are similar triangles. What is the value of n?
(31)

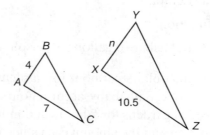

*12. **Predict** An estimate of the number of tagged foxes to the total number of foxes in a forest is 3:13. A forest warden recorded 21 tagged foxes. About how many foxes are in the forest?
(31)

13. **Multi-Step** A skydiver falls at a rate given by $s = 1.05\sqrt{w}$, where s is the falling speed in feet per second and w is the weight of the skydiver with gear in pounds. What is the approximate falling speed of a 170-pound man with 40 pounds of gear? (Round to the nearest whole number.)
(13)

Using Rates, Ratios, and Proportions

LESSON 31

14. Copy and complete the table for $y = x^2 + 2$. Then use the table to graph the equation.
(30)

x	−3	−1	0	1	3
y	11	3	2	3	11

*15. (**Shopping**) Glenn buys 4 computers for $2800. How much will 6 computers cost?
(31)

16. Probability A spinner is divided into 5 sections labeled *A* through *E*. The bar graph shows the results of 50 spins. What is the experimental probability that the next spin will land on *A* or *D*?
(22)

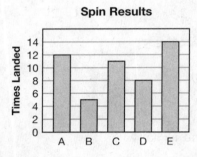

Spin Results

17. (**Mileage**) The table shows how far a car travels for each gallon of gasoline it uses.
(30)

Number of Gallons, x	1	2	3	4
Miles Traveled, f(x)	33	66	99	132

a. Use the table to make a graph.

b. Write a rule for the function.

c. How far will the car travel using 10 gallons of gasoline?

18. Multi-Step Students are paid *d* dollars per hour for gardening and *g* dollars per hour for babysitting and housework. Sally babysat for 6 hours and mowed lawns for 3 hours. Her brother weeded gardens for 5 hours and mopped floors for 1 hour.
(18)

a. Write an expression to represent the amount each student earned.

b. Write expressions for the total amount they earned together.

c. If they are paid $5 an hour for gardening and $4 an hour for babysitting and housework, how much did they earn together?

Using Rates, Ratios, and Proportions

LESSON 31

***19.** (Carpentry) A carpenter has propped a board up against a wall. The wall, board, and ground form a right triangle. What will be the measures of the three angles?
(26)

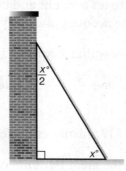

20. Give the domain and range of the relation.
(25)
$$\{(12, 2); (11, 10); (18, 0); (19, 1); (13, 4)\}$$

 ***21.** Use a graphing calculator to make a table of values for $f(x) = x^2 - 1$. Graph the function and determine the domain and range.
(30)

***22. Write** Why is there no conclusive value for x in the equation
(28) $\frac{3}{2}x + 5 = 2x - \frac{1}{2}x + 5$? Explain.

23. Multi-Step Amy works in a kitchen appliance store. She earns $65 daily and a
(17) commission worth $10 less than one-fifth of the value of each appliance she sells. Let m equal the value of an appliance.

   a. Write an expression for Amy's daily salary if she sells one appliance every day.

   b. Write an expression for Amy's daily salary if she sells n appliances every day.

   c. If you know the value of m and n, which part of the expression would you solve first?

24. Multiple-Choice Which expression is equivalent to $4(x^2 - 4) + 3z^3(4z^7)$?
(15)
   A $4x^2 - 16 + 3z^{10}$ **B** $4x^2 - 16 + 12z^{10}$

   C $4x^2 - 16 + 12z^{21}$ **D** $4x - 16 + 12z^{10}$

***25. Error Analysis** Two students solved $\frac{3z}{2} - \frac{4q}{3} = 6$ for z. Which student is correct?
(29) Explain the error.

Student A	Student B
$\frac{3z}{2} - \frac{4q}{3} = 6$	$\frac{3z}{2} - \frac{4q}{3} = 6$
$3(3z) - 2(4q) = 6$	$3(3z) - 2(4q) = 6(6)$
$9z - 8q = 6$	$9z - 8q = 36$
$9z = 6 + 8q$	$9z = 36 + 8q$
$z = \frac{6 + 8q}{9}$	$z = 4 + \frac{8}{9}q$

Using Rates, Ratios, and Proportions

LESSON 31

26. **(Finance)** Compound interest is calculated using the formula $A = P\left(1 + \frac{r}{1}\right)^t$, where P = principal (amount originally deposited), r = the interest rate, and t = time in years. If $1500 is deposited into an account and compounded annually at 5.5%, how much money will be in the account after 10 years?

*27. **Verify** Suppose that $\frac{3}{4} = \frac{x}{100}$. Show that x equals 75.

28. **Verify** Is $m = \frac{2}{3}$ a solution for $\frac{1}{3}m + \frac{5}{6} = \frac{11}{18}$? Explain. If false, provide a correct solution and check.

29. **Justify** Simplify $20 \cdot \$9 + 10 \cdot \$13$. Justify each step.

30. **(The Great Pyramid)** The base of the great Pyramid of Giza is almost a perfect square. The perimeter of the base measures about 916 meters. What is the length of each side of the pyramid's base?

Name _____ Date _____ Class _____

Simplifying and Evaluating Expressions with Integer and Zero Exponents
LESSON 32

Simplify.

*1. (32) $y^0 \dfrac{y^6}{y^5}$

*2. (32) $\dfrac{m^3 p^2 q^{10}}{m^{-2} p^4 q^{-6}}$

Solve.

3. (28) $9x - 2 = 2x + 12$

*4. (28) $3y - y + 2y - 5 = 7 - 2y + 5$

5. (28) $2y + 3 = 3(y + 7)$

6. (28) $5(r - 1) = 2(r - 4) - 6$

 *7. (32) **Geometry** Express the ratio of the area of the circle to the area of the square.

8. (17) The sum of twice a number and 17 is 55. Find the number.

*9. (31) **Error Analysis** Two students solved the proportion $\dfrac{3}{8} = \dfrac{x}{4}$. Which student is correct? Explain the error.

Student A	Student B
$\dfrac{3}{8} = \dfrac{x}{4}$	$\dfrac{3}{8} = \dfrac{x}{4}$
$3 \cdot x = 8 \cdot 4$	$8 \cdot x = 3 \cdot 4$
$x = 10\dfrac{2}{3}$	$x = 1\dfrac{1}{2}$

*10. (27) **(Health)** The circle graph shows the prevalence of all listed types of allergies among people who suffer from allergies. What about the graph may lead someone to an inaccurate conclusion?

Allergy Prevalence

- Eye 4%
- Insect 4%
- Latex 4%
- Food and Drug 6%
- Skin 7%
- Indoor and Outdoor 75%

11. (26) **Write** Why is it best to combine like terms in an equation, such as $3n + 9 - 2n = 6 - 2n + 12$, before attempting to isolate the variable?

12. (23) **Verify** Is $n = 9$ a solution for $-28 = -4n + 8$? Explain. If false, provide a correct solution and check.

© Saxon. All rights reserved.   Saxon Algebra 1

Simplifying and Evaluating Expressions with Integer and Zero Exponents

LESSON 32

13. The table lists the ordered pairs from a relation. Determine whether the relation represents a function. Explain why or why not
(25)

Domain (x)	Range (y)
1	5
0	6
2	4
1	8
3	3

14. If there are 60 dozen pencils in 12 cartons, how many are in 1 carton?
(31)

15. Multi-Step How many seconds are in 1 day?
(31)

16. (Roller Coasters) The table shows the number of roller coasters in several countries. Suppose one student displays the data in a bar graph, and another student makes a circle graph of the data. Compare the information that each type of display shows.
(22)

Roller Coasters Worldwide

Country	Japan	United Kingdom	Germany	France	China	South Korea	Canada	United States
Number	240	160	108	65	60	54	51	624

17. If there are 720 pencils in 6 cartons, how many dozen pencils are in 10 cartons?
(31)

18. Multi-Step How many centimeters are in 1 kilometer?
(31)

***19. (Geography)** On a map, Brownsville and Evanstown are 2.5 inches apart. The scale on the map is 1 inch:25 miles. How far apart are the two towns?
(31)

20. Copy and complete the table for $y = |x| + 10$. Then use the table to graph the equation. Is the graph of the equation a function?
(30)

x	−3	−2	−1	0	1	2
y						

***21. Multiple Choice** Which expression is simplified?
(32)

A $\dfrac{6xy^2}{z^0}$ B $\dfrac{6x^3y^{-2}}{z}$ C $\dfrac{6x^3y^2}{z}$ D $\dfrac{6x^3y^2z}{z}$

***22. (Chemistry)** An electron has a mass of 10^{-28} grams and a proton has a mass of 10^{-24} grams. How many times greater is the mass of a proton than the mass of an electron?
(32)

Simplifying and Evaluating Expressions with Integer and Zero Exponents

LESSON 32

23. Multi-Step A border is being built along two sides of a triangular garden. The third side is next to the house.
(17)

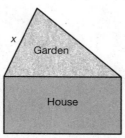

a. The second side of the garden is 4 feet longer than first side. Write an expression for the length of the second side.

b. If the total amount of border is 28 feet, how long are the sides of the garden that are not next to the house?

24. Multi-Step The temperature of a liquid is 72°F. The first step of a set of instructions requires that a scientist cools the liquid by 15°F. The second step requires that she warms it until it reaches 85°F. By how many degrees will she warm the liquid in the second step?
(19)

25. Analyze Megan and Molly have an age gap of 6 years. Megan is older. If Molly is 8 years old, then how old is Megan?
(29)

26. (Fuel Costs) It cost Rayna $73.25 to fill her truck with gas, not including tax. The gasoline tax is $0.32 per gallon. If the price for gasoline including tax is $3.25 per gallon, how many gallons of gas did she buy?
(24)
a. Write an equation to represent the problem.
b. How many gallons of gas did she buy?

27. Expand the expression $(5p - 2c)4xy$ by using the Distributive Property.
(15)

Solve each proportion.

28. $\dfrac{7}{x} = \dfrac{1}{0.5}$
(31)

*29. $\dfrac{1}{x} = \dfrac{-3}{x + 2}$
(31)

30. Multi-Step How far Sam bikes in two hours depends on the rate at which he rides. His distance is represented by the equation $y = 25x$, where x is the time in hours and y is the distance in miles.
(20)

a. Copy and complete the table and graph the solutions.

x	1	2	3	4	5
y					

b. Connect the points. What do you notice?

c. **Predict** If Sam rides at the same rate, how long will it take him to ride 80 miles?

Name _____ Date _____ Class _____

Finding the Probability of Independent and Dependent Events

LESSON 33

Solve.

1. (28) $-5v = 6v + 5 - v$
2. (26) $-3(b + 9) = -6$
3. (26) $-22 = -p - 12$
4. (26) $-\frac{2}{5} = -\frac{1}{3}m + \frac{3}{5}$
5. (31) $\frac{2}{x} = \frac{30}{-6}$
6. (31) $\frac{x-4}{6} = \frac{x+2}{12}$

Simplify.

7. (32) $\frac{y^6 x^5}{y^5 x^7}$
*8. (32) $\frac{w^{-5} z^{-3}}{w^{-3} z^2}$
9. (32) $\frac{4x^2 z^0}{2x^3 z}$

*10. (33) **Model** There are 10 little marbles and 4 big marbles in a bag. A big marble is drawn and not replaced. Draw a picture that represents how the contents of the bag change between the first draw and the second draw.

Before draw

*11. (33) **Write** Explain the difference between probability and odds.

*12. (33) True or False: Two rolls of a number cube are independent events.

*13. (1) Is the set of whole numbers closed under subtraction? Explain.

*14. (33) **Multiple Choice** A bag contains 3 blue stones, 5 red stones, and 2 white stones. What is the probability of picking a blue stone, keeping it, and then picking a white stone?

   A $\frac{3}{50}$ B $\frac{1}{15}$ C $\frac{3}{28}$ D $\frac{1}{2}$

15. (11) **Stock Market** The value of an investor's stock changed by $-1\frac{3}{4}$ points last week. This week the value changed by 3 times as much. How much did the value of the investor's stock change this week?

*16. (33) **Predict** What is the probability of rolling a 3 twice in a row on a six-sided number cube?

17. (27) **Write** Give an example of a situation in which someone may want to use large intervals on a graph to persuade people to come to a certain conclusion.

*18. (32) **Analyze** Simplify $x^3 \cdot x^{-3}$. What is the mathematical relationship between x^n and x^{-n}?

19. (32) **Time** A nanosecond is 10^{-9} times as fast as 1 second and a microsecond is 10^{-6} times as fast as 1 second. How much faster is the nanosecond than the microsecond?

Finding the Probability of Independent and Dependent Events

LESSON 33

20. Convert 30 quarts per mile to gallons per mile.
(31)

*21. Error Analysis Two students solved the proportion $\frac{5}{9} = \frac{c}{45}$. Which student is correct?
(31) Explain the error.

Student A	Student B
$\frac{5}{9} = \frac{c}{45}$	$\frac{5}{9} = \frac{c}{45}$
$9c = 225$	$9c = 225$
$c = 25$	$c = 2025$

*22. Vehicle Rental One moving company charges $19.85 plus $0.20 per mile to rent a van.
(28) The company also rents trucks for $24.95 plus $0.17 per mile. At how many miles is the price the same for renting the vehicles?

23. If a set of ordered pairs is not a relation, can the set still be a function? Explain.
(25)

24. Write Explain how to find the solution of $0.09n + 0.2 = 2.9$.
(24)

25. Keeping Cool The British thermal unit (BTU) is a unit of energy used globally in air
(21) conditioning industries. The number of BTUs needed to cool a room depends on the area of the room. To find the number of BTUs recommended for any size room, use the formula $B = 377lw$, where B is the number of BTUs, l is the length of the room, and w is the width of the room. The room you want to cool uses the recomended number of 12,252.5 BTUs and is 5 meters wide. Find the area of the room.

26. Multi-Step A quarterback throws the ball approximately 30 times per game. He has
(20) already thrown the ball 125 times this season. The equation $y = 30x + 125$ predicts how many times he will have thrown the ball after x more games.

x	y
2	
3	
4	
5	

a. Copy and complete the table using a graphing calculator and then graph the solutions.

b. When will he have thrown the ball more than 300 times?

27. Marathons A marathon is 26.2 miles long. In order to qualify for the Boston
(21) Marathon, Jill must first complete a different marathon within $3\frac{2}{3}$ hours. Her average speed in the last marathon she completed was 7.8 miles per hour. Did she qualify for the Boston Marathon? Explain.

Finding the Probability of Independent and Dependent Events

LESSON 33

28. Geometry The rectangles shown are similar.

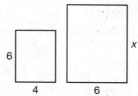

a. Find the ratio of the side lengths of the smaller rectangle to the larger rectangle.

b. Find the longer side length of the larger rectangle using proportions.

29. Jack is building a square pen for his dog. If he wants the area of the pen to be 144 square feet, how long should he make each side of the pen?

30. True or False: Whole numbers include negative numbers.

Recognizing and Extending Arithmetic Sequences

LESSON 34

Solve each proportion

1. $\frac{2}{10} = \frac{x}{-20}$
(31)

2. $\frac{32}{4} = \frac{x+4}{3}$
(31)

*3. (Construction) An amphitheater with tiered rows is being constructed. The first row
(34) will have 24 seats and each row after that will have an additional 2 seats. If there will be a total of 15 rows, how many seats will be in the last row?

*4. Use a graphing calculator to complete the table of values
(30) for the function $f(x) = 2x^2 - 5$. Graph the function.

x	y
−2	
−1	
0	
1	
2	

*5. Solve $y = x + \frac{z}{3}$ for z.
(29)

Solve each equation. Check your answer.

6. $4x + 2 = 5(x + 10)$
(28)

7. $2\left(n + \frac{1}{3}\right) = \frac{3}{2}n + 1 + \frac{1}{2}n - \frac{1}{3}$
(28)

8. A bead is drawn from a bag, kept, and then a second bead is drawn. Identify these
(33) events as independent or dependent.

*9. Justify Is the sequence 0.3, −0.5, −1.3, −2.1, ... an arithmetic sequence? Justify your
(34) answer.

*10. Write Explain why the sequence 2, 4, 8, 12, 16, ... is not an arithmetic sequence.
(34)

*11. Multiple Choice In the rule for the nth term of an arithmetic sequence
(34) $a_n = a_1 + (n - 1)d$, what does d represent?

    A the number of terms   **B** the first term

    C the nth term   **D** the common difference

*12. Is the sequence 7, 14, 21, 28, ... an arithmetic sequence? If it is, then find the
(34) common difference and the next two terms. If it is not, then find the next two terms.

13. Statistics A poll is taken and each person is asked two questions. The results are shown in the table. What is the probability that someone answered "yes" to both questions?
(33)

	Question 1	Question 2
Yes	55	30
No	45	70

14. Predict A number cube labeled 1–6 is rolled two times. What is the probability of
(33) rolling a 2 and then a 3?

Recognizing and Extending Arithmetic Sequences

LESSON 34

★15. Geometry A rectangle with perimeter 10 units has a length of 3 units and a width of 2
(34)   units. Additional rectangles are added as shown below.

P = 10 units

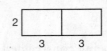
P = 16 units

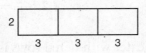
P = 22 units

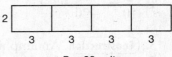
P = 28 units

   a. Write a rule for the perimeter of *n* rectangles.

   b. Use the rule to find the perimeter of 12 rectangles.

★16. Work uniforms include pants or a skirt, a shirt, and a tie or a vest. There are 3 pairs
(33) of pants, 5 skirts, 10 shirts, 2 ties, and 1 vest in a wardrobe.

   a. What is the probability of choosing a pair of pants, a shirt, and a tie?

   b. The pants and shirt from the previous day must be washed, but the tie returns to
      the wardrobe. What is the probability of choosing pants, a shirt, and a tie the next
      day?

17. Evaluate the expression $d = 6 \cdot \frac{1}{c^{-2}}$ for $c = 2$.
(32)

18. (**Physical Science**) The wavelengths of microwaves can range from 10^{-3} m to 10^{-1} m.
(32)   Express the range of wavelengths using positive exponents.

19. Multiple Choice Ms. Markelsden baked 36 cookies in 45 minutes. How many cookies
(31)   can she bake in 3 hours?

   A 45 cookies   **B** 81 cookies

   C 64 cookies   **D** 144 cookies

20. Does $y = x^2 + 2$ represent a function? Explain how you know.
(30)

21. (**Architecture**) A model of a building is 15 inches tall. In the scale drawing, 1 inch
(31)   represents 20 feet. How tall is the building?

★22. Generalize Given $\frac{2}{b} = \frac{1}{a}$, where *a* and *b* are positive numbers, write an equation
(31)   that shows how to find *a*.

23. The line graph at right shows the costs of
(27)   tuition at a university over the past 5 years.
       How might this graph be misleading?

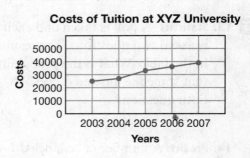

24. Write The equation $x + 5 = x - 5$ has
(28)   no solution. Explain why it has no solution.

25. Verify Solve $16x + 4(2x - 6) = 60$ for *x*.
(26)   Check your answer.

26. True or False: Any irrational number divided by an irrational number will be an
(1)   irrational number. Explain your answer.

120   Saxon Algebra 1

Recognizing and Extending Arithmetic Sequences

LESSON 34

27. (**Savings**) Hector has $400 in his savings account. Each week he deposits his $612.50 paycheck and takes out $250 to live on for the week. If he wants to buy a car for $5500, about how many months will it take him to save up for the car?
(23)

28. Multi-Step Jamal is riding his bike at a rate of about 8 miles per hour. How many hours will it take Jamal to ride 50 miles?
(31)

29. (**Car Rental**) A family rented a car that cost $45 per day plus $0.23 per mile. If the family rented the car for 7 days and paid $395.50 altogether, how many miles did they drive?
(24)

30. Error Analysis Two students determine whether the ordered pairs in the table represent a function. Which student is correct? Explain the error.
(25)

x	y
7	10
−2	−3
7	12
5	4

Student A	Student B
(7, 12) and (7, 10) The *x*-values are the same, so it is not a function.	All the *y*-values are different, so it is not a function.

© Saxon. All rights reserved.   Saxon Algebra 1

Locating and Using Intercepts

LESSON 35

Solve.

1. $\dfrac{-2.25}{x} = \dfrac{9}{6}$
   (31)

*2. $\dfrac{y+2}{y+7} = \dfrac{11}{31}$
   (31)

3. $2(f+3) + 4f = 6 + 6f$
   (28)

4. $3x + 7 - 2x = 4x + 10$
   (28)

Evaluate each expression for the given value of the variable.

5. $(m+6) \div (2-5)$ for $m = 9$
   (9)

6. $-3(x + 12 \cdot 2)$ for $x = -8$
   (9)

Simplify by combining like terms.

7. $10y^3 + 5y - 4y^3$
   (18)

8. $10xy^2 - 5x^2y + 3y^2x$
   (18)

9. Identify the subsets of real numbers to which the number $\sqrt{7}$ belongs.
   (1)

*10. Find the x- and y-intercepts for $5x + 10y = -20$.
   (35)

*11. Find the x- and y-intercepts for $-8x + 20y = 40$.
   (35)

*12. **Write** Explain how knowing the x- and y-intercepts is helpful in graphing a linear equation.
   (35)

*13. **Multiple Choice** What is the x-intercept for the equation $15x + 9y = 45$?
   (35)
       A (0, 3)   **B** (3, 0)   **C** (5, 0)   **D** (0, 5)

*14. (**Fishery**) A Pacific salmon can swim at a maximum speed of 8 miles per hour. The function $y = 8x$ describes how many miles y the fish swims in x hours. Graph the function. Use the graph to estimate the number of miles the fish swims in 2.5 hours.
   (35)

15. Determine if the sequence 34, 29, 24, 19, ... is an arithmetic sequence or not. If yes, find the common difference and the next two terms. If no, find the next two terms.
    (34)

16. **Error Analysis** Two students are finding the common difference for the arithmetic sequence 18, 15, 12, 9, ... Which student is correct? Explain the error.
    (34)

Student A	Student B
$18 - 15 = 3$	$15 - 18 = -3$
$15 - 12 = 3$	$12 - 15 = -3$
$12 - 9 = 3$	$9 - 12 = -3$

*17. **Geometry** A right triangle is formed by the origin and the x- and y-intercepts of $14x + 7y = 56$. Find the area of the triangle.
   (35)

Saxon Algebra 1

Locating and Using Intercepts
LESSON 35

 18. Data Analysis The table shows the weights of a
(34) newborn baby who was 7.5 lb at birth.
  a. Write a recursive formula for the baby's weight gain.
  b. If the pattern continues, how much will the baby weigh after 7 weeks?

Week Number	Weight (lb)
1	9
2	10.5
3	12
4	13.5

*19. **Multi-Step** Use the arithmetic sequence $-65, -72, -79, -86, \ldots$.
(34)
  a. What is the value of a_1?
  b. What is the common difference d?
  c. Write a rule for the nth term of the sequence.

 20. Write A coin is flipped and lands on heads. It is flipped again and lands on tails.
(33) Identify these events as independent or dependent.

21. (Economics) You have agreed to a babysitting job that will last 14 days. On the first day,
(34) you earn $25, but on each day after that you will earn $15. How much will you earn if you babysit for 7 days?

 22. Probability A bag holds 5 red marbles, 3 white marbles, and 2 green marbles.
(33) A marble is drawn, kept out, and then another marble is drawn. What is the probability of drawing two white marbles?

*23. **Multiple Choice** Which of the following expressions is the simplified solution
(32) of $\frac{m^3 n^{-10} p^5}{m n^0 p^{-2}}$?

  A $\frac{m^3 p^3}{n^{10}}$ B $\frac{m^3 p^7}{n^{10}}$ C $\frac{m^2 p^3}{n^9}$ D $\frac{m^2 p^7}{n^{10}}$

*24. **Verify** Is the statement $4^{-2} = -16$ correct? Explain your reasoning.
(32)

25. Convert 45 miles per hour to miles per minute.
(31)

26. Rewrite the following question so it is not biased: Would you rather buy a brand new
(Inv 3) luxury SUV or a cheap used car?

27. Identify the independent variable and the dependent variable: money earned, hours
(20) worked.

28. (Temperature) Use the formula $F = \frac{9}{5}C + 32$ to find an equivalent Fahrenheit
(23) temperature when the temperature is $-12°C$

Locating and Using Intercepts

29. (Homework) A student has to write a book report on a book that contains 1440 pages. Suppose she plans to read 32 pages per day. Using function notation, express how many pages remain after reading for *d* days.

30. (Soccer) For every hour a player practices soccer, he must drink 8 fluid ounces of liquid to stay hydrated. Write an equation describing this relation and determine whether it is a function.

Name _____ Date _____ Class _____

Writing and Solving Proportions — LESSON 36

Solve each proportion.

*1. $\dfrac{3}{4} = \dfrac{x}{100}$

*2. $\dfrac{5.5}{x} = \dfrac{1.375}{11}$

Simplify each expression.

3. $2^2 + 6(8 - 5) \div 2$

4. $\dfrac{(3 + 2)(4 + 3) + 5^2}{6 - 2^2}$

5. $\dfrac{14 - 8}{-2^2 + 1}$

6. The point (3, 5) is graphed in which quadrant of a coordinate plane?

7. True or False: The set of ordered pairs below defines a function.

$$\{(1, 3), (2, 3), (3, 3), (4, 3)\}$$

*8. The triangles at right are similar. Find the missing length.

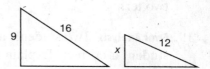

*9. **Multiple Choice** One triangle has side lengths 3, 5, and 6. A similar triangle has side lengths 18, 15, and 9. Which of the following ratios is the scale factor of the triangles?

A $\dfrac{1}{6}$ B $\dfrac{1}{3}$ C $\dfrac{1}{5}$ D $\dfrac{2}{3}$

*10. **Landscaping** A landscaping company needs to measure the height of a tree. The tree casts a shadow that is 6 feet long. A person who is 5 feet tall casts a shadow that is 2 feet long.
 a. Draw a picture to represent the problem.
 b. Use your picture to find the height of the tree.

*11. A real estate company sells small models of houses. The scale factor of the models to the actual houses is 0.5 ft:10 ft. Find the ratio of the areas of the model to the actual house.

12. **Entrepreneurship** A small child decided to sell his artwork. He sold black-and-white drawings for $2 and colored drawings for $3. The equation $2x + 3y = 24$ shows that he earned $24. Find the x- and y-intercepts.

*13. **Verify** A tree casts a shadow 14 feet long. A flagpole that is 20 feet tall casts a shadow 8 feet long. The triangle formed by the tree and its shadow is similar to the triangle formed by the flagpole and its shadow. Verify that the tree is 35 feet tall.

14. Find the x-intercept for $11x - 33y = 99$.

15. Find the y-intercept for $-7x - 8y = 56$.

Writing and Solving Proportions

LESSON 36

*16. Geometry A right triangle is formed by the origin and the *x*- and *y*-intercepts
(35) of the line $11x - 4y = 22$. Find the area of the triangle.

17. Multi-Step A car wash is held as a school fundraiser to earn $280 for a field trip. The
(35) charge is $7 for a car and $10 for an SUV. Let *x* be the number of cars and
y be the number of SUVs washed. The profits are calculated using the equation
$7x = -10y + 280$.
   a. Rewrite the profit equation in standard form.
   b. Calculate the *y*-intercept and explain its real-world meaning.
   c. Calculate the *x*-intercept and explain its real-world meaning.

18. Determine if the sequence 0.4, 0.1, −0.2, −0.5, … is an arithmetic sequence or
(34) not. If yes, find the common difference and the next two terms. If no, find the next
two terms.

*19. Error Analysis Two students are writing an example of an arithmetic sequence. Which
(34) student is correct? Explain the error.

Student A	Student B
5, 1, −3, −7, …	1, 4, 16, 44, …

*20. Verify At a raffle, 5 students' names are in a hat. There are 3 prizes in a bag: 2 books
(33) and a free lunch. Once the name and a prize are drawn, they are not replaced. After
giving out a book in the first drawing, a remaining student quickly calculates her
probability of winning a book in the second drawing as $\frac{1}{8}$. Show that she is correct.

21. (Astronomy) The force of gravity on the moon is about one-sixth of that on Earth.
(31) If an object on Earth weighs 200 pounds, about how much does it weigh on the moon?

22. Multiple Choice What are the odds against spinning a B on the
(33) spinner?
   A 1:2   **B** 1:3
   C 2:1   **D** 3:1

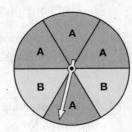

23. Multi-Step Steve has $300 in the bank. Each week he spends $10. Mario has $100 in
(18) the bank and deposits $5 each week.
   a. Write expressions to represent how many dollars each person has in the bank
after *w* weeks.
   b. Write an expression that represents how many dollars they have altogether.
   c. After 6 weeks, how much money do they have?

Writing and Solving Proportions

LESSON 36

24. Identify the independent variable and the dependent variable in the following statement: The fire was very large, so many firefighters were there..
(20)

25. **Multi-Step** The measure of angle B is three times the measure of angle A. The sum of the angle measures is 128°. Find the value of x.
(23)

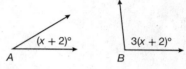

26. (**Piano Lessons**) A piano student has a $250 scholarship and an additional $422 saved for piano lessons. If each lesson costs $42, how many lessons will he have?
(23)

27. Simplify $\dfrac{4x^2z^0}{2x^3z}$.
(32)

28. (**Meteorology**) Meteorologists sometimes use a measure known as virtual temperature (T_v) in kelvins (K) to compare dry and moist air. It can be calculated as $T_v = T(1 + 0.61r)$, where T is the temperature of the air and r is the mixing ratio of water vapor and dry air. For temperatures of $T = 282.5$ K and $T_v = 285$ K, find the mixing ratio to the nearest thousandth.
(26)

29. **Measurement** For a science project, Joe must measure out 5 samples of a liquid. The graph shows the size of his samples. Why might the data require smaller intervals on the graph?
(27)

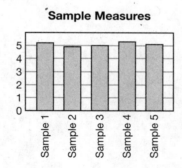

30. **Justify** Solve the equation $34 - 2(x + 17) = 23x - 15 - 3x$. Write out and justify each step.
(28)

Name _____ Date _____ Class _____

Using Scientific Notation

LESSON 37

Simplify each expression.

1. $18 \div 3^2 - 5 + 2$
2. $7^2 + 4^2 + 3$
3. $3[-2(8 - 13)]$

Simplify each expression by combining like terms.

4. $13b^2 + 5b - b^2$
5. $-3(8x + 4) + \frac{1}{2}(6x - 24)$

*6. Write 7.4×10^{-9} in standard notation.

*7. **Write** Explain how to recognize if a number is in scientific notation.

*8. **Write** Explain why anyone would want to use scientific notation.

*9. **Multiple Choice** What is $(3.4 \times 10^{10})(4.8 \times 10^5)$ in scientific notation?

    A 1.632×10^{15} **B** 1.632×10^{16} **C** 16.32×10^{15} **D** 16.32×10^{16}

*10. **(Physiology)** The diameter of a red blood cell is about 4×10^{-5} inches. Write this number in standard notation.

11. The triangles shown are similar. Find the missing length.

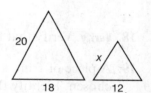

12. A student's final grade is determined by adding four test grades and dividing by 4. The student's first three test grades are 79, 88, and 94. What must the student make on the last test to get a final grade of 90?

*13. Graph $50x - 100y = 300$ using the x- and y-intercepts.

 14. **Geometry** A square has side lengths of 3 centimeters. Another square has side lengths of 6 centimeters.

    a. What is the scale factor of the sides of the smaller square to the larger square?

    b. What is the perimeter of each square?

    c. What is the ratio of the perimeter of the smaller square to the perimeter of the larger square?

    d. What is the area of each square?

    e. What is the ratio of the area of the smaller square to the area of the larger square?

Using Scientific Notation

LESSON 37

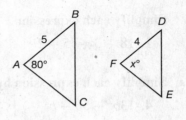

*15. Error Analysis In the figures at right, $\angle A$ and $\angle F$
(36) correspond. Two students are finding the measure of
$\angle F$. Which student is correct? Explain the error.

Student A	Student B
$\dfrac{5}{4} = \dfrac{80}{x}$	$m\angle A = m\angle F$
$5x = 320$	$80° = m\angle F$
$x = 64$	
$m\angle F = 64°$	

16. **Architecture** A room is 10 feet by 12 feet. If the scale of the blueprints to the room is
(36) 1 inch to 2 feet, find the dimensions of the room on the blueprints.

*17. Measurement What is the ratio of the area of the smaller circle to
(36) the area of the larger circle?

18. Verify Verify that the sequence 4, $2\tfrac{2}{3}$, $1\tfrac{1}{3}$, 0, … is an arithmetic sequence.
(34)

19. A piece of fruit is chosen from a box, eaten, and then a second piece of fruit is
(33) chosen. Identify these events as independent or dependent.

20. Predict An estimate of the number of tagged foxes to the total number of foxes in
(31) a forest is 3 to 13. A forest warden noted 21 tagged foxes during a trip in the forest.
Write a proportion to indicate the total number of foxes that might be in the forest.

21. Multiple Choice In the rule for the nth term of arithmetic sequence
(34) $a_n = a_1 + (n-1)d$, what does a_1 represent?

   A the number of terms   **B** the first term

   C the nth term   **D** the common difference

*22. **Physical Science** The wavelengths of ultraviolet light can range from 10^{-9} meter to
(32) 10^{-7} meter. Express the range of wavelengths using positive exponents.

23. Estimate Between which two whole numbers is the solution to $\dfrac{13}{14} = \dfrac{x}{10}$?
(31)

Using Scientific Notation
LESSON 37

24. **Dog Breeds** The table shows the number of dogs of the top five breeds registered with
(27) the American Kennel Club in 2006. Describe how the data could be displayed in a potentially misleading way.

Breed	Labrador Retriever	Yorkshire Terrier	German Shepherd	Golden Retriever	Beagle
Number	123,760	48,346	43,575	42,962	39,484

25. Choose an appropriate graph to display a survey showing what type of sport most
(22) people like. Explain your answer.

26. **Exercising** A weightlifter averages 2 minutes on each exercise. Each workout includes
(25) a 20-minute swim. Write a rule in function notation to describe the time it takes to complete w exercises and the swim.

27. **Estimate** Using the order of magnitude, estimate the value of 89,678 multiplied
(3) by 11,004,734.

28. **Justify** Solve for x: $7x + 9 = 2(4x + 2)$. Justify each step.
(29)

*29. An oceanographer wants to convert measurements that are above and below sea level
(30) from yards to feet. He takes measurements of depths and heights in yards and feet.

yards	−679	−125	32	79
feet	−2037	−375	96	237

   a. Formulate Use the table to write a formula to convert from yards to feet.

   b. Predict Use the formula to convert 27.5 yards to feet.

   c. Write a formula to convert yards to inches.

30. **Multi-Step** A rectangle has a perimeter of $38 + x$ centimeters. The rectangle has a
(28) length of $3x - 2$ centimeters and a width of x centimeters. What is the length of the rectangle?

   a. Substitute the dimensions of the rectangle into the perimeter formula $P = 2w + 2l$.

   b. Solve for x.

   c. Find the length of the rectangle.

Simplifying Expressions Using the GCF

LESSON 38

Solve each equation for the variable indicated.

1. $6 = hj + k$ for j
2. $\dfrac{a+3}{b} = c$ for a

Draw a graph that represents each situation.

3. A tomato plant grows taller at a steady pace.

4. A tomato plant grows at a slow pace, and then grows rapidly with more sun and water.

5. A tomato plant grows slowly at first, remains a constant height during a dry spell, and then grows rapidly with more sun and water.

Find each unit rate.

6. Thirty textbooks weigh 144 pounds.

7. Doug makes $43.45 in 5.5 hours.

8. Write 2×10^6 in standard notation.

9. Solve $\dfrac{p}{3} = \dfrac{18}{21}$.

*10. Find the prime factorization of 140.

*11. **Multiple Choice** Which of the following expressions is the correct simplification of $\dfrac{10x+5}{5}$?

   A $2x + 5$ B $2x + 1$ C $10x + 1$ D $5x$

*12. (**Free Fall**) The function $h = 40 - 16t^2$ can be used to find the height of an object as it falls to the ground after being dropped from 40 feet in the air. Rewrite the equation by factoring the right side.

*13. **Write** Explain how the Distributive Property and factoring a polynomial are related.

*14. **Generalize** Explain why the algebraic fraction $\dfrac{6(x-1)}{6}$ can be reduced, and why the fraction $\dfrac{6x-1}{6}$ cannot be reduced.

15. (**Biology**) The approximate diameter of a DNA helix is 0.000000002 meters. Write this number in scientific notation.

16. **Measurement** A nanosecond is one-billionth of a second. Write this number in scientific notation.

17. Write 78,000,000 in scientific notation.

*18. **Geometry** A square has side length 6.04×10^{-5} meters. What is its area?

*19. The triangles are similar. Find the missing length.

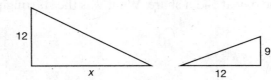

Simplifying Expressions Using the GCF

LESSON 38

20. Multi-Step An adult brain weighs about 3 pounds.
(37)
a. There are about 100 billion brain cells in the brain. Write this number in scientific notation.

b. Divide the weight of an adult brain by the number of cells and find how many pounds one brain cell weighs. Write the answer in scientific notation.

*21. Analyze Find the x- and y-intercepts for $y = 12x$ and explain how they relate to the graph of the equation.
(35)

22. (Fundraising) The math club has a carwash to raise money. Out of the first 40 vehicles, 22 are SUVs and 18 are cars. What are the odds against the next one being a car?
(33)

*23. Justify Explain why the statement $3^{-2} = -6$ is false.
(32)

24. Analysis A bookstore wants to show the number of different types of books that were sold on a given day. Why is this graph misleading?
(Inv 3)

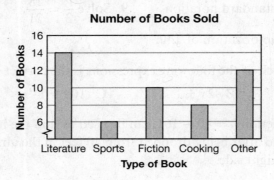

*25. Determine if the sequence $\frac{5}{4}, 2, \frac{11}{4}, \frac{7}{2}, \ldots$ is an arithmetic sequence. If yes, find the common difference and the next two terms.
(34)

26. Multiple Choice Which equation is in standard form?
(35)
A $y - 6 = 3(x + 4)$ **B** $y = -6x + 13$
C $10y = 12y + 25$ **D** $9x + 11y = 65$

27. How is the value 30 represented in a stem-and-leaf plot?
(22)

28. (Pool Charges) Barton Springs Pool charges $2 a visit plus a membership fee of $20.90 a month. Blue Danube Pool charges $2.95 a visit, with no membership fee. At what number of visits per month will the total fees for each pool be the same?
(28)

29. (Stock Market) On a day of heavy trading, one share of ABC Industries' stock originally decreased by $5 only to increase later in the day by twice the original value. The stock ended the day at $43 a share. What was the starting price of one share?
(26)

Simplifying Expressions Using the GCF

LESSON 38

30. Multi-Step The table shows the total number of shrubs a gardener planted after each half hour.
₍₃₀₎

Time (hr)	0.5	1	1.5	2
Number of Shrubs	1	3	7	8

a. Plot this data on a coordinate grid.

b. Determine if the graph is a function. Explain.

c. **Predict** Can you predict the number of shrubs the gardener will plant in 3 hours? Why or why not?

Name _____ Date _____ Class _____

Using the Distributive Property to Simplify Rational Expressions
LESSON 39

Solve each equation. Check your answer.

1. $4\left(y + \frac{3}{2}\right) = -18$
(26)

2. $x - 4 + 2x = 14$
(26)

3. True or False: The set of integers is closed under division. If false, give a counterexample.
(1)

Translate words into algebraic expressions.

4. the sum of a and 3
(17)

5. 2.5 more than k
(17)

6. 3 less than x
(17)

7. 2 more than the product of 3 and y
(17)

*8. Simplify $\frac{d^2}{s^2}\left(\frac{d^2}{s} + \frac{9s^3}{h}\right)$.
(39)

*9. **Write** Why isn't division by zero allowed?
(39)

*10. **Justify** Simplify $\frac{x^{-2}}{n^{-1}}(2x^{-4} + n^{-3})$ and explain each step.
(39)

*11. **Multiple Choice** Simplify $\frac{g^{-2}s}{b^2}\left(\frac{g^{-3}s^{-1}}{b^{-1}} + \frac{4}{b^3}\right)$.
(39)

A $\frac{4g^{-5}s^{-1}}{b^4}$

B $\frac{g^{-5}s^{-1}}{b} + \frac{4g^{-2}s}{b^5}$

C $\frac{g^{-5}}{b} + \frac{4g^{-2}s}{b^5}$

D $\frac{1}{bg^5} + \frac{4s}{b^5g^2}$

*12. Simplify the expression $\frac{w^2p}{t}\left(\frac{4}{w^4} - \frac{t^2}{p^5}\right)$.
(39)

*13. Find the prime factorization of 918.
(38)

*14. **Error Analysis** Two students factor the polynomial $16x^4y^2z + 28x^3y^4z^2 + 4x^3y^2z$ as shown below. Which student is correct? Explain the error.
(38)

Student A	Student B
$4x^3y^2z(4x + 7y^2z)$	$4x^3y^2z(4x + 7y^2z + 1)$

15. **Geometry** The area of a rectangle is represented by the polynomial $6a^2b + 15ab$. Find two factors that could be used to represent the length and width of the rectangle.
(38)

16. **Multiple Choice** Complete the following statement: The side lengths of similar figures _____.
(36)

A must be congruent

B cannot be congruent

C are in proportion

D must be whole numbers

Using the Distributive Property to Simplify Rational Expressions

LESSON 39

17. Multi-Step Use the expression $24x^2y^3 + 18xy^2 + 6xy$.
(38)
  a. What is the GCF of the polynomial?
  b. Use the GCF to factor the polynomial completely.

*18. Probability The probability that a point selected at random is in the shaded region of the figure is represented by the fraction $\frac{\text{area of shaded rectangle}}{\text{area of entire rectangle}}$. Find the probability. Write your answer in simplest form.
(38)

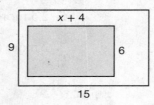

*19. Analyze In order to double the volume of water in a fish tank, is it necessary to double the length, width, and height of the tank? If yes, explain why. If no, explain how to double the volume of water.
(36)

*20. Graph $27x + 9y = 54$ using the x- and y-intercepts.
(35)

*21. (Fundraising) For a fundraiser, the science club sold posters for $5 and mugs for $8. The equation $5x + 8y = 480$ shows that they made $480. Find the x- and y-intercepts.
(35)

22. (Entertainment) A contestant is in the bonus round of a game show where she can win $1500 for answering the first question correctly and then an additional $500 for each correct response to each of the next five questions. If she answers all of the questions correctly, how much money will she receive when she answers the sixth question?
(34)

23. Write 0.00608 in scientific notation.
(37)

24. Error Analysis Two students write 1.32×10^{-5} in standard form. Which student is correct? Explain the error.
(37)

Student A	Student B
0.00000132	0.0000132

25. Verify Show that $4\frac{3}{4}$ is the solution to $\frac{1}{n-1} = \frac{4}{15}$.
(31)

26. Evaluate the expression $\frac{x^2y^{-2}}{z^2}$ if $x = 3$, $y = 4$, and $z = -2$.
(32)

27. Analyze The odds of winning a CD in a raffle are 3:7. Explain how to find the probability of not winning a CD.
(33)

Using the Distributive Property to Simplify Rational Expressions

LESSON 39

28. **(Stamp Collecting)** The table shows some collectible stamps with their estimated values. Explain whether the ordered pairs, such as (2, $2) and (2, $3), will be a function.

Number	Stamp	Value (low)	Value (high)
1	11¢ President Hayes (1931)	$2	$4
2	14¢ American Indian (1931)	$2	$3
3	4¢ President Taft (1930)	$1	$3
4	1¢ Benjamin Franklin (1911)	$5	$50

29. **(Salaries)** In an interview with a potential employee, an employer shows a line graph displaying the average salary of employees over several years. Explain why the graph is potentially misleading and why the employer might have shown this graph.

30. Formulate Write a rule for the table in function notation.

g	2	4	6	8	10
$f(g)$	1.5	2.5	3.5	4.5	5.5

Name _____ Date _____ Class _____

Simplifying and Evaluating Expressions Using the Power Property of Exponents
LESSON 40

Solve each proportion.

1. $\dfrac{3}{12} = \dfrac{-24}{m}$ (31)

2. $\dfrac{-4}{0.8} = \dfrac{2}{x-1}$ (31)

3. $\dfrac{5}{12} = \dfrac{1.25}{k}$ (31)

4. True or False: All whole numbers are integers. If false, give a counterexample. (1)

*5. Simplify $(4^4)^5$ using exponents. (40)

*6. **Multiple Choice** Which expression simplifies to $-24x^4y^3$? (40)
   A $(-2x^2y)^2(6y)$
   B $-2(x^2y)^2(6y)$
   C $-(2x^2y)^2(6y)$
   D $(-2xy)^3(3)$

7. Simplify $\dfrac{e^3}{r^5}\left(\dfrac{e^2}{4r} + \dfrac{r^9}{k}\right)$. (39)

*8. (**Cooking**) Use the formula $A = \pi r^2$ for the area of a circle. A 6-inch pizza covers an area of $\pi(6)^2 = 36\pi$ square inches. What happens to the area of the pizza if you double the radius and make a 12-inch pizza? (40)

*9. **Verify** Is the statement $(a + b)^n = a^n + b^n$ true? Verify your answer with a numeric example. (40)

*10. **Generalize** When do you know to add exponents and when to multiply exponents? (40)

11. (**Painting**) A rectangular top on a bench is to be painted. Its area is $\dfrac{wd^{-3}}{c}\left(\dfrac{d}{w^{-4}} + \dfrac{c^{-2}}{wd}\right)$. Simplify. (39)

12. Simplify $\dfrac{a^2}{d^2}\left(\dfrac{a^{-2}x}{d^{-1}} - \dfrac{2x}{d^{-3}}\right)$. (39)

*13. **Geometry** The equation of an ellipse is $\dfrac{wx^2}{g^2} + \dfrac{gy^2}{w^2} = 1$. To enlarge the ellipse, the left side is multiplied by $\dfrac{g^5}{w^{-2}}$. This expression is $\dfrac{g^5}{w^{-2}}\left(\dfrac{wx^2}{g^2} + \dfrac{gy^2}{w^2}\right)$. Simplify. (39)

14. The trim around a window has a total length of $\dfrac{rt}{w^3}\left(\dfrac{rty}{w} + 2ty - \dfrac{8}{w^2}\right)$. (39)
   a. Simplify the expression.
   b. Identify the variables that cannot equal zero.

*15. Find the GCF of $4xy^2z^4 - 2x^2y^3z^2 + 6x^3y^4z$. (38)

Simplifying and Evaluating Expressions Using the Power Property of Exponents

LESSON 40

16. Error Analysis Two students are simplifying the fraction $\frac{3x-6}{9}$ as shown below.
(38) Which student is correct? Explain the error.

Student A	Student B
$\frac{3x-6}{9} = \frac{\cancel{3}(x-2)}{\cancel{9}_3} = \frac{x-2}{3}$	$\frac{\cancel{3}x-6}{\cancel{9}_3} = \frac{x-\cancel{6}^2}{\cancel{3}} = x-2$

*17. (Shipping) A shipping container is in the shape of a rectangular box that has
(38) a length of $10x + 15$ units, a width of $5x$ units, and a height of 2 units.
  a. Write an expression that can be used to find the volume of the box.
  b. Factor the expression completely.

18. 0.78 of 250 is what number?
(24)

19. Give the domain and range of $\{(4, 9); (4, 7); (2, 4); (5, 12); (9, 4)\}$.
(25)

20. The heights of 8 trees were 250, 190, 225, 205, 180, 240, 210, and 220 feet. How
(27) could a misleading graph make you think the trees are all very similar in size?
  a. Make a bar graph of the data using a broken axis.
  b. Make a bar graph of the data using large increments.
  c. Compare the two graphs.

21. Justify Without changing the number to standard form, explain how you can tell that
(37) $-10 < 1 \times 10^{-4}$.

*22. Multiple Choice What is $\frac{1.6 \times 10^7}{6.4 \times 10^2}$ in scientific notation?
(37)
   A 2.5×10^4 **B** 0.25×10^5 **C** 2.5×10^6 **D** 4×10^5

23. (Astronomy) The diameter of the moon is approximately 3,480,000 meters.
(37) Write this distance in scientific notation.

24. The rectangles below are similar. Find the missing length.
(36)

*25. (Drama) The cost of presenting a play was $110. Each ticket was sold for $5.50. The
(35) equation $11x - 2y = 110$ shows how much money was made after ticket sales. Graph
this equation using the intercepts.

26. Justify Is the sequence 0.2, 2, 20, 200, ... an arithmetic sequence? Justify your answer.
(34)

27. There are 2 yellow stickers and 4 purple stickers. Make a tree diagram showing all
(33) possible outcomes of drawing two stickers. How many possible ways are there to
draw a purple sticker, keep it, and then draw another purple sticker?

Simplifying and Evaluating Expressions Using the Power Property of Exponents

LESSON 40

28. In a stem-and-leaf plot, which digit of the number 65 would be a leaf?
(22)

29. Measurement How many inches are there in 18 yards?
(31)

30. Analyze The rule for negative exponents states that for every nonzero number x, $x^{-n} = \frac{1}{x^n}$. Explain why the base, x, cannot be zero.
(32)

Name _____ Date _____ Class _____

Finding Rates of Change and Slope — LESSON 41

Solve each equation. Check your answer.

1. (26) $-2(b + 5) = -6$
2. (26) $4(y + 1) = -8$
3. (26) $\dfrac{5}{8} = 2m + \dfrac{3}{8}$

Find the slope of the line.

*4. (41) line a
*5. (41) line b
*6. (41) line c
*7. (41) line d

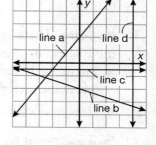

8. (37) Write 110,400 in scientific notation.

*9. (38) **Multiple Choice** Which of the following expressions is the GCF of $45a^3b^4c^2 + 30a^2bc^3$?

   A $15a^2bc^2$
   B $5a^2bc^2$
   C $3a^3b^4c^3$
   D $15a^3b^4c^3$

10. (38) **Verify** Is $2x(5x^3 + 6x^2 - 3x)$ completely factored? Explain why or why not. If no, factor the polynomial completely.

*11. (41) **Finances** The graph shows the amount of money Siobhan has in her bank account over time. What is the rate of change in her balance over time?

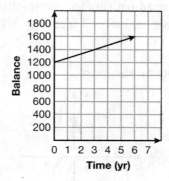

*12. (41) **Analyze** At some point on Mindy's mountain climbing trip, a graph relating her distance up the mountain over her climbing time shows a line with a negative slope. Why might this be?

13. (40) Simplify $(b^3)^5$ using exponents.

14. (40) **Error Analysis** Two students simplify the expression $(-2x^7)^5$ as shown below. Which student is correct? Explain the error.

Student A	Student B
$(-2x^7)^5 = -10x^{35}$	$(-2x^7)^5 = -32x^{35}$

Saxon Algebra 1

Finding Rates of Change and Slope

LESSON 41

***15. Geometry** Use the formula $A = s^2$ to find the area of the square.
₍₄₀₎

16. Multi-Step There are 10 millimeters in a centimeter. Therefore, there are 10^3 cubic
₍₄₀₎ millimeters in one cubic centimeter.
   a. There are 10^2 centimeters in a meter. How many cubic centimeters are in one cubic meter?
   b. How many cubic millimeters are in one cubic meter?

17. (Packing) The sides of a box in the shape of a cube are $5ab$ inches long. What is the
₍₄₀₎ volume of the box?

Simplify.

18. $\dfrac{fr}{d^3}\left(\dfrac{fsr}{d^2} + 3fs - \dfrac{8}{d}\right)$
₍₃₉₎

*19. $\dfrac{rt^{-2}}{g^{-3}h}\left(\dfrac{tg^4}{r^3h^{-2}} - \dfrac{r^3h}{g^{-2}r^{-2}}\right)$
₍₃₉₎

*20. (Woodworking) You are making furniture for a miniature dollhouse with a scale of
₍₃₆₎ 1 in.:12 in. If an actual chair measures 3 feet high, what will be the height of the chair in the dollhouse?

21. Find the x- and y-intercepts on the graph.
₍₃₅₎

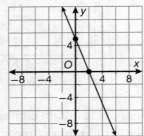

22. Measurement The formula $F = \frac{9}{5}C + 32$ calculates the number of degrees Fahrenheit
₍₃₅₎ from the degrees Celcius. Write this formula in standard form.

23. Write a recursive formula for the arithmetic sequence 8, 19, 30, 41,…. Then find
₍₃₄₎ the next two terms of the sequence.

24. Write Explain the difference between dependent and independent events.
₍₃₃₎

25. (Cooking) To make two batches of cookies, Ralph needs $\frac{3}{4}$ cup molasses. How many
₍₃₁₎ cups would he need to make 6 batches of cookies?

Finding Rates of Change and Slope

LESSON 41

26. (**Painting**) It will cost $2.50 per square foot to paint the walls of a storage shed. Write a rule in function notation to describe the cost of painting x square feet of outside wall space.

27. The two parabolas have the same shape but different orientations. Determine whether each parabola is a function.

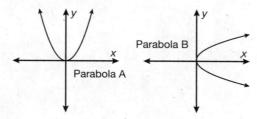

28. Multi-Step A physicist wants to find the kinetic energy of a ball that has a mass of 2.5 kg and travels a distance of 5.8 meters in 2.5 seconds. The formula for kinetic energy is $E_K = \frac{1}{2}mv^2$, where m is the mass of the object and v is the object's velocity. The formula for velocity is $v = \frac{d}{t}$, where d is distance traveled and t is the amount of time. What are the velocity and the kinetic energy of the ball?

29. Multi-Step Two cars travel the same distance. Car 1 travels at 50 mph and Car 2 at 65 mph. If it takes Car 1 one more hour to travel than Car 2, how far did the cars travel? Remember, $d = rt$.
a. Write an expression to represent the distance Car 2 travels.
b. Write an expression to represent the distance Car 1 travels.
c. Use these expressions to find the time Car 2 traveled.
d. Use this information to determine the distance traveled.

30. Multi-Step The length of a rectangle is 5 inches less than 3 times the width. The perimeter of the rectangle is 14 inches. Find the length and width of the rectangle.

Name _____ Date _____ Class _____

Solving Percent Problems — LESSON 42

Solve each equation. Check your answer.

1. (29) $5d - 8 = 3 + 7d$

2. (29) $9 + 2.7t = -4.8t - 6$

Solve each equation for the indicated variable.

3. (29) $V = \frac{1}{3}lwh$ for w

4. (29) $d = rt$ for t

5. (13) $\sqrt{42}$ is between which two integers?

6. (42) What is 18% of 340?

*7. (42) What is 270% of 93?

8. (40) Simplify $(6mn^3)^2$.

*9. (42) **Multiple Choice** What is 54% of 1200?
   A 64.8   **B** 600   **C** 648   **D** 1254

*10. (42) **(Gas Prices)** Shoshannah researched gasoline prices and found that gas was 224% more expensive than it had been during the same week 5 years previously, when the average price of regular gasoline was $1.36 per gallon. What was the new price?

*11. (42) **Write** Explain how fractions, percentages, and decimals are all related.

12. (22) **Multi-Step** The graph shows the number of pitches thrown in each inning of a baseball game.

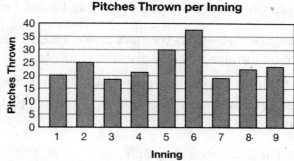

a. Make a stem-and-leaf plot of the data.

b. Find the greatest and least number of pitches thrown during the game.

13. (24) **(Wages)** Maria earned $10.50 per hour working at an ice cream shop. She earned $147 each week before taxes. Find the number of hours she worked each week.

14. (32) **(Computers)** One megabyte holds 10^6 pieces of information and one terabyte holds 10^{12} pieces of information. A terabyte holds how many times more information than a megabyte?

15. (29) **Verify** For $6x + \frac{1}{4}(y) = 44$, is $y = 88 - 24x$? If the soulution for y is incorrect, give the correct solution for y.

Solving Percent Problems

LESSON 42

16. **(Pricing)** A baker sells 2 pies for $5, 4 pies for $10, and 6 pies for $15. What is the price
    (30) of 14 pies?
    a. Create a table comparing the number of pies sold to the total cost.
    b. Illustrate the data using a graph. Does the graph show a function?
    c. Determine a rule for the function.
    d. Graph the function using a graphing calculator and determine the price of 14 pies.

17. How could a company use a line graph to make it appear as though they had a
    (27) larger profit?

18. **Multi-Step** A family plans to take a road trip. They leave Town A, go to Town B, and
    (33) then end at Town C. On the way back, they decide to travel different roads for a change of scenery.

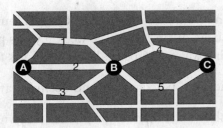

   a. What is the probability they took Road 1 and Road 4 on the way to Town C?
   b. What is the probability they will take Road 2 and Road 5 on the way back?

19. **Generalize** During the past 4 years you have grown 3 inches per year.
    (34)
    a. Write a recursive formula that describes the sequence.
    b. If you were 52 inches tall 4 years ago, how tall are you now?
    c. Would you expect this pattern of growth to continue for a long time? Explain your reasoning.

*20. Find the x- and y-intercepts on the graph.
    (35)

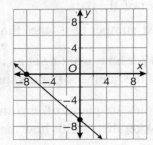

*21. **Generalize** A gerbil cage measures 24 inches by 10 inches by 18 inches. If you double
    (36) the lengths of each side of the cage, how will this increase affect the volume of the cage?

Solving Percent Problems

LESSON 42

*22. Find the GCF of $14p^5qr^2 - 28p^2q^2r^3$.
(38)

23. **Verify** Show that $\frac{r^{-2}}{s^{-3}}\left(\frac{rs^{-2}}{sr^{-1}} - \frac{s^{-3}r^{-1}}{r^{-3}}\right) = 0$.
(39)

24. **Earth Science** A grain of sand is approximately 0.002 meters in diameter. Write this number in scientific notation.
(37)

*25. **Multiple Choice** In the expression $\frac{pg^3}{s}\left(\frac{g^2p^3}{s^4} + \frac{9p}{s^2} - n\right)$, which variable cannot equal 0?
(39)
  A p **B** g **C** s **D** n

26. **Fuel Efficiency** The graph represents the amount of gasoline in a car's tank and the number of miles driven. What is the rate of change?
(41)

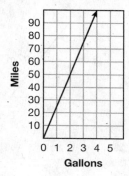

27. **Error Analysis** Two students simplify the expression $(3^4)^2$ as shown below. Which student is correct? Explain the error.
(40)

Student A	Student B
$(3^4)^2 = 3^{4 \cdot 2} = 3^8$	$(3^4)^2 = 9^{4 \cdot 2} = 9^8$

*28. Find the rate of change from the graph.
(41)

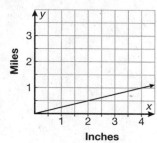

Solving Percent Problems

LESSON 42

*29. Multi-Step A mountain resort charges a set fee for cabin rentals as well as for each day
(41) of snowboard rentals.

Days of Cabin and Snowboard Rentals	1	2	3	4
Cost	$405	$427	$449	$471

   a. What is the rate of change in the rental costs?

   b. What will the mountain resort charge for 7 days?

30. Geometry A right triangle has side lengths 6 cm, 8 cm, and 10 cm. Use the shorter leg
(41) as the base of the triangle and find the slope of the hypotenuse. Then use the longer
leg as the base of the triangle and find the slope of the hypotenuse. Assume the
slopes are positive.

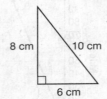

Name _____ Date _____ Class _____

Simplifying Rational Expressions   LESSON 43

Evaluate each function for the given input value.

1. $f(x) = -2x$ for $x = -5$
(30)

2. $h(x) = 3x - 1$ for $x = 7$
(30)

Find the indicated term of each arithmetic sequence.

3. $a_n = 16 + (n-1)(-0.5)$, 15th term
(34)

4. $-8, -6, -4, -2, \ldots$, 100th term
(34)

Identify the values that make the rational expressions below undefined.

*5. $\dfrac{25}{x+10}$
(43)

*6. $\dfrac{12 + 3x}{5 - x}$
(43)

7. What is 14% of 120?
(42)

8. What is 75% of 60?
(42)

9. A school needs to raise $2700. A nearby store promises to donate twice the amount the students raise. How much money do the students need to raise?
(23)

10. Ann-Marie has $2.55 in nickels and dimes. She has $1.80 in dimes. How many nickels does Ann-Marie have?
(24)
    a. Write an equation to represent the situation.
    b. How many nickels does she have?

11. **Error Analysis** Two students use the diagram to determine whether the relation is a function. Which student is correct? Explain the error.
(25)

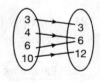

Student A	Student B
{(3, 3), (4, 6), (6, 6), (10, 12)} The y-values are not all different; y equals 6 twice. So, it is not a function.	{(3, 3), (4, 6), (6, 6), (10, 12)} All the x-values are different, so it is a function.

12. **Write** Explain how you can use a graph to tell whether a relation is a function.
(30)

13. (**Geography**) A map shows a 2.5-inch distance between Brownsville and Evanstown. The scale on the map is 1 inch:25 miles. How far apart are the two towns?
(31)

*14. **Multiple Choice** Which of the following rational expressions is undefined at $x = -6$?
(43)

A $\dfrac{x-6}{12x + 72}$

B $\dfrac{x}{2(x+12)}$

C $\dfrac{x+6}{72 - 12x}$

D $\dfrac{2x + 12}{x}$

© Saxon. All rights reserved.   155   Saxon Algebra 1

Simplifying Rational Expressions

LESSON 43

***15.** **(Digital Signal Processing)** In digital signal processing, electronic signals are often represented as rational expressions. The rational expression $\frac{3z^2 + 2.7z}{(z + 0.9)(z - 0.9)}$ models a digital signal. What is the simplified form of this signal representation?
(43)

*16. **Analyze** Given the rational expression $\frac{8x}{2x + 16}$, for what values of x would its reciprocal be undefined?
(43)

17. **(Phone Numbers)** A student forgets the last two digits of her friend's phone number. She does remember that the digits are different. What is the probability of guessing the last two digits correctly on the first try?
(33)

18. **Multi-Step** In an arithmetic sequence, $a_1 = 17$ and $d = 10$.
(34)
  a. Write a rule for the nth term of the sequence.
  b. Use the rule to find the fourth and eleventh terms of the sequence.

19. **Verify** Verify that $5x + 6y = -12$ is the standard form of the equation $y = -\frac{5}{6}x - 2$.
(35)

20. Name the corresponding sides and angles of the two similar triangles shown at right.
(36)

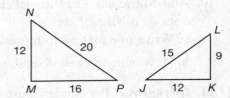

*21. Multiply $(1.6 \times 10^{-5})(2.2 \times 10^3)$ and write the answer in scientific notation.
(37)

22. **(Packaging)** The bottom of a rectangular box has a length of $6x + 4$ inches and a width of $8x$ inches.
(38)
  a. Write an expression that can be used to find the area of the bottom of the box.
  b. Factor the expression completely.

*23. **Multiple Choice** Simplify $(10g^3h^{-4})^2 (3gh^6)^3$.
(40)
  A $2700g^9h^{10}$ **B** $180g^9h^{10}$ **C** $180g^5h^7$ **D** $2700g^{18}h^{16}$

24. **Formulate** Multiply $(8x^3)(2x)^{-3}$. What is true about $(8x^3)$ and $(2x)^{-3}$? (Hint: Simplify the expression.)
(40)

25. Simplify $\frac{k}{g}\left(\frac{rtw}{nk} - 5k^2w^6\right)$.
(39)

*26. **Probability** $P(A) = \frac{r^2}{t}$, $P(B) = \frac{t}{s}$, and $P(C) = \frac{s^2}{rt}$. $P(A$ and $(B$ or $C))$ is represented by the expression $\frac{r^2}{t}\left(\frac{t}{s} + \frac{s^2}{rt}\right)$. Simplify.
(39)

Simplifying Rational Expressions

LESSON 43

27. Use the table to find the rate of change.
(41)

Tables	3	5	7	9
Guests	36	60	84	108

*28. Multi-Step A plant in Selena's garden will eventually reach a height of 134.4
(42) centimeters. Today the plant measures 42 centimeters in height. What percent of the present height of Selena's plant is the final height of the plant?
   a. Write a proportion to solve the problem.
   b. Find the solution.
   c. Choose a method with which to check your work and determine the reasonableness of your solution.

29. (**Restoration**) At the beginning of 1999, the Leaning Tower of Pisa leaned 14.5 feet
(42) past center and was in danger of falling over. Engineers tried removing dirt from the side of the tower opposite from the direction in which it leans, and were able to reduce the lean by about 10%. About how far past center did the tower lean after the dirt was removed?

*30. Geometry Two similar triangles are shown
(42) at right. The sides on the larger triangle are 130% of those on the smaller triangle. What is the length of side x?

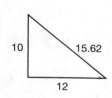
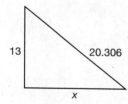

Finding Slope Using the Slope Formula

LESSON 44

Write each number in standard notation.

1. 8.2×10^{-9}
2. 0.23×10^6

Write each number in scientific notation.

3. 112,500
4. 0.00058

List the domain and range of the following relations.

5. (1, 2), (3, 4), (5, 6) (7, 8)
6. (3, 4), (3, 5), (4, 4) (4, 5)
7. **Exercising** A gymnast spends 20 minutes on each stretching exercise. She completes the workout with a 3-hour practice session. Write a function to describe the time it takes to complete x exercises and the practice session.
8. What percent of 520 is 26?
9. Find 35% more than 90.
10. Find the value that makes the expression $\frac{22 + x}{16 + 8x}$ undefined.

Use the formula for slope to determine the slope of the line connecting the given points.

*11. $(-5, 1)$ and $(5, 7)$

*12. $(-3, -4)$ and $(3, -6)$

*13. Find the slope of a line containing $(3, -6)$ and $(3, 4)$.

*14. **Zoology** If a red-tailed boa, a western hemisphere constrictor snake, grows from 22 inches at 2 months of age to 42 inches at 1 year of age, what is its average growth rate?

15. **Predict** The points in the table all lie on the same line. What is the missing y-value?

x	y
7	66
5	46
3	26
1	6
0	

Finding Slope Using the Slope Formula

LESSON 44

16. (Planting) An orchard has enough space to plant 21 rows with y trees in each row,
(28) or 18 rows with $y + 7$ trees in each row. If each orchard plan contains the same
number of trees, how many trees can each orchard contain?

17. Multi-Step The price of 4 pounds of hamburger is $13.80. The price of a pound of
(31) potato salad is $3.75. How much would it cost to buy 12 pounds of hamburger and
3 pounds of potato salad?

18. (Architecture) Plans for the construction of an eight-floor hotel call for the first floor
(34) to be 16 feet high and each additional floor to be 10.5 feet high. What is the height of
the building?

19. Multi-Step Pencils and T-shirts were bought to sell at the school bookstore. The
(35) bookstore's profits are calculated using the equation $2y - 500 = -20x$, where
y is the number of boxes of pencils sold and x is the number of T-shirts sold.
  a. Rewrite the profit equation in standard form.
  b. Calculate the y-intercept and explain its real-world meaning.
  c. Calculate the x-intercept and explain its real-world meaning.

20. Multi-Step A swimmer's average time per lap is $\frac{3}{4}$ minute. If he swims for $10\frac{1}{2}$ minutes,
(24) how many laps does he swim?
  a. Write an equation to find the number of laps.
  b. Write Explain how to solve the problem in two different ways.

21. Multiply $(4.2 \times 10^{12})(3.14 \times 10^{-4})$ and write the answer in scientific notation.
(37)

22. Factor the polynomial $3x^2y^2 + 3xy^3 - 6x^3y^6$ completely.
(38)

23. Simplify $(-2^3)^3$ using exponents.
(40)

*24. Probability A student takes a quiz that has 5 true/false questions and 5 multiple-choice
(40) questions with four choices for each question. This means there are 2^5 possible ways to
answer the true/false questions and 2^{10} ways to answer the multiple-choice questions.
The student guesses on every question.
  a. How many ways are there to answer all of the questions?
  b. What is the probability that the student will get every question right?

*25. Multiple Choice What is the slope of the line that passes through the points $(-5, 8)$
(41) and $(5, 2)$?

  A $-1\frac{2}{3}$ **B** $-\frac{3}{5}$

  C $\frac{3}{5}$ **D** $1\frac{2}{3}$

26. Generalize Give an example of a real-world situation that could be represented by a
(41) graph with a negative slope.

Finding Slope Using the Slope Formula

LESSON 44

*27. **Taxi Fares** A taxi company charges $0.75 per mile ($0.75m) on the weekends. During
(43) the week, the company charges a base fee of $2.50 plus $0.50 a mile
($2.50 + $0.50m). Write a simplified rational expression for the ratio of the weekend
charge to the weekday charge.

*28. **Error Analysis** Two students found the value at which $\frac{x+5}{x+3}$ is undefined. Which student
(43) is correct? Explain the error.

Student A	Student B
$\frac{x+5}{x+3}$ denominator $= x + 3$ $x + 3 = 0$ when $x = -3$ $\frac{x+5}{x+3}, x \neq -3$	$\frac{x+5}{x+3}$ denominator $= x + 3$ $x + 3 = 0$ when $x = 3$ $\frac{x+5}{x+3}, x \neq 3$

29. **Multi-Step** To make a rope that is x^2 units long requires that 4 strands be twisted
(43) together. Each strand must be $x^2 + 6x$ units in length.
  a. What is the total length of the 4 strands?
  b. Write a rational expression for the length of the rope divided by the total length of the strands.
  c. Simplify the rational expression.

30. **Geometry** The right circular cylinder shown has a height that
(43) is 2 units more than the radius of its bases.
The ratio $\frac{2\pi r^2 + 4\pi r}{2\pi r^2}$ gives the surface area around the cylinder to the
surface area of both bases. Simplify.

Name _____ Date _____ Class _____

Translating Between Words and Inequalities — LESSON 45

Use unit analysis to convert each measurement.

1. 42 feet to centimeters
(8)

2. 2 miles to inches
(8)

Simplify.

3. $-2(-3-3)(-2-4) - (-3-2) + 3(4-2)$
(4)

4. $\dfrac{5(-5+3) + 7(-5+9) + 2}{(4-2) + 3 + 5}$
(4)

Translate each sentence into an inequality.

*5. The product of 6 and an unknown number is less than or equal to 15.
(45)

*6. Josephine sleeps more than 7 hours each night.
(45)

Translate each inequality into a sentence.

*7. $-4b \geq 7$
(45)

*8. $\dfrac{t}{7} - 4 < 8$
(45)

9. **Geometry** On a separate sheet of paper, plot the points (1, 3), (3, 2), (5, −1),
(41)   and (2, 1). Connect the points, in order, to form a quadrilateral. Find the slope of
each side.

10. **Multiple Choice** What percent of 160 is 24?
(42)
    A 6.67%   **B** 15%   **C** 136%   **D** 184%

*11. **Write** Explain how to translate the phrase "4 more than the quotient of an unknown
(45)   and 9 is no less than 15."

*12. (Sports) Harriet's current score after two rounds in a freestyle skating contest is
(45)   45.7 points. She needs to have a score of 83.2 or better to win first place. Write
an inequality that expresses the possible scores she can achieve to win first place.

*13. **Multiple Choice** Which of the following inequalities represents "Five less than an
(45)   unknown is at most seven"?
    A $x + 5 \leq 7$ **B** $x + 5 > 7$ **C** $x - 5 \leq 7$ **D** $x - 5 > 7$

14. Find the slope of a line containing the two points (−5, −6) and (8, 7).
(44)

Translating Between Words and Inequalities

LESSON 45

***15. Multi-Step** Stock A and Stock B have both experienced rapid growth on the market over the past several days. The graph shows the growth of both stocks over 9 days.

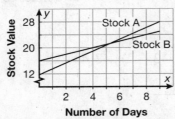

a. After 9 days, what was the average rate of increase for Stock A?

b. What was the average rate of increase for Stock B?

c. **Write** Explain which stock was the better buy based on the graphs shown.

16. (Atmospheric Science) Scientists at Mauna Loa Observatory in Hawaii have measured the levels of carbon dioxide present in the atmosphere for many years. In 1962 the level was measured at 315 parts per million. It had increased to 320 parts per million by 1969. What was the average rate of increase for carbon dioxide between 1962 and 1969 to two decimal places?

17. Geometry A student draws a right triangle on a coordinate plane. What is the slope of the triangle's hypotenuse?

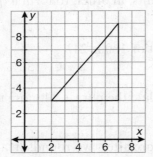

18. Factor the polynomial $5m^2n^4 + 10m^3n$ completely.

19. Simplify $\dfrac{x^{-3}}{w^2}\left(\dfrac{4x^2}{w} - \dfrac{j^{-3}w}{x}\right)$.

20. (Membership Dues) A local gym charges a monthly fee of $12 plus $4 for each class attended. The equation $y = 4x + 12$ can be used to calculate the monthly bill. Find the intercepts.

Translating Between Words and Inequalities

LESSON 45

21. Multi-Step You and a friend are taking a trip. The scale on your map is
(36) 1 inch:16 miles.

a. If the first leg of the trip measures 5 inches on the map, how many miles will you be driving?

b. If the second leg of the trip measures 17.5 inches on the map, how many miles will you be driving on this leg?

c. If you find a shorter route for the two legs that totals 20 inches on the map, how many miles shorter will the shorter route be?

d. Will the shorter route always be the faster route? Explain your reasoning.

22. (Astronomy) The radius of the moon is about 1.75×10^3 kilometers. Use the formula
(40) $A = 4\pi r^2$ to find the surface area of the moon. Write your answer in terms of π.

23. Use the table to find the rate of change.
(41)

Hours	2	4	6	8
Earnings	$13.50	$27	$40.50	$54

*24. Model A dog digs a conical hole that has a volume of 803 in³. The next day the dog
(42) increases the volume by another 25%. Write an equation or proportion to find the new volume and draw a model to represent the problem.

25. Simplify $\frac{24 + 9x}{x}$. Identify any excluded values.
(43)

*26. Error Analysis Two students simplified the following rational expression: $\frac{x^2 - 11x}{2x^2 - 6x}$.
(43) Which student is correct? Explain the error.

Student A	Student B
$\frac{x^2 - 11x}{2x^2 - 6x}$ numerator GCF = x denominator GCF = $2x$ $\frac{x^2 - 11x}{2x^2 - 6x} = \frac{2x(x - 11)}{2x(x - 3)} = \frac{x - 11}{x - 3}$ $\frac{x^2 - 11x}{2x^2 - 6x} = \frac{x - 11}{x - 3}, x \neq 3$	$\frac{x^2 - 11x}{2x^2 - 6x}$ numerator GCF = x denominator GCF = $2x$ $\frac{x^2 - 11x}{2x^2 - 6x} = \frac{x(x - 11)}{2x(x - 3)} = \frac{x - 11}{2(x - 3)}$ $\frac{x^2 - 11x}{2x^2 - 6x} = \frac{x - 11}{2(x - 3)}, x \neq 3$

27. Use the information in the graph to complete the table below.
(30)

x	0	1	−1	2	4
y	1	0.5	1.5	0	−1

Translating Between Words and Inequalities

LESSON 45

28. Multi-step A centimeter is 10^{-2} times the length of a meter and a kilometer is 10^3 times the length of a meter.
   a. Using the Quotient Property of Exponents, how many times longer is the length of a kilometer than the length of a centimeter?
   b. First, using exponents, express how many times longer a meter is then a centimeter. Then use that answer to find how many times longer a kilometer is than a centimeter. Are the results the same as for part **a**?

29. (Names) A new baby girl arrives. The parents have 6 names they really like, including Maria and Gail. They are going to pick one of the 6 names for a first name and one for a middle name. What is the probability that the baby will be named Maria Gail?

30. Generalize Is there a pattern in the number of zeros in standard notation and the exponent in scientific notation? Consider 0.0204, 8,000,000, and 6700.

Simplifying Expressions with Square Roots and Higher-Order Roots

LESSON 46

Evaluate each expression for the indicated values.

1. $(p - x)(a - px)$ for $a = -3$, $p = 3$, and $x = -4$
   (9)

2. $-a[-a(x - a)]$ for $a = -2$ and $x = 3$
   (9)

Evaluate each expression.

*3. $\sqrt{-10{,}000}$
(46)

*4. $-\sqrt[4]{10{,}000}$
(46)

5. Simplify the following expression by adding like terms:
   (18)
   $xym^2 + 3xy^2m - 4m^2xy + 5mxy^2$.

*6. **Multiple Choice** Evaluate $-\sqrt[3]{-\dfrac{27}{64}}$.
(46)

    A $-\dfrac{3}{4}$ **B** $\dfrac{9}{16}$ **C** $\dfrac{3}{4}$ **D** no real solution

Translate each sentence into an inequality.

7. The opposite of 2 is less than or equal to the difference of a number and 7.
   (45)

8. A U.S. citizen must be at least 35 years old in order to run for President.
   (45)

*9. **Estimate** Find the two whole numbers that $\sqrt[3]{1500}$ is between.
(46)

10. (**Farming**) For every 15 cows, a farmer puts out 1 mineral block for them to eat as a source of additional nutrients. What are the dependent and independent variables?
    (20)

11. Use the table of ordered pairs below to make a graph.
    (30)

x	0	1	3	4	5
y	-1	2	8	11	14

12. Solve $\dfrac{5}{7} = \dfrac{h}{49}$.
    (31)

13. **Multi-Step** Work uniforms include pants or a skirt, a shirt, and a tie or a vest. There are 3 pairs of pants, 5 skirts, 10 shirts, 2 ties, and 1 vest in a wardrobe.
    (33)
    a. What is the probability of choosing a skirt, a shirt, and a vest?
    b. The skirt and shirt from the previous day must be washed, but the vest returns to the wardrobe. What is the probability of choosing pants, a shirt, and a tie the next day?

*14. **Multi-Step** A student earns $10 for every yard he mows in his neighborhood. He wants to buy a new MP3 player for $180 including tax. At least how many yards will he need to mow in order to earn the money for the MP3 player?
(45)
   a. Write an inequality that could be used to solve the problem.
   b. Solve the inequality.

Simplifying Expressions with Square Roots and Higher-Order Roots

LESSON 46

Simplify.

15. $\dfrac{gt}{d^2}\left(\dfrac{gth}{d} - 3th + \dfrac{t}{5}\right)$
(39)

16. $\left(\dfrac{-8x^4}{3}\right)^2$
(40)

*17. **Measurement** Write an inequality to represent the value of the diameter of a circle
(45) that has a circumference greater than 5 inches.

18. (**Scale Models**) A designer is making a model of a room. The model is 9 inches by
(36) 15 inches. If the scale factor is 1 in.:10 in., what are the dimensions of the room?

19. **Multi-Step** At the end of 2006, the U.S. debt was $4.9 trillion. The population was
(37) approximately 300 million.
   a. Write these amounts in scientific notation.
   b. Divide the debt by the population to find the approximate debt per person in the United States.

20. **Justify** Show that the GCF of the polynomial $27x^2y^3z + 12xy^2z$ is $3xy^2z$. Then factor
(38) the polynomial.

21. (**Fill Rate**) Ailani places a hose in a partially filled swimming pool to fill it. The table
(41) shows the number of gallons in the pool over time. Find the rate at which the hose adds water.

Time (min)	10	15	30	60
Water (gal)	15,048.5	15,072.75	15,145.51	15,291

22. What is 120% of 250?
(42)

23. **Multiple Choice** Which one of the following rational expressions cannot be further
(43) simplified?

   A $\dfrac{7x+1}{5x^2 - x}$ B $\dfrac{x^2}{15 - x}$

   C $\dfrac{x^2+1}{x^2 - 3x}$ D $\dfrac{8(2-x)}{6 - 3x}$

*24. **Justify** Simplify the rational expression $\dfrac{6-6x}{9-9x}$. Explain your reasoning.
(43)

Simplifying Expressions with Square Roots and Higher-Order Roots

LESSON 46

25. Find the slope of the line containing the points in the table.
(44)

x	y
−8	−1
−4	−2
0	−3
4	−4
8	−5

26. Find the slope of a line containing the points (1, 7) and (5, 8).
(44)

27. Error Analysis Two students translate the sentence "The product of 25 and an unknown number is at most 150." Which student is correct? Explain the error.
(45)

Student A	Student B
$25x \leq 150$	$25x \geq 150$

*28. Geometry Gretchen wants to design a rectangular flower bed so that the length is twice the width. She has 45 feet of edging. Write an inequality that represents the greatest possible width of the flower bed.
(45)

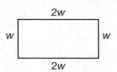

*29. Scheduling Marcy needs to read a 400-page book for her literature class. She has already read 255 pages. If she has 8 more days to read the book, about how many pages does she need to read each day to complete the reading assignment on or before time?
(45)

*30. Sports A can containing 3 tennis balls has a volume of $\frac{128}{9}\pi$ cubic inches. What is the diameter of each tennis ball? (Hint: The formula for the volume of a cylinder is $V = \pi r^2 h$.)
(46)

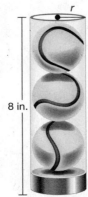

Name _____ Date _____ Class _____

Solving Problems Involving the Percent of Change
LESSON 47

Solve the equations. Check your answers.

1. $-(x-3) - 2(x-4) = 7$
   (26)

2. $3p - 4 - 6 = -2(p-5)$
   (28)

3. $k + 4 - 5(k+2) = 3k - 2$
   (28)

4. Write an inequality for the product of 3.6 and an unknown number that is greater than 18.
   (45)

5. Simplify $a^3x - |x^3|$ for $a = -3$ and $x = -2$.
   (9)

6. If 15 boxes of cereal costs $67.50, what is the unit price?
   (31)

7. Evaluate $\sqrt[3]{-512}$.
   (46)

*8. Find the percentage of increase or decrease from an original price of $35 to a new price of $49.
   (47)

*9. **Multiple Choice** 78 changes to 88. Find the percent of increase or decrease.
   (47)
   A 10%   **B** 12%   **C** 13%   **D** 11%

*10. **Analyze** Is it possible to have a percent of increase of more than 100%? Is it possible to have a percent of decrease of more than 100%? Explain.
   (47)

*11. The first 3 numbers in a sequence are 30,000, 12,000, and 4800. Use percent to describe the pattern of the sequence. List the next 2 terms.
   (47)

12. **Multiple Choice** Which points are connected by a line with an undefined slope?
   (44)
   A $(-16, 8)$ and $(-1, 10)$ **B** $(-5, -15)$ and $(18, 6)$
   C $(1, 1)$ and $(-1, -1)$ **D** $(-1, -4)$ and $(-1, 15)$

13. **Write** Based on the slope formula, why is the slope often described as rise over run $\left(m = \frac{\text{rise}}{\text{run}}\right)$?
   (44)

*14. **Sale Prices** A holiday weekend sale advertised a 25% discount on a TV that regularly sells for $500. After the holiday weekend, the sale price was increased by 25%. The management said this should put the new price back to its original price of $500. Is the management correct? If not, which price is higher? Explain.
   (47)

15. **Error Analysis** Two students evaluate the expression $16^{\frac{1}{4}}$. Which student is correct? Explain the error.
   (46)

Student A	Student B
$16^{\frac{1}{4}} = 2$	$16^{\frac{1}{4}} = \frac{1}{64}$

© Saxon. All rights reserved.   Saxon Algebra 1

Solving Problems Involving the Percent of Change

LESSON 47

***16. (Basketball Rules)** In the NBA, the inner part of a basketball hoop has an area of 81π square inches. What is the circumference of the hoop? (Hint: The formula for the area of a circle is $A = \pi r^2$.)
(46)

***17. Data Analysis** The geometric mean is the nth root of the product of n numbers. For example, the geometric mean of 1, 2, and 4 is $\sqrt[3]{1 \cdot 2 \cdot 4} = 2$. Find the geometric mean of 2, 4, 25, and 50.
(46)

18. On average, the ratio of sparrows to doves who come to Jenn's birdbath is 7 to 8. On Tuesday 45 birds came to bathe. How many sparrows and how many doves came to bathe that day?
(31)

19. Evaluate the expression $\dfrac{3n^0}{m^{-2}}$ if $m = -3$ and $n = 8$.
(32)

20. Multi-Step In an arithmetic sequence, $a_1 = 9$ and $d = 13$.
(34)
  a. Write a recursive formula for the sequence.
  b. Using that formula, find the first four terms of the sequence.

21. Error Analysis Two students translate the sentence "twice a number is less than or equal to 5" as shown below. Which student is correct? Explain the error.
(45)

Student A	Student B
$2 + x \leq 5$	$2x \leq 5$

***22. Coordinate Geometry** A diamond is centered on a coordinate axis. What are the slopes of its two steepest sides?
(44)

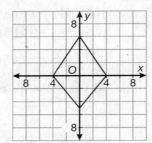

***23.** Determine any values where the rational expression $\dfrac{x + 4x^2}{16}$ is undefined.
(43)

***24. (Budgeting)** A family wants to write a budget where they put a total of 15% of their income each month into retirement savings. They are already putting 6% of their income into a retirement account. If the family's income each month is $3484.00, how much more do they need to put into retirement each month?
(42)

172 Saxon Algebra 1

Solving Problems Involving the Percent of Change

LESSON 47

25. Use the graph to find the rate of change.
₍₄₁₎

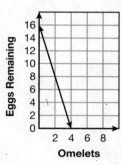

26. Simplify $(-4ab^2c^2)^3$.
₍₄₀₎

27. Analyze For the expression $\dfrac{3x^{-2}}{f}$, why can x not equal zero?
₍₃₂₎

28. (Chemistry) The approximate radius of an electron is 2.817939×10^{-15} meter. Write this number in standard notation.
₍₃₇₎

29. Multi-Step Use the expression $4a + 4ab + 4bc$.
₍₃₈₎
  a. Is the expression $2(2a + 2ab + 2bc)$ a correct factorization of the polynomial? Explain.
  b. Does the expression $2(2a + 2ab + 2bc)$ factor the polynomial completely?

30. (Finance) Over the past six months, Shari made four deposits for a total of $854.71. She wrote down the first two deposits of $220.25 and $318.12 in her register. The final two deposits were of equal amounts that she forgot to write down. What amount should Shari write in her register for each of those deposits?
₍₂₆₎

Name _____ Date _____ Class _____

Analyzing Measures of Central Tendency

LESSON 48

Find the greatest common factor.

1. $4ab^2c^4 - 2a^2b^3c^2 + 6a^3b^4c$
   (38)

2. $5m^2x^2y^5 - 10m^2xy^2 + 15m^2x^2y^4$
   (38)

Simplify.

3. $4x^2(ax - 2)$
   (15)

4. $\dfrac{6a^{-3}c^{-3}}{a^{-2}cd^0}$
   (32)

5. A deck of cards contains 6 green cards and 2 yellow cards. What is the probability of drawing a green card, keeping it, and then drawing another green card?
   (33)

*6. Find the mean, median, and mode of the following data set:
   (48)
   number of kittens in 11 litters: 3, 5, 4, 2, 6, 6, 2, 1, 4, 6, 5.

7. **Statistics** The numbers below have a mean of 17.5. What four identical values could be added to this set of numbers to increase the mean to 19?
   (26)

   {19, 18, 17, 17, 19, 15}

8. **Multi-Step** The area of a square game board is represented by the expression $\left(\dfrac{x}{y} - \dfrac{r}{x}\right)^2$.
   (39)
   a. Write this expression as a product of two terms.
   b. Distribute both fractions in the first factor, $\dfrac{x}{y}$ and $-\dfrac{r}{x}$, one at a time, as if it were two separate problems: $\dfrac{x}{y}\left(\dfrac{x}{y} - \dfrac{r}{x}\right)$ and $-\dfrac{r}{x}\left(\dfrac{x}{y} - \dfrac{r}{x}\right)$.
   c. Simplify the result of part b.

9. **Write** When is the statement $(-2)^n = 2^n$ true for a positive integer n? When is it false? Explain.
   (40)

10. **Data Analysis** If $x + 12$ represents the number of data values in a set of data, the mean of this data is 9, and the sum of all the data values is 216, how many data values are in the set?
    (48)
    a. Write an equation to find the value of x. Start with the formula for finding the mean.
    b. Solve for x.
    c. Find the number of data values in the set.

*11. **Multiple Choice** The following data set shows the number of minutes 8 students spent studying last night: 30, 90, 60, 30, 90, 40, 90, 50. Which measure has a value of 55?
    (48)
    A mean   **B** median   **C** mode   **D** range

*12. **Write** Can 7 be the mode of 1, 2, 3, 4, 7, 7 even though it fails to occur in the middle of the data set? Explain.
    (48)

Analyzing Measures of Central Tendency

LESSON 48

*13. **Multi-Step** Matthew opened a checking account last year for $150. Since then, the balance has increased by 15%, decreased by 8%, and then increased again by 25%.
(47)
a. What is the account balance now?
b. Did the balance increase or decrease over the original amount? By what percent?

*14. (**Travel**) A round-trip plane fare from Orlando, Florida, to Manila, Philippines, during autumn costs $1689. However, the same round-trip flight during winter costs $3474. Find the percent of change.
(47)

*15. Provide a counterexample for this statement: The sum of an even number and an odd number is an even number.
(Inv 4)

*16. **Analyze** Juan recorded the number of points he scored individually in the last 13 basketball games his team played: 33, 12, 18, 21, 10, 18, 14, 20, 11, 24, 0, 0, 0. Tell why the mode is not the best measure of central tendency to describe the data set.
(48)

*17. (**Hockey**) In hockey, a goalie gets a shutout if he or she can prevent the opposing team from scoring for the entire game. Study the data in the table. Identify the outlier. What would this outlier represent in the data?
(48)

Shutouts for the 2006–2007 Regular Hockey Season													
2	5	8	3	5	2	4	7	5	4	6	2	5	4	12
5	3	5	2	7	4	5	3	3	5	4	2	5	2	2

18. **Multi-Step** A cell phone bill can be calculated using the equation $10y = 3x + 360$, where x is the number of minutes used and y is the amount of the bill.
(35)
a. Rewrite this equation in standard form.
b. Calculate the y-intercept and explain its real-world meaning.
c. Calculate the x-intercept and explain its real-world meaning.

19. (**Recreation**) The formula $h = -16t^2 + 12t + 2$ can be used to find the height of a ball that is kicked into the air from 2 feet off the ground with an initial velocity of 12 feet/second. Rewrite the formula by factoring the right side of the equation using the GCF and making the t^2-term positive.
(38)

20. Find 93% of 24 using a proportion.
(42)

21. (**Transporting Freight**) Ramps are often used to load freight on to trucks for transporting. The figure at right shows one possible structure of a ramp. The ratio comparing the area of the top of the ramp to the area of one of its sides is $\dfrac{x\ell}{0.5(x^2 + 4x)}$. Simplify this expression.
(43)

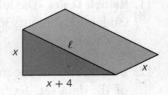

22. Determine the slope of the line containing the two points $(2, -9)$ and $(4, -25)$.
(44)

Analyzing Measures of Central Tendency

LESSON 48

23. Formulate Translate "the sum of $\frac{1}{2}$ an unknown and the opposite of 4 is less than 6."
(45)

24. Multiple Choice Marshall deposits $45 into his savings account every month. His current balance is $215. After how many months will his balance exceed $500? Write an inequality to represent the situation.
(45)

A $45w + 215 > 500$ B $45w + 500 \leq 215$

C $45w + 215 \leq 500$ D $45w + 500 > 215$

25. Write $\sqrt[6]{m}$ with a fractional exponent.
(46)

*26. Multiple Choice Which expression is equal to $-15^{\frac{1}{4}}$?
(46)

A $\sqrt[4]{-15}$ B $-\sqrt[4]{15}$

C $\dfrac{1}{\sqrt[4]{-15}}$ D $-\dfrac{1}{\sqrt[4]{15}}$

27. Find the percent of increase or decrease from an original price of $500 to a new price of $400.
(47)

*28. Error Analysis Student A and Student B computed the percent of increase of the price of gasoline in Idaho at $3.25 in June 2007, from the nationwide average of $3.07. Which student is correct? Explain the error.
(47)

Student A	Student B
$\dfrac{\$3.25 - \$3.07}{\$3.07} = \dfrac{\$0.18}{\$3.07}$ ≈ 0.06 $= 6\%$	$\dfrac{\$3.07 - \$3.25}{\$3.25} = -\dfrac{\$0.18}{\$3.25}$ ≈ -0.06 $= -6\%$

*29. Geometry Draw a similar figure whose perimeter is 75% more than the perimeter of the square below.
(47)

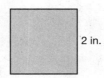
2 in.

30. Analyze Determine if the premise and the conclusion use inductive or deductive reasoning. Explain your choice.
(Inv 4)

Premise: The light turned on the last 50 times the switch was flipped.

Conclusion: The light will turn on the next time the switch is flipped.

Writing Equations in Slope-Intercept Form

LESSON 49

Expand each expression by using the Distributive Property.

1. $x^2y^3(3xy - 5y)$
2. $-2x^3y^3(4x^2y - 3xy)$

Evaluate each expression for the given values.

3. x^2y^3z if $x = 3$, $y = -2$, and $z = 4$
4. $-x^2 - y^3$ if $x = -3$ and $y = -2$

5. Find the x-intercept of the line $3x + 2y - 10 = 0$.

*6. Identify the slope and y-intercept of the line $2x - 5y - 6 = 0$.

*7. **Multiple Choice** What is the equation of the graphed line?

A $y = -\frac{1}{3}x + 3$
B $y = -\frac{1}{3}x - 3$
C $y = -3x + 3$
D $y = -3x - 3$

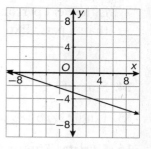

*8. **Multi-Step** The directions on a box of frozen biscuits state to cook one biscuit for 90 seconds on high and to add 15 seconds of cooking time for each additional biscuit.
   a. Write a linear equation in slope-intercept form to represent the situation. Identify what the variables y and x represent.
   b. **Analyze** What is the y-intercept of the graph? Does it have any meaning? Why or why not?

*9. (**Measurement Conversion**) The formula for converting from Celsius to Fahrenheit is $F = \frac{9}{5}C + 32$, where C is the temperature in degrees Celsius and F is the temperature in degrees Fahrenheit. Identify the slope and the y-intercept of the equation and then graph it.

*10. (**Pet Care**) The table shows a feeding chart on the bag of a certain brand of cat food.

Maximum Weight of Cat	Amount of Food Per Day
6 lb	$\frac{3}{4}$ cup
9 lb	$1\frac{1}{4}$ cups
12 lb	$1\frac{3}{4}$ cups

   a. Write an equation in slope-intercept form.
   b. **Write** The three points in the table show that this is a linear equation. Do you think this relationship would be a linear equation? Explain.

11. Find the mean, median, and mode of the set of data below.
    number of pockets in 8 pairs of pants: 4, 2, 0, 4, 5, 4, 2, 3

Writing Equations in Slope-Intercept Form

LESSON 49

12. Error Analysis Two students studied the table of data. Student A says that the median lowest temperature per month for San Diego, California, is 38.5°F. Student B says that the median is 52.5°F. Which student is correct? Explain the error.

Lowest Recorded Temperature in San Diego (in °F)			
July 2006	68	Jan 2007	35
Aug 2006	63	Feb 2007	45
Sep 2006	61	Mar 2007	45
Oct 2006	55	Apr 2007	50
Nov 2006	42	May 2007	55
Dec 2006	42	Jun 2007	58

*13. Probability Hari surveyed the first 15 students who came to class about the number of pets they owned. According to the data he collected, is the next person he surveys likely to have more than 4 pets? Explain.

2, 1, 3, 1, 1, 0, 2, 2, 3, 7, 0, 4, 2, 1, 1

14. Simplify $\dfrac{x^2 - 12x}{2x^2 - x}$.

15. Find 22% of 80 using a proportion.

16. Estimate Approximate the slope of the line shown in the graph.

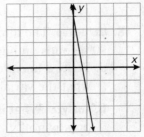

17. (Travel) Quick Cab charges a $5 fee for a ride, plus $0.15 for every block. Speedy Cab charges $7 for a ride, but only $0.05 every block. For what number of blocks is the cost of travel the same?

18. A deck of cards contains 15 cards, five of each number 1, 2, and 3. What are the odds of getting a 1? What are the odds against getting a 3?

19. Write a recursive formula for the arithmetic sequence with $a_1 = -3$ and common difference $d = 9$. Then find the first four terms of the sequence.

20. Multi-Step Use the similar triangles shown.
   a. What angle corresponds with $\angle M$?
   b. Find $m\angle Q$
   c. What is the scale factor of the triangles?
   d. Use the scale factor to find the value of x.

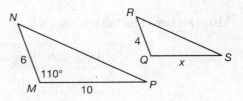

21. (Tourism) The number of visitors to national parks in the United States each year for the years 1990 to 2005 (in millions) are as follows: 258.7, 267.8, 274.7, 273.1, 268.6, 269.6, 265.8, 275.3, 286.7, 287.1, 285.9, 279.9, 277.3, 266.1, 276.9, 273.5. Find the range of the data.

Writing Equations in Slope-Intercept Form

LESSON 49

*22. Geometry Ferdinand found the areas of seven triangles. This was his data set: 7 cm², 8 cm², 12 cm², 6 cm², 7 cm², 10 cm², 6 cm². If the triangle at right has the same area as the mean value in the data set, what is the height (h) of the triangle?
(48)

23. Error Analysis A box of chocolates sells for $12.50. The price of the chocolates decreased by 15% from last year. Student A and Student B tried to find the price from last year. Which student is correct? Explain the error.
(47)

Student A	Student B
$0.15n = \$12.50$	$0.85n = \$12.50$
$\dfrac{0.15n}{0.15} = \dfrac{\$12.50}{0.15}$	$\dfrac{0.85n}{0.85} = \dfrac{\$12.50}{0.85}$
$n = \$83.33$	$n = \$14.71$

*24. Multi-Step A square table has an area of 2116 square inches. A tablecloth hangs 3 inches over each end of the table.
(46)

a. What is the length of one side of the table?

b. What is the area of the tablecloth?

c. What is the area of the part of the tablecloth that hangs over the table?

*25. Analyze Copy and complete the statement with $\sqrt[4]{\sqrt{256}}$ ___?___ $\sqrt{\sqrt[4]{256}}$ >, <, or =.
(46)

26. Find the percent of increase or decrease to the nearest percent from an original price of $20 to a new price of $23.
(47)

27. Write an inequality for the following sentence: The sum of twice a number and $\frac{1}{3}$ is less than $1\frac{2}{3}$.
(45)

*28. Remote Sensing Satellites perform sophisticated data collection, but over time their orbits decay, affecting their measurements. In 1980 the NOAA-06 satellite was holding an orbit of 826 km. By 1996 the orbit had decayed to 804 km. What was the average rate of orbital decay for NOAA-06?
(44)

29. Analyze Determine if the following statement uses inductive or deductive reasoning. Explain your choice.
(Inv 4)

"If Sharon made at least one goal at each of her last soccer games, then she will make at least one goal at her next soccer game."

30. Multi-Step A leather footstool in the shape of a cube is being made for a living room. The length of one side of the footstool is $2x$ inches.
(40)

a. What is the area of one side of the footstool?

b. What is the surface area of leather needed to make the footstool?

181

Saxon Algebra 1

Name _____ Date _____ Class _____

Graphing Inequalities — LESSON 50

Factor the greatest common factor.

1. $6k^5m^2 - 2k^3m - km$
   (38)

2. $mx^4y^2 - m^2x^3y^3 + 5m^2x^6y^2$
   (38)

Simplify.

3. $\left(\dfrac{2x}{3y^4}\right)^3$
   (40)

4. $(2x^3y^2)^4$
   (40)

5. True or False: The set of whole numbers is closed under subtraction. If false, give a counterexample.
   (1)

6. Determine the slope of the line shown in the graph.
   (44)

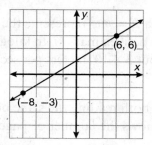

7. Write $5x - 2 = 6y$ in standard form.
   (35)

8. Write the expression $\sqrt[4]{y}$ with a fractional exponent.
   (46)

9. **Probability** 7 cards labeled P, E, R, C, E, N, and T are in a jar. Find the probability of picking a P and then an E if the card drawn is not replaced.
   (33)

10. (**Vacation**) At an amusement park, there are 5000 families. Ten families are chosen to have a meal with the park's owner each day. What are the odds that a family will be chosen?
    (33)

11. Write a rule for the nth term of the sequence with $a_1 = 32$ and common difference $d = -6$. Then find the fifth and twelfth terms of the sequence.
    (34)

*12. **Error Analysis** Two students wrote the equation $3x + 5y = 15$ in slope-intercept form. Which student is correct? Explain the error.
   (49)

Student A	Student B
$y = -\dfrac{3}{5}x + 3$	$x = -\dfrac{5}{3}y + 5$

*13. Determine which values in the set $\{-6, 0, 1, 6\}$ are solutions to the inequality $-2y + 3 < 0$.
   (50)

Graphing Inequalities
LESSON 50

*14. **Multiple Choice** Which of the following inequalities describes the graph?
(50)

A $x \leq 7$ B $7 < x$

C $x \geq 7$ D $x > 7$

*15. **Write** Explain how to graph the inequality $n \geq 12$.
(50)

16. **Formulate** How could Yalda write the inequality $x \neq 2$ as two separate inequalities?
(50)

17. **Multi-Step** The approximate diameter of Earth is 12,756,000 meters. The diameter of the Sun is approximately 695,900,000 meters.
(37)
   a. Write these distances in scientific notation.
   b. About how many times larger is the Sun than Earth in diameter?

18. (**Physical Science**) The formula $E = mc^2$ represents the amount of energy E, in joules, contained in an object with a given mass m, in kilograms. The variable c represents the speed of light, which is about 3×10^8 meters per second. Find the amount of energy in an object weighing 2 kg.
(40)

19. **Multi-Step** Ivan and Jed are both reading the same book for a school report. The graph shows the number of pages they read over time.
(41)
   a. What is Ivan's reading rate?
   b. What is Jed's reading rate?

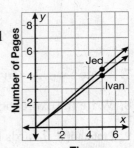

*20. **Justify** Janis is trying to find 36% of 212 using mental math. She knows how to find 25% of 212, 10% of 212, and 1% of 212. Can she use the sum of these percentages to find 36% of 212? Why or why not?
(42)

21. Simplify the rational expression $\frac{6x + 30}{36x + 6}$. State any excluded values.
(43)

*22. (**Personal Finance**) Ricardo babysits his brother for $5 per hour. He wants to earn enough money to buy a new game console that costs $280. If he has already saved $135, write an inequality that represents the least number of hours of babysitting he needs to do to be able to buy the game console.
(45)

23. **Multiple Choice** 49 changes to 45. Find the percent of increase or decrease.
(47)
   A 4%   B 8%   C 2%   D 6%

24. **Write** Kwami's video game collection increased from 6 video games to 9 video games. Lisa's CD collection increased from 20 CDs to 28 CDs. Compare the amount of increase and the percents of increase of both collections.
(47)

Graphing Inequalities

LESSON 50

*25. Troy recorded the number of minutes 11 people spent jogging: 25, 26, 18, 28, 20, 15, 27, 70, 32, 15, 21. Find the mean, median, and mode for the data.
(48)

26. Two students looked at the following data that show the number of hospitals in each state. Find the range of the data.
(48)

108, 19, 62, 87, 361, 70, 35, 6, 203, 146, 24, 39, 191, 113, 115, 134, 105,
131, 37, 50, 78, 144, 132, 93, 119, 54, 85, 30, 28, 80, 37, 206, 115, 40,
166, 109, 58, 197, 11, 62, 51, 127, 418, 43, 14, 88, 85, 57, 121, 24

27. Write Doyle wrote the linear equation of a graphed line for a homework problem. Explain how a table of values for the graph of a linear equation can be used to verify his answer.
(49)

*28. Geometry Write an equation in slope-intercept form that shows the relationship between the perimeter of a square (y) and the length (x) of a side. Graph the equation and find the coordinates of 3 points on the line.
(49)

*29. (Hobbies) Jean Claude is making painted candles to sell at a local craft market. He purchased $185 in materials and sells each candle for $7.50.
(49)
 a. Write a linear equation in slope-intercept form that can be used to calculate his profit (y) based on the number of candles sold (x).
 b. Use a graphing calcuator to graph the equation.
 c. How many candles must Jean Claude sell to make a profit?

*30. (Chemistry) The freezing temperature of water at normal atmospheric conditions is 32°F. Write and graph an inequality to model the temperatures at which water freezes.
(50)

Name _____ Date _____ Class _____

Simplifying Rational Expressions with Like Denominators — LESSON 51

Simplify.

1. $\sqrt[4]{81}$
(46)

2. $\sqrt[3]{-27}$
(46)

3. $\sqrt[3]{64}$
(46)

4. $\sqrt[3]{-64}$
(46)

5. Write $7y = \frac{3}{8}x - 1$ in standard form.
(35)

6. Simplify $\frac{12x^2 - 16x}{16xy}$.
(38)

Add.

*7. $\frac{d\,m^{-2}}{3} + \frac{5d}{m^2}$
(51)

*8. $\frac{8h^{-6}}{y^2} + \frac{y^{-2}}{h^6}$
(51)

*9. **Multiple Choice** What is the excluded value for
(51)

    A $h \neq -2$ **B** $h \neq 0$

    C $h \neq 2$ **D** $h \neq 3$

*10. **(Football)** A NCAA football field has a width of $\frac{160 + 160f}{f+1}$ feet and a length of $\frac{360 + 360f}{f+1}$ feet. A person walks the outside boundary line to mark it with chalk. How far does the person walk?
(51)

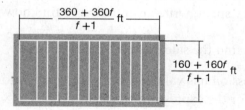

*11. **Write** Find the excluded value of the rational expression $\frac{5p}{p-6}$. Explain.
(51)

12. **(Travel)** You begin a long-distance trip along the highway at mile marker 21. After 5 minutes, you pass mile marker 32. After another 5 minutes, you pass mile marker 43. You have been traveling at a constant rate. If you continue to travel at this constant rate, what mile marker will you be at after 60 minutes? (Hint: Consider marker 21 (after 0 minutes) to be a_1, marker 32 (5 minutes) to be a_2, marker 43 (after 10 minutes) to be a_3, and so on.)
(34)

13. Write the converse of the following statement: If a number is a whole number, then the number is a natural number. Then determine whether the new statement is true or false. If false, give a counterexample.
(Inv 5)

Simplifying Rational Expressions with Like Denominators

LESSON 51

14. Multiple Choice The following data set shows the height in inches of 9 eighth graders:
(48) 66, 62, 56, 64, 60, 62, 58, 57, 59. What is the range of the data?
A 7   B 10
C 60   D 62

15. The scale factor of two similar triangles is 4:5. If one angle of the smaller triangle
(36) measures 60°, what is the measure of its corresponding angle in the larger triangle?

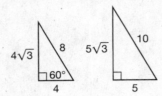

16. (Produce Cost) Find the unit price of apples at the grocery store.
(41)

Apples (lb)	1.5	2	3	3.5
Cost	$2.40	$3.20	$4.80	$5.60

*17. Multi-Step Luca and Paolo buy old bikes for a few dollars each, and then fix them up
(42) so that they can sell them for a profit. If they bought a rare tandem bike for $8, spent
an additional $150 on the bike for paint and parts, and sold it at a 285% profit, then
how much money did they sell the bike for?
  a. Analyze If profit means the amount of money Luca and Paolo made after taking
    into account what they spent, write an equation to find how much the bike
    sold for.
  b. Solve the equation to find the sale price of the bike.

*18. Write If a rational expression has values at which the numerator equals 0, are
(43) these points undefined for the expression? Explain.

19. Determine the slope of the line shown on the graph.
(44)

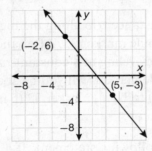

20. Find the percent of increase or decrease if the original price was $7000 and the new
(47) price is $10,200.

Simplifying Rational Expressions with Like Denominators

LESSON 51

21. Analyze Determine if the premise and the conclusion use inductive or deductive
(Inv 4) reasoning. Explain your choice.

Premise: The measures of two angles in a triangle add up to 80°.

Conclusion: The third angle is 100°.

*22. (Astronomy) The surface temperature of the sun is about 5880 kelvins. The sun's core is
(50) much hotter. Create a graph of an inequality that represents the temperature of the sun's core.

*23. Geometry According to a city bylaw, the area of a store's sign cannot exceed
(50) 900 square decimeters. Write and graph an inequality to represent the situation.

24. Justify The numbers of students who bought lunch each day this month are
(48) listed below.

$$176, 134, 208, 170, 149, 153, 136,$$
$$200, 168, 150, 157, 141, 211, 176,$$
$$145, 155, 128, 199, 182, 148.$$

a. How would you find the median of the set of data above?

b. Find the median.

25. Write the equation $5x + 3y = 9$ in slope-intercept form.
(49)

26. Multiple Choice Which equation has a slope of $-\frac{1}{2}$ and a y-intercept of -3?
(49)
A $4x + 2y + 3 = 0$

B $3x + 6y + 6 = 0$

C $5x + 10y + 30 = 0$

D $6x + 2y + 1 = 0$

*27. Determine which values in the set $\{-1, 0, 1, 2\}$ are solutions to the inequality
(50) $x - 1 \geq 4$.

28. Error Analysis Two students graphed the statement "A number is at least 9" as shown
(50) below. Which student is correct? Explain the error.

Student A

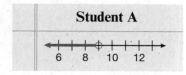

Student B

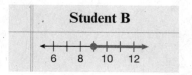

Simplifying Rational Expressions with Like Denominators

LESSON 51

29. Multi-Step The graph shows that the temperature is now at least the temperature it was this morning. How can the current temperature be expressed with a sentence?

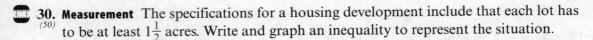

  a. Write an inequality for the graph.
  b. Translate the inequality into a sentence.

30. Measurement The specifications for a housing development include that each lot has to be at least $1\frac{1}{2}$ acres. Write and graph an inequality to represent the situation.

Determining the Equation of a Line Given Two Points

LESSON 52

Given the domain of a function, find the range.

1. $f(x) = 3x - 5$; domain: $\{0, 1, 2, 3\}$

2. $f(x) = \frac{1}{2}x + 3$; domain: $\{-2, 0, 2, 4\}$

Write an inequality for each situation.

3. To qualify for the job, the applicants must have more than 3 years of experience in the field.

4. In 2005 the minimum wage in the United States was $5.15 per hour.

5. Divide $\frac{(3 \times 10^{-9})}{(4.8 \times 10^{-1})}$ and write the answer in scientific notation.

*6. Graph the line that has a slope of -1 and passes through the point $(3, 1)$.

7. The scale of a map is 1 in.:20 mi. Find the actual distance that corresponds to a map distance of 4 inches.

8. (**Carpentry**) You are constructing a picnic table and are using a scale of 1 inch to 6 inches. If the length of the table on the drawing is 7 inches, what will the actual length of the table be?

9. A box with a volume of $\frac{t^3}{y^4}\left(\frac{t^2}{y^2} + \frac{5y}{m}\right)$ receives a fixed postage rate.
   a. Simplify the expression above.
   b. Identify the variables that cannot equal zero.

10. (**Simple Interest**) Simple interest is the amount of money the borrower pays based on the amount borrowed (the principal) for a given period of time (months or years). It is calculated this way: $I = prt$. If a person borrows $20,000 ($p$) to buy a car, pays 6.95% interest (r), and takes 5 years (t) to repay the loan, how much will the borrower pay in simple interest?

11. **Analyze** The rational expression $\frac{x-5}{15x^2 - 75x}$ is not defined for all real numbers.
    a. When is the denominator equal to 0?
    b. When is the numerator equal to 0?
    c. When is the rational expression undefined?

12. **Analyze** The slope of a line is 3. Two points on this line are $(-1, -2)$ and $(4, ?)$. Using the formula for slope, determine the missing y-value for the second point.

13. Translate the following sentence into an inequality:

    The difference of -4 and an unknown number is less than or equal to 0.

Determining the Equation of a Line Given Two Points

LESSON 52

14. Evaluate the expression for the given value.
(46)
$$-\sqrt{x} \text{ when } x = \frac{25}{16}$$

*15. (Business) Arminda owns a ladies' bag business. The total cost of producing a group
(28) of bags is $1100. In addition, each bag costs $18 in materials. If Arminda sells each bag for $40, how many bags should Arminda sell to gain a profit of at least $2200?

16. The data below are the weights of 10 newborns (to the nearest pound). Find the
(48) mean, median, and mode of the data.

5, 8, 6, 5, 7, 8, 10, 7, 8, 6.

*17. **Data Analysis** Study the graph. It shows the cost of playing
(49) x number of games after renting equipment.
   a. Write an equation that represents the line of the graph in slope-intercept form.
   b. What is the cost of playing 5 games?

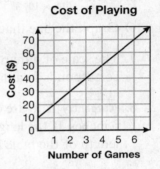
Cost of Playing

18. **Multi-Step** Murietta is planning a party at a bowling alley. She wants to rent two
(49) lanes. The rental fee for the two lanes is $40 plus the rental of $2 per pair of bowling shoes.
   a. Write an equation in slope-intercept form to represent this situation.
   b. Murietta is renting two lanes and inviting 9 people. If everyone including herself rents shoes, what will the cost of the party be?

19. Write an inequality for the graph below.
(50)

*20. **Error Analysis** Two students graphed the inequality $2.5 > b$ as shown below.
(50) Which student is correct? Explain the error.

Student A

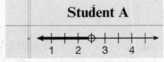

Student B

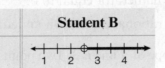

21. Find the excluded value in the expression $\frac{b^5}{b^3}$.
(51)

*22. (Soccer) In international play, the maximum length of a soccer field is $\frac{525x + 100}{5x + 1}$
(51) meters and the maximum width is $\frac{400x + 85}{5x + 1}$ meters. Each linesman is responsible for watching half of the field's perimeter. How many meters along the border of the field must the linesman watch during the game?

Determining the Equation of a Line Given Two Points

LESSON 52

***23.** Write the inverse of the statement: "If a polygon is has four sides, then it is a quadrilateral." Then determine whether the new statement is true or false. If false, give a counterexample.
(Inv 5)

24. Error Analysis Two students simplify the expression $-r^3s^{-4} + \frac{2s^{-4}}{r^{-3}}$. Which student is correct? Explain the error.
(51)

Student A	Student B
$-r^3s^{-4} + \frac{2s^{-4}}{r^{-3}}$	$-r^3s^{-4} + \frac{2s^{-4}}{r^{-3}}$
cannot simplify	$= \frac{-r^3}{s^4} + \frac{2r^3}{s^4}$
	$= \frac{r^3}{s^4}$

25. Geometry Two sides of a triangle measure $4x + 8$ and $x^2 + 2x$. Find the ratio of the first side to the second side. Simplify, if possible.
(43)

26. A rectangular window has a length of $\frac{6a}{3a-1}$ meters and a width of $\frac{4}{3a-1}$ meters. How many feet of trim will be painted if the trim goes around three sides of the window (both lengths and one width)?
(51)

*27. Multiple Choice On the first day of the school play, Carlos sold four tickets. On the second day, he sold seven tickets. Which equation of a line represents the line that passes through the two points that represent the data in this problem?
(52)

A $y = 3x + 1$ **B** $y = \frac{1}{3}x + 1$ **C** $y = -3x + 1$ **D** $y = 3x - 1$

*28. Write Find the slope of the line that passes through $(-1, 2)$ and $(3, 2)$. Explain.
(52)

*29. Rachel and a Michelle are crocheting a baby blanket that will be 72 inches long. Rachel crochets the first 24 inches and then gives the blanket to Michelle to finish. Michelle expects to crochet at a rate of 8 inches per day. How many days will it take Michelle to finish the blanket?
(52)

   a. Write an equation giving the length y of the blanket (in inches) when crocheting for x days.

   b. Graph the equation using a graphing calculator.

   c. How long it will take Michelle to finish the blanket?

Determining the Equation of a Line Given Two Points

LESSON 52

*30. **Transportation** John is 500 miles from home. He is traveling toward home at a constant rate of 65 miles per hour. The distance d (in miles) away from home after t (in hours) is given by the equation $d = 500 - 65t$.
(52)

a. Use a graphing calculator to make a table of values with the values of t from 0 to 5 in increments of 1.

b. How far away from home John is after 4 hours?

Adding and Subtracting Polynomials

LESSON 53

Simplify each expression.

1. $18 - 12 + 4^2$
   (4)

2. $-2[7 + 6(3 - 5)]$
   (4)

3. Graph the inequality $x \leq 8$.
   (50)

4. Write an inequality for the graph shown below.
   (50)

5. Compare: $0.00304 \bigcirc 3.04 \times 10^{-4}$.
   (37)

6. Factor the polynomial $18a^2b^3c - 45ab^6c$ completely.
   (38)

7. **Rockets** The formula $h = -16t^2 + 80t + 8$ can be used to find the height of a rocket
   (38) that is launched into the air from 4 feet off the ground with an initial velocity of
   80 feet/second. Write the formula by factoring the right side of the equation using
   the GCF.

8. Simplify the expression $(3a)(6a^2b)^3$.
   (40)

9. **Automotive Performance** The ratio $\frac{40 + 6t}{20 + 10t}$ compares the speeds of a car starting from
   (43) two different cruising speeds after an acceleration lasting t seconds. What is the
   simplified form of the expression?

10. **Multi-Step** The inflation rate, the rate at which the things people buy and use increase
    (44) in price, is constantly changing. During the first month of 2007, the inflation rate
    was 2.08%. By the sixth month, the rate had changed to 2.69%.
    a. Graph this information on a coordinate plane as 2 points connected by a line segment.
    b. What is the slope of the line segment?
    c. **Predict** Use the slope to predict the probable inflation rate for July of 2007.

11. **Write** What phrases can be used to indicate the inequality $\geq$?
    (45)

12. Write the expression $-\sqrt{b}$ with a fractional exponent.
    (46)

Adding and Subtracting Polynomials

LESSON 53

13. **Astronomy** The table shows the sidereal periods (the time it takes for a planet to orbit the sun) for the 8 planets of our solar system. What is the mean time it takes a planet in our solar system to revolve around the sun? What is the median time?
(48)

Planet	Sidereal Period (in days)
Mercury	88
Venus	225
Earth	365
Mars	687
Jupiter	4,329
Saturn	10,753
Uranus	30,660
Neptune	60,150

14. **Multiple Choice** Which of the following graphs is correct for $-3 \leq y$?
(50)

A [number line from -6 to 0, closed dot at -2, shaded left]
B [number line from -6 to 0, open dot at -2, shaded right]
C [number line from -6 to 0, closed dot at -2, shaded right]
D [number line from -4 to 0, open dot at -2, shaded left]

15. **Generalize** How can the graph of an inequality show if a number is in the solution set?
(50)

16. Find the percent of increase or decrease to the nearest percent if the original price was $48,763 and the new price is $39,400.
(47)

17. Identify the slope and y-intercept in the equation of a line given below.
(49)

$$1.5x + 3y - 6 = 0$$

*18. Simplify the rational expression, if possible. Identify any excluded values.
(51)

$$\frac{2k + 6}{k + 2}$$

*19. **Water Polo** A water-polo pool is $\frac{50x + 150}{3x + 5}$ meters long and $\frac{40x}{3x + 5}$ meters wide. Find the perimeter of the pool.
(51)

Adding and Subtracting Polynomials

20. Error Analysis Two students simplify the expression $\frac{y+4}{y-2}$. Which student is correct? Explain the error.
(51)

Student A	Student B
$\frac{y+4}{y-2}$ cannot be simplified	$\frac{y+4}{y-2}$ $\frac{y+4}{y-2}$ $\frac{4}{-2}$ -2

21. Geometry Points A, B, and C are three vertices of a rectangle. Plot the three
(52) points. Then find the coordinates of the fourth point, D, to complete the rectangle. Finally, write the equation of the line that passes through points B and D and forms a diagonal of the rectangle.

$$A\,(-2,\,3),\ B\,(4,\,3),\ C\,(4,\,-1)$$

*22. Measurement Two farmers each harvested 50 acres of tomatoes per day from their
(52) fields. The area of one farmer's field is 800 acres and the area of the other farmer's field is 600 acres.

   a. Write an equation giving the unharvested area y of the larger field (in acres) after x days.

   b. Write an equation giving the unharvested area y of the smaller field (in acres) after x days.

   c. What is the unharvested area of each field (in acres) after 10 days?

*23. Write an equation of a line that passes through points $(6, -3)$ and has a slope
(52) of -2.

*24. Multiple Choice Which expression is not a polynomial?
(53)
   A $-12b$ **B** $x^2 + x^{-1}$

   C $y^2 - y + 6$ **D** -60

*25. (Advertisement) The polynomials below approximate the amount of dollars (in
(53) millions) one company spent on advertising children's books on the national and local level for each year during a certain period. In each polynomial, x represents the years since the company began its campaign. Write a polynomial that gives a combined amount spent each year on national and local advertising.

$$\text{national} = 59x^2 - 262x + 3888$$
$$\text{local} = -33x^3 + 611x^2 - 1433x + 28{,}060$$

Adding and Subtracting Polynomials

LESSON 53

*26. What is the degree of $-a^2b^2c^3 + 5x^5$?
(53)

*27. A line passes through the points (4, 5) and (6, 4).
(52)
  a. Write the equation of the line in point-slope form.
  b. Write the equation of the line in slope-intercept form.
  c. Graph the line using a graphing calculator.
  d. Fill in the missing coordinates of the following points: $(x, 4)$ and $(3, y)$.

28. Write an equation of the line in point-slope form that has a slope of -1
(52) and passes through the point (3, 1).

*29. **Write** Is 4 equal to $4x^0$? Explain.
(53)

*30. **Verify** What polynomial can be subtracted from $3x^2 + 7x - 6$ to get -6?
(53)

Name _____ Date _____ Class _____

Displaying Data in a Box-and-Whisker Plot

LESSON 54

Solve each equation. Check your answer.

1. (26) $\frac{1}{2} + \frac{3}{8}x - 5 = 10\frac{1}{2}$

2. (24) $0.02x - 4 - 0.01x - 2 = -6.3$

3. (28) $x - 5x + 4(x - 2) = 3x - 8$

Simplify.

4. (38) $\dfrac{2x^2 - 10x}{2x}$

5. (39) $\dfrac{b^2}{d^{-3}}\left(\dfrac{db^{-2}}{4} - \dfrac{3f^{-3}d^2}{b^{-2}}\right)$

*6. (40) **Chemistry** Avogadro's number is represented by 6.02×10^{23}. Writing this number as a product, which value would have to be in the exponent of the expression $(6.02 \times 10^{15})(10^{-})^2$?

7. (41) **Multi-Step** The formula to convert degrees Fahrenheit to degrees Celsius is $C = \frac{5}{9}(F - 32)$.
   a. On a separate sheet of paper, make a table of the equivalent Celsius temperature to -4, 32, 50, and 77 degrees Fahrenheit.
   b. Use the table to make a graph of the relationship.
   c. Find the slope of the graph.

*8. (54) A class makes a box-and-whisker plot to show how many children are in each family. Identify the median, upper and lower quartiles, upper and lower extremes, and the interquartile range.

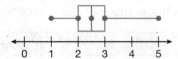
Children per Family

*9. (54) A doctor makes a box-and-whisker plot to show the number of patients she sees each day. Identify the median, upper and lower quartiles, upper and lower extremes, and the interquartile range.

Patients per Day

*10. (54) **Formulate** Create a data set that meets the following criteria: lower extreme 62, lower quartile 70, median 84, upper quartile 86, and upper extreme 95.

Displaying Data in a Box-and-Whisker Plot

LESSON 54

*11. **Multiple Choice** Using a box-and-whisker plot, which information can you gather?
(54)

    A the mode   **B** the range

    C the mean   **D** the number of data values

*12. **Astronomy** The planets' distances (in millions of miles) from the sun are as follows:
(54)

    36, 67, 93, 142, 484, 887, 1765, and 2791

Make a box-and-whisker plot of these distances and determine if any planet's distance is an outlier.

13. Find the percent of increase or decrease to the nearest percent from the original price
(47) of $2175.00 to the new price of $2392.50.

*14. Choose an appropriate measure of central tendency to represent the data set. Justify
(48) your answer.

    12 quiz scores (in percents): 86, 92, 88, 100, 86, 94, 92, 78, 90, 96, 94, 84.

*15. **Manufacturing** A skateboard factory has 467 skateboards in stock. The factory can
(49) produce 115 skateboards per hour. Write a linear equation in slope-intercept form to represent the number of skateboards in inventory after so many hours if no shipments are made.

16. Write an inequality for the graph below.
(50)

17. **Automotive Maintenance** The following chart shows the wear on a particular brand of
(44) tires every 10,000 miles. What is the average rate of wear for this brand of tires?

Mileage	Tread Depth
10,000	20 mm
20,000	16 mm
30,000	12 mm
40,000	8 mm

Displaying Data in a Box-and-Whisker Plot

LESSON 54

18. The diagram shows types of transportation. Use the diagram to determine if each
(Inv 4) statement is true or false. If the statement is false, provide a counterexample.

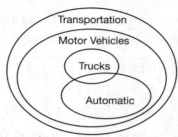

   a. If a vehicle is a truck, then the vehicle is an automatic.

   b. All trucks are motor vehicles.

19. **Write** Explain the difference between $\sqrt{-1}$ and $\sqrt{1}$.
(46)

20. **Write** Explain why $\frac{2g}{2g+6}$ cannot be simplified to $\frac{1}{6}$.
(51)

21. **Multiple Choice** Which expression is not equivalent to $3rd^{-1} - \frac{6}{r^{-1}d}$?
(51)

   A $-3rd^{-1}$ **B** $\frac{-3r}{d}$ **C** $\frac{-3d}{r}$ **D** $\frac{-3}{r^{-1}d}$

*22. **Telecommunications** Jane bought a prepaid phone card that had 500 minutes. She
(52) used about 25 minutes of calling time per week. Write and graph an equation
to approximate her remaining calling time y (in minutes) after 9 weeks.

23. Find the slope of the line that passes through $(1, 6)$ and $(3, -4)$.
(52)

24. Describe a line that has a slope of 0 and passes through the point $(-1, 1)$.
(52)

25. Write an equation in slope-intercept form of a line that passes through the
(52) points $(14, -3)$ and $(-6, 9)$.

26. **Error Analysis** Students were asked to find the sum of the polynomials vertically.
(53) Which student is correct? Explain the error.

Student A	Student B
$-6x^3 - 3x^2 + 5$	$-6x^3 - 3x^2 + 5$
$+\ \ 2x^3 - x\ \ - 7$	$+\ \ \ \ 2x^3\ \ \ \ - x - 7$
$-4x^3 - 4x^2 - 2$	$-4x^3 - 3x^2 - x - 2$

27. **Geometry** Write a polynomial expression for the perimeter of the triangle. Simplify the
(53) polynomial and give your answer in standard form.

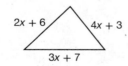

201

Saxon Algebra 1

Displaying Data in a Box-and-Whisker Plot

LESSON 54

28. Measurement The length of the sidewalk that runs in front of Trina's house is $3x - 16$ and the width is $5x + 21$. Find the perimeter of the sidewalk.
(53)

*29. Multi-Step The table shows the amounts that Doug and Jane plan to deposit in their
(53) savings account. Their savings account has the same annual growth rate g.

Date	1/1/04	1/1/05	1/1/06	1/1/07
Doug	$300	$400	$200	$25
Jane	$375	$410	$50	$200

a. On January 1, 2007, the value of Doug's account D can be modeled by $D = 300g^3 + 400g^2 + 200g + 25$, where g is the annual growth rate. Find a model for Jane's account J on January 1, 2007.

b. Find a model for the combined amounts of Doug and Jane's account on January 1, 2007.

30. Find the sum of $(9x^3 + 12) + (16x^3 - 4x + 2)$ using a horizontal format.
(53)

Solving Systems of Linear Equations by Graphing

LESSON 55

Solve each equation. Check your answer.

1. $39.95 + 0.99d = 55.79$
(24)

2. $12.6 = 4p + 1$
(24)

3. $2(b - 4) = 8b - 11$
(28)

4. $1.8r + 9 = -5.7r - 6$
(28)

Simplify.

5. $\dfrac{k^{-4}}{2}$
(32)

6. $10r^{-3}t^4$
(32)

7. $\dfrac{p^{-9}q^{-4}}{r^2 s^{-3}}$
(32)

8. **Write** A student is conducting a class survey. Write each question so
(58) that it is not biased.
   a. Doesn't milk taste better than the juice for lunch?
   b. Isn't math the best class?
   c. Shouldn't the class exercise in the morning?

9. Simplify $\dfrac{rs}{z^2}\left(\dfrac{pr^{-5}s^4}{z^{-7}} - 7p^{-2}s^{-1} + \dfrac{5}{z^{-3}}\right)$.
(39)

10. Simplify $(2ab^2)^2(-2b^2)^2$.
(40)

11. **(Handicap Ramps)** Building codes require that handicap ramps rise no more than 1 foot
(41) for every 12 feet of horizontal distance. What is the maximum allowable slope of a handicap ramp?

12. **Multi-Step** Majid is hiking a total distance of 620 yards. He has already hiked
(42) 272.8 yards. What percentage of the total has Majid already completed?
   a. Write an equation to solve the problem.
   b. Find the solution.
   c. Determine the reasonability of your solution.

*13. **(Charity)** Gaby has $250 in donations, and she receives pledges of $40 for every mile
(45) she walks in a charity walk-a-thon. She wants to have total donations of at least $500. Write an inequality to represent this situation.

Solving Systems of Linear Equations by Graphing

LESSON 55

14. Multi-Step A storage box in the shape of a cube is filled halfway with packing material. There are 5324 cubic inches of packing material in the box.
 a. What is the volume of the box?
 b. What is the length of one side of the box?

15. Verify Vickie's grade on her history report was 48 out of 60. Annie's grade on her history report was 47 out of 60. How much greater was Vickie's percent grade than Annie's?

16. The data represents the number of clocks in 9 households: 7, 5, 13, 5, 7, 10, 5, 5, 8. What measure of central tendency best describes the data?

*17. Estimate In 1997 the enrollment in a preschool was approximately 500 students. After that, the enrollment increased by approximately 20 students per year.
 a. Write an equation to model the school's enrollment y in terms of x, the number of years since 1997.
 b. Estimate the school's enrollment in 2005.

18. Multiple Choice A line through which of the following pairs of points has a slope of 2?
 A (0, 0) (0, 4)
 B (0, 4) (6, 4)
 C (4, 0) (6, 4)
 D (6, 1) (2, 4)

*19. Graph the linear equation $2x - 7y + 5 = 0$ using slope-intercept form. Check your graph with a graphing calculator.

20. (Leisure Time) The current top score on Danielle's favorite arcade game is 13,468 points. Write an inequality expressing all the possible scores Danielle could score to not score lower than her top score. Graph the inequality.

*21. Simplify the rational expression $\dfrac{5f^9}{20f^4}$, if possible. Identify any excluded values.

22. Error Analysis Students were asked to find the degree of the polynomial $2x^3 + 5x^2 - 3x + 1$. Which student is correct? Explain the error

Student A	Student B
The degree is 3.	The degree is 6.

*23. (Golf) The height (in feet) of a golf ball t seconds after it is hit is given by the polynomial $-16t^2 + 100t$. What is the height of the golf ball 3 seconds after it is hit? 5 seconds after it is hit?

Solving Systems of Linear Equations by Graphing

LESSON 55

24. Find the difference of the following expressions using a vertical form:
$(3n^3 + 2n - 7) - (n^3 - n - 2)$.

*25. A box-and-whisker plot is made to show the gas prices per gallon. Identify the median, upper and lower quartiles, upper and lower extremes, and the interquartile range.

Gasoline Prices

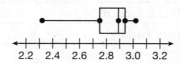

26. Probability What is the probability that a data value will fall into the interquartile range?

*27. Verify Use a graphing calculator to find the solution to the linear system. Check your answer.

$$y = -\frac{1}{2}x + 4$$
$$y = \frac{1}{4}x - 2$$

*28. Multiple Choice What is the solution of the linear system below?

$$y = \frac{3}{4}x + 1$$
$$y = -\frac{1}{4}x - 3$$

A $(4, -2)$ **B** $(-2, 4)$

C $(-4, -2)$ **D** $(4, 0)$

*29. (Phone Service) A cell phone company named Talk-A-Lot charges $1.25 for a connection fee and $0.25 for each minute. Another company named Save-N-Talk charges $0.50 for each minute but has no connection fee.

a. Write a system of linear equations to represent the total cost for a phone call through each company.

b. Graph the system of equations using a graphing calculator and calculate the point of intersection.

c. What does the point of intersection represent?

Solving Systems of Linear Equations by Graphing

*30. A dietician encourages students to eat lean meats. The table shows the fat grams of different types of meat.

Type of Meat (1 oz)	Fat Grams
White Meat Chicken	1
White Meat Turkey	1
Fresh Fish	1
Dark Meat Chicken	3
Dark Meat Turkey	3
Fish	3
Ground Beef	5
Beef	5
Pork	5
Lamb	5
Pork Sausage	8
Spare Ribs	8
Lunch Meat	8
Hot Dog (Turkey or Chicken)	8

a. Make a box-and-whisker plot for the set of data.

b. Why is this graph missing a whisker?

Name _____ Date _____ Class _____

Identifying, Writing, and Graphing Direct Variation

LESSON 56

Solve.

1. (26) $-4\frac{3}{4} + 3\frac{3}{5}x = 13\frac{1}{4}$

2. (26) $0.3x - 0.02x + 0.2 = 1.18$

3. (26) Solve the following equation for p: $7p + 3w = w - 12 - 3p$.

4. (38) Write the prime factorization of 315.

5. (42) What percent of 160 is 88?

6. (42) What number is 140 percent of 70?

7. (42) Twenty percent of what number is 18?

*8. (56) Tell whether the equation $\frac{y}{11} = x$ represents a direct variation.

*9. (56) Tell whether the equation $3y = x$ represents a direct variation.

10. (40) Simplify $\left(\frac{10x^3}{y}\right)^2 (-2x^2y^2)^3$.

11. (42) **Market Research** An automobile manufacturer sent out 22,000 questionnaires to a sample of their customers who bought a new car from them this year. 63% of the customers who received the questionnaires filled them out and returned them. How many customers returned their questionnaires?

12. (43) **Multi-Step** A circle with radius $10x$ is inscribed in a square.
    a. Write an expression for the area of the circle.
    b. Write an expression for the area of the square.
    c. Form a simplified rational expression from the ratio of the two areas. Use the circle area as the numerator.

*13. (56) **Generalize** Explain why the graph of a direct variation will always go through the origin (0, 0).

*14. (56) **Multiple Choice** Which equation represents a direct variation?
    A $y = 2 - x$ B $x + 3 = y$ C $xy = -2$ D $3y = -2x$

*15. (56) **Currency** The number of euros received varies directly with the number of dollars exchanged. If $5.00 is exchanged for 3.65 euros, how many euros would be received for $8.00?

16. (49) Write the equation of the line in slope-intercept form.
$$-4y + x = 2$$

17. (50) Write an inequality for the graph at right.

Identifying, Writing, and Graphing Direct Variation

LESSON 56

18. Camping A tent is in the shape of a triangular pyramid with a square footprint (base). The tent is 6 feet high and has a volume of 128 cubic feet. What is the length of the base?
(46)

19. Model Formulate a data set that has a mean value of 9, a median of 9, and a mode of 12.
(48)

*20. Tennis An avid tennis player is painting lines to make a court at home. The tennis court measures $\frac{150x + 200}{2x + 4}$ feet long and $\frac{60x + 220}{2x + 4}$ feet wide. How many feet did the tennis player paint to mark the outside lines of the court?
(51)

*21. Write the equation of the line that has a slope of $\frac{1}{2}$ and passes through the point (2, 8) in point-slope form.
(52)

22. Write How is the degree of a polynomial determined?
(53)

23. Multiple Choice What is the degree of $a^2b^3 - a^3b^4 + 2ab$?
(53)
   A 14   **B** 2   **C** 7   **D** 5

24. Make a box-and-whisker plot to display the data titled "Average High Temperature in Phoenix, AZ".
(54)

        66, 70, 75, 84, 93, 103, 105, 103, 99, 88, 75, 66

25. Running A group of friends run a marathon together. Their times in minutes are shown below.
(54)

        201, 240, 236, 270, 245, 252, 239, 267, 241, 229, 360

Make a box-and-whisker plot of these times and determine the outliers, if any.

*26. Error Analysis Two students decide whether there is an outlier. Which student is correct? Explain the error.
(54)

grades on a math test: 100, 75, 80, 96, 77, 82, 88, 89, 79, 92, 100, 78, 95, 60

Student A	Student B
$78 - 1.5(95 - 78) = 52.5$	$78 - 1.5(85 - 78) = 67.5$
There is no outlier.	Yes, 60 is an outlier.

*27. Measurement A square has side length s, while an equilateral triangle has length $s + 1$. Write and solve a system of linear equalities to determine the length of the side at which the perimeters of the triangle and the rectangle would be equal.
(55)

$$P = 4s$$
$$P = 3(s + 1)$$

Identifying, Writing, and Graphing Direct Variation

LESSON 56

28. Error Analysis Two students used a graphing method to solve the system of equations.
$$-4x + 3y = 5$$
$$2x + 3y = -16$$
They used substitution to verify their solutions. Which student is correct? Explain the error.

Student A	Student B
$-4(-3.5) + 3(-3) = 5$	$-4(4) + 3(7) = 5$
$2(-3.5) + 3(-3) = -16$	$2(4) + 3(7) = -16$

*29. Geometry The measure of the angles in a parallelogram total 360°. Write and solve a system of equations to determine the values of x and y.

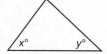 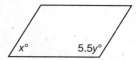

30. Multi-Step Thomas has a younger brother named Miguel. Thomas is 5 years older than Miguel. If Thomas's age is decreased by three years, it is equivalent to double Miguel's age.

a. **Formulate** Write a system of equations to represent the age of Thomas. Let t represent Thomas's age and m represent Miguel's age.

b. Solve the system to determine the ages of Thomas and Miguel.

c. **Verify** Explain why your answer is reasonable.

Name _____ Date _____ Class _____

Finding the Least Common Multiple — LESSON 57

Solve each equation for the variable indicated.

1. (29) $s + 4t = r$ for s
2. (29) $3m - 7n = p$ for m

Solve each proportion.

3. (36) $\dfrac{3}{4} = \dfrac{a+5}{21}$
4. (36) $\dfrac{3}{y-3} = \dfrac{1}{9}$

Find each percent change. Tell whether it is a percent increase or decrease.

5. (47) 50 to 20
6. (47) 12 to 96

*7. (57) **Write** Explain how the processes for finding the LCM and the GCF of two numbers differ.

*8. (57) Find the LCM of 24 and 84.

9. (38) Find the GCF of 24 and 84.

10. (41) Draw the horizontal and vertical lines passing through the point (6, 5).

11. (44) **Multi-Step** The graph at right shows the temperature measured at different times during a summer day. A weak cold front passed through in the early afternoon.

    a. What was the rate of increase in temperature from 6:00 in the morning until noon?

    b. What was the rate of change in temperature from noon until 2:00 p.m.?

    c. What would have been the rate of increase in temperature from 6:00 a.m. to 4:00 p.m. if those were the only times at which anyone had measured the temperature?

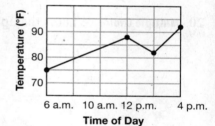

12. (48) **Multi-Step** Tenecia timed her classmates to see how many seconds they could jump up and down on one leg, and got these data results: 40, 33, 41, 36, 26, 38, 44, 46, 44, 35, 40, 37, 40, 31, 38, 41, 33, 51, 38, 40, 43, 35, 37, 38.

    a. Show Tenecia's data on a line plot.

    b. **Analyze** Describe the overall shape of the line plot.

    c. **Generalize** Make a statement about this data based on the line plot.

13. (47) (Zoo) The number of visitors at the petting zoo jumped from 600 in May to 900 in July. What is the percent of increase?

14. (53) Find the sum of $(-9z^3 - 3z) + (13z - 8z^2)$ using a vertical form.

© Saxon. All rights reserved. Saxon Algebra 1

Finding the Least Common Multiple

LESSON 57

15. (**Phone Rates**) Ronnie is considering switching cell phone plans. The new plan
(49) would cost $5.95 per month plus $0.04 cents per minute.
   a. Analyze What are the restrictions on the domain and range of this situation?
   b. Write an equation in slope-intercept form to represent this situation.

16. Simplify the rational expression, if possible. Identify any excluded values.
(51) $\dfrac{k-3}{7k-21}$

17. Write an inequality for the graph.
(50)

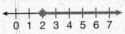

18. (**Fitness**) After six weeks of participating in a fitness program, Joyce jogs 40 miles
(49) per week. Her average mile gain has been 2 miles per week. Write an equation that
models Joyce's weekly mileage m in terms of the number of weeks w that she stays in
the program. When will Joyce jog over 60 miles?

*19. Verify Verify that 221 is an outlier for the following data set that shows the number
(54) of dollars individuals have in a savings account.

143, 95, 116, 169, 146, 131, 144, 150, 191, 127, 162, 221, 150

20. Multiple Choice What is the upper extreme in the box-and-whisker plot?
(54)

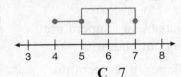

A 5   **B** 6   **C** 7   **D** There is none.

*21. (**Basketball**) At the beginning of the third quarter of a basketball game, Stephen had
(55) 16 points and Robert had 14 points. During this quarter Stephen only attempted
2-point shots and Robert only attempted 3-point shots. For every shot Stephen made,
Robert made a shot. At the end of the third quarter, they were tied.
   a. Write a system of linear equations to represent the total points earned by each
   player by the end of the third quarter. Let x be the number of shots that were
   made and y be the total number of points.
   b. How many shots would each player have to make so that the game would be tied?
   c. How many points had each player earned by the end of the third quarter?

*22. Use a graphing calculator to solve the system of linear equations.
(55)
$2x + 3y = 12$
$-3x + 2y = 8$

23. Tell whether the equation $7 + y = x$ represents a direct variation.
(56)

Finding the Least Common Multiple

LESSON 57

24. Measurement The number of centimeters varies directly with the number of inches. A table is 25 inches, or 63.5 centimeters, long. How many centimeters long is a book that is 9 inches?

25. Error Analysis Two students solve the following problem:

A person's age in months is directly proportional to the number of years the person has been alive. A 3-year-old is 36 months old. How many months old is a 24-year-old?

Which student is correct? Explain the error.

Student A	Student B
$y = kx$	$y = kx$
$36 = k \cdot 3$	$36 = k \cdot 3$
$12 = k$	$12 = k$
$24 = 12x$	$y = 12 \cdot 24$
$x = 2$	$y = 288$

*26. Geometry The circumference of a circle is directly proportional to the length of the radius. If the radius is 10 centimeters, the circumference is 20π centimeters. What is the circumference of a circle with a radius of 29 feet?

*27. Multi-Step The fluid pressure on an object submerged in water varies directly with the depth of the object. At 2 feet below the surface of the water, there is a fluid pressure of 124.8 pounds per square feet.
a. Write the equation that represents this relationship.
b. Graph this relationship using a graphing calculator, and estimate the amount of pressure exerted on an object 9 feet below the surface.
c. Use the graph to estimate how deep an object is that experiences 312 pounds per square foot of pressure.

*28. Write Explain how the LCM is used when adding and subtracting fractions.

*29. Multiple Choice What is the LCM of $15k^{11}$ and $36k^6$?
A $3k^6$ B $180k^6$ C $180k^{11}$ D $540k^{17}$

*30. (Clocks) One clock chimes every $4x^3y^6$ minutes. Another clock cuckoos every $2xy^2$ minutes. A third clock rings every $6x^4y$ minutes. All the clocks just sounded together. How many more minutes will it be until they all go off together again?

Name _____ Date _____ Class _____

Multiplying Polynomials
LESSON 58

Graph the inequalities on a number line.

1. $x \leq -2$
(50)

2. $x > 2$
(50)

Solve the equations. Check your answers.

3. $4(2x - 3) = 3 + 8x - 11$
(28)

4. $-5m + 2 + 8m = 2m + 11$
(28)

*5. Multiple Choice Which product equals $6x^2 - 7x + 2$?
(58)
A $(6x + 1)(x - 7)$
B $(2x - 1)(3x + 2)$
C $(-6x + 1)(-x + 7)$
D $(-2x + 1)(-3x + 2)$

*6. (Web Design) Luke is designing a web page. At the top is a banner ad that is
(58) $3x^2 + 6x + 4$ inches high. The length is $x + 5$ inches. What is the area of the banner ad on Luke's web page?

*7. $(x - 2)(x + 3)$
(58)

*8. $(2x - 3)(2x + 3)$
(58)

*9. Analyze Monroe says the product of x^2 and $x^3 + 5x^2 + 1$ is $x^6 + 5x^4 + x^2$. Do you
(58) agree? If not, explain your reasoning.

*10. Find the product of $5(x^2 + 3x - 7)$.
(58)

11. Find the LCM of 35, 60, and 100.
(57)

*12. Measurement On a piece of plywood, there is a seam every 8 feet and a stud every
(57) 16 inches. After how many feet or how many inches is there a seam and a stud?

13. Error Analysis Two students are asked to find the LCM of $24x^5y^2$ and $32x^3y^6$. Which
(57) student is correct? Explain the error.

Student A
$24x^5y^2 = 2 \cdot 2 \cdot 2 \cdot 3 \cdot x \cdot x \cdot x \cdot x \cdot x \cdot y \cdot y$
$32x^3y^6 = 2 \cdot 2 \cdot 2 \cdot 2 \cdot 2 \cdot x \cdot x \cdot x \cdot y \cdot y \cdot y \cdot y \cdot y \cdot y$
LCM $= 2 \cdot 2 \cdot 2 \cdot x \cdot x \cdot x \cdot y \cdot y$
LCM $= 8x^3y^2$

Student B
$24x^5y^2 = 2 \cdot 2 \cdot 2 \cdot 3 \cdot x \cdot x \cdot x \cdot x \cdot x \cdot y \cdot y$
$32x^3y^6 = 2 \cdot 2 \cdot 2 \cdot 2 \cdot 2 \cdot x \cdot x \cdot x \cdot y \cdot y \cdot y \cdot y \cdot y \cdot y$
LCM $= 2 \cdot 2 \cdot 2 \cdot 2 \cdot 2 \cdot 3 \cdot x^5 \cdot y^6$
LCM $= 96x^5y^6$

*14. Geometry Write a counterexample for the statement: If the measures of the angles in
(Inv 4) a polygon add up to 360 degrees, then the polygon is a rectangle.

© Saxon. All rights reserved. Saxon Algebra 1

Multiplying Polynomials

LESSON 58

*15. **Multi-Step** A line of children's shoes are designed so that every 10th pair has blinking lights, every 24th pair glows in the dark, and every 15th pair is waterproof.
   a. How many pairs of shoes are made before one pair blinks, glows, and is waterproof?
   b. If 2000 pairs of shoes are made, how many blink, glow, and are waterproof?

16. Tell whether the set of ordered pairs (9, 6), (11, 8), and (22, 19) represents a direct variation.

*17. **Meteorology** The number of seconds between seeing a lightning flash and hearing the thunder is directly proportional to the number of kilometers between you and the lightning. When there are 27 seconds between the flash and the sound, you are 9 kilometers from the lightning. How far is the lightning if the time is 51 seconds?

18. **Error Analysis** Two students are writing an equation for a direct variation that includes the point (15, −5). Which student is correct? Explain the error.

Student A	Student B
$y = kx$	$y = kx$
$-5 = k(15)$	$15 = k(-5)$
$-\frac{1}{3} = k$	$-3 = k$
$y = -\frac{1}{3}k$	$y = -3k$

19. **Data Analysis** A box-and-whisker plot shows the number of signatures in yearbooks at the end of the school year. Identify the median, upper and lower quartiles, upper and lower extremes, and the interquartile range.

Signatures in Yearbooks

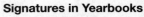

20. Use a proportion to find what percent of 108 is 81.

21. Simplify $\dfrac{9x - 81}{4x^2 - 36x}$.

22. Simplify $\dfrac{(x-4)(2x-3)}{7x - 28}$.

23. **Physics** According to the Theory of Relativity, and as represented by the equation below, the mass of a particle changes with its velocity. When is the rational expression of the equation undefined? (Hint: Think of the fraction in the denominator as a single variable.)

$$m^2 = \dfrac{m_0^2}{1 - \dfrac{v^2}{c^2}}, \quad m_0 = \text{rest mass}, \; v = \text{velocity}, \; c = \text{speed of light}$$

Multiplying Polynomials   LESSON 58

24. Multi-Step The $5\frac{1}{2}$ cups of orange juice and half of the apricot juice left in the
(45) refrigerator added up to no more than 10 cups when combined in the pitcher.
   a. Translate the sentence into an inequality.
   b. Translate the inequality into words.

25. (Weather) The table shows the total
(48) snowfall amounts for five months
in Fargo, North Dakota, and
St. Paul, Minnesota. Based on the
measures of central tendency,
which city receives less snow over
the winter on average?

Date	Fargo, North Dakota	St. Paul, Minnesota
November 2006	0.2 in.	0.2 in.
December 2006	5.4 in.	4.3 in.
January 2007	2.9 in.	5.5 in.
February 2007	10.9 in.	12.6 in.
March 2007	9.4 in.	3.6 in.

26. Write Create a table of values and a graph for the equation $4x - 2y = 10$. Explain
(49) how the three representations are related.

27. (Energy) The amount (in billions of dollars) spent on natural gas N and electricity E
(53) by a randomly selected country in a survey can be modeled by the equations below,
where t is the number of years since the survey began. Write an expression that
models the total amount A (in billions of dollars) spent on natural gas and electricity
by this country's residents.

$N = 1.521t^2 - 2.304t + 56.659$
$E = -0.104t^2 + 5.879t + 196.432$

28. Formulate Kami challenged her classmate to find the pattern in a logic puzzle. When
(52) she says 8, she means 9. When she says 16, she means 8. Find the equation of the line
that passes through the two points that represent the data in the problem. Use the
equation to predict what Kami means when she says -8.

29. Simplify the rational expression $\frac{4m - 8}{2m - 4}$, if possible. Identify any excluded values.
(51)

30. Analyze Grace is making applesauce for a party. She has enough apples for
(50) 10 servings. If each guest gets 1 serving, how many guests could come to the party?
Would a graph of this situation be helpful? Explain.

Name _____ Date _____ Class _____

Solving Systems of Linear Equations by Substitution
LESSON 59

Solve.

1. (28) $-[-(-k)] - (-2)(-2 + k) = -k - (4k + 3)$

2. (26) $\frac{1}{3} + 5\frac{1}{3}k + 3\frac{2}{9} = 0$

3. (13) Simplify $\sqrt{9} + \sqrt{16} - \sqrt{225}$.

4. (1) Give an example of a rational number that is not an integer.

5. (13) True or False: The number $\sqrt{49}$ is a rational number.

*6. (59) Solve by substitution: $y = 3x - 5$
$y = -2x + 15$.

*7. (59) Solve by substitution: $y = -8x + 21$
$y = -3x + 6$.

*8. (59) **Write** How do you know if a point is a solution to a system of equations?

*9. (59) **Multiple Choice** Which ordered pair is a solution to the system of equations?

$$4x + 9y = 75$$
$$8x + 6y = 66$$

A (3, 7)

B (0, 11)

C (10, 4)

D (12, −5)

*10. (59) The sum of two numbers is 64. Their difference is 14. Find each of the numbers.

11. (58) **Error Analysis** Students were asked to find the product of $2b^2(b^3 + 4)$. Which student is correct? Explain the error.

Student A	Student B
$2b^2(b^3 + 4)$	$2b^2(b^3 + 4)$
$2b^2(b^3) + 2b^2(4)$	$2b^2(b^3) + 2b^2(4)$
$2b^5 + 8b^2$	$2b^6 + 8b^2$

Solving Systems of Linear Equations by Substitution

LESSON 59

*12. **Geometry** The length of a rectangular pool is four times the width. A four-foot-wide deck surrounds the pool. Write a polynomial expression for the area of the pool and deck. Use the Distributive Property and write your answer in standard form.
(58)

*13. **Multi-Step** Henry has a game that includes a number cube. The side length of the cube is $(5x + 1)$ inches. Find the volume of the number cube.
(58)
a. Find the area of the base. Multiply the length times the width.
b. Find the volume by multiplying the product found in part **a** by the height.

*14. **Measurement** Tim has a garden in the shape of a right triangle. The triangle has a base of $6x^2 + 8x + 12$ feet and a height of $(x - 1)$ feet. What is the area of Tim's garden?
(58)

$(x - 1)$ ft
$(6x^2 + 8x + 12)$ ft

15. Find the product of $4x(x^2 + 2x - 9)$.
(58)

16. Write an equation of the line in slope-intercept form. The slope is 2; the y-intercept is -1.
(49)

17. What is the degree of $12x^4x^3 + 6xy + 41x^2y^3$?
(53)

18. **Baseball** The percentage of games won is used to determine a team's standing in the league. In the American League, the following percentages were recorded:
(54)

0.604, 0.540, 0.505, 0.465, 0.380, 0.600, 0.584, 0.505, 0.446,
0.430, 0.580, 0.545, 0.475, 0.451

Make a box-and-whisker plot of these percentages and determine if there is an outlier.

*19. **Estimate** The cost of a chain is directly proportional to its length. Use the graph to estimate the cost of a chain that is 18 inches long.
(56)

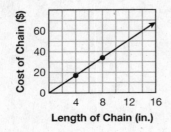

Solving Systems of Linear Equations by Substitution

LESSON 59

20. Multiple Choice Which point represents the same direct variation as $(3, -9)$?
  A $(4, -8)$ **B** $(4, -7)$ **C** $(4, -12)$ **D** $(4, -16)$

21. Find the LCM of $16c^6$ and $24c^3$.

*22. Games Every $20x^3y$ turns, you win \$500. Every $12xy^3c$ turns, you get to roll again. How many turns do you take before you win \$500 and get to roll again on the same turn?

23. Find the LCM of $300d^2$ and $90d^4$.

24. Formulate The equations $y = 3x - 24$ and $y = 24x + 9$ define two lines. Write a rational expression that represents the ratio of the first line to the second line. Simplify the expression, if possible.

25. Determine the slope of the line that goes through the points $(-1, 1)$ and $(1, -1)$.

26. Biology Write a counterexample for the following statement: If an animal has wings, then the animal is an insect.

27. Multi-Step A canister of oatmeal has a height of 7 inches. Its volume is 28π cubic inches.
  a. Write an equation you can use to find the radius of the cylinder.
  b. Find the radius of the canister.

28. Astronomy The relative gravity on Jupiter is 2.34. This means that the weight of an object on Jupiter is 2.34 times greater than its weight on Earth. Identify the slope and the y-intercept of the equation representing this relationship and then write an equation for the situation in slope-intercept form.

29. Verify Write the converse of the following statement: If a number is an integer, then it is a rational number. Give an example to show that the converse is false.

30. Write What is an excluded value for a rational expression? Why is it excluded?

Finding Special Products of Binomials

LESSON 60

Simplify.

1. $(3k + 2k^2 - 4) - (k^2 + k - 6)$
2. $(-2m + 1) + (6m^2 - m - 2)$
3. $(x + 4)(x - 5)$
4. $(x + 2)(6x^2 + 4x + 5)$

Find the square of each binomial.

*5. $(3t - 1)^2$
*6. $(3t + 1)^2$

*7. **Painting** Dat is painting a picture for his grandmother. He wants a 3-inch-wide blue border around the square painting. Write a special product and simplify to find the area of the picture including the border.

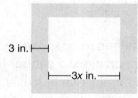

8. Solve the linear system by substitution.

$y = 2x - 9$
$8x - 6y = 34$

*9. **Multiple Choice** Which of the following quadratic expressions is the product of $(6x + 7)^2$?

A $36x^2 + 42x + 49$
B $12x^2 + 84x + 49$
C $36x^2 + 84x + 49$
D $36x + 84x + 49$

10. Solve the linear system by substitution.

$y = 2x - 4$
$y = x + 5$

*11. **Write** How can you check your work when finding a product of two binomials using special-product patterns?

*12. **Verify** Tell whether the statement $(9x + 8)(9x + 8) = 81x^2 + 64$ is true or false. If false, explain why.

*13. **Measurement** The perimeter of a rectangle is 78 feet. The length is 3 feet more than 3 times the width. Find the dimensions of the rectangle.

Finding Special Products of Binomials

LESSON 60

14. Error Analysis Two students are in the process of finding the solution to the system of
(59) equations. Which student is correct? Explain the error.

$$x + 4y = 19$$
$$6x + 5y = 38$$

Student A	Student B
$x = -4y + 19$	$x = -4y + 19$
$6(-4y + 19) + 5y = 38$	$6(-4y + 19) + 5y = 38$
$-24y + 114 + 5y = 38$	$-24y + 19 + 5y = 38$
$-19y = -76$	$-19y = 19$
$y = \frac{76}{19} = 4$	$y = -1$

15. (Installation) A contractor is installing some special light bulbs for two floors in a
(55) building. For the first floor, she purchased 5 natural-light bulbs and 2 ceiling bulbs at
a cost of $23. For the second floor, she purchased 3 natural-light bulbs and 4 ceiling
bulbs at a cost of $25. How much does each type of bulb cost?

*16. Multi-Step The sum of 4 times a girl's age and 7 times a boy's age is 169. The boy
(59) is 1 year older than twice the age of the girl. Find how old each will be 10 years
from now.

*17. Geometry The perimeter of a rectangle is 24 centimeters. The length is 4 centimeters
(59) less than 7 times the width. Find the dimensions of the rectangle.

18. Data Analysis Display the data using a box-and-whisker plot titled "Average Monthly
(54) Rainfall in Cloudcroft, NM (in inches)". Identify any outliers.

1.68, 1.90, 1.54, 0.84, 1.35, 2.21, 6.10, 6.04, 3.11, 1.78, 1.58, 2.33

19. Verify Show that $12f^4$ is the LCM of $6f^4$ and $4f^2$.
(57)

20. Multiple Choice What is the LCM of $(4x^4 - 14x^3)$ and $(6x^2 - 21x)$?
(57)
  A $6x^3(2x - 7)$ **B** $6x^4(2x - 7)$
  C $6(2x^4 - 7x^3)(2x^2 - 7x)$ **D** $(4x^4 - 14x^3)(6x^2 - 21x)$

21. Tell whether the set of ordered pairs (2, 8), (4, 16), and (7, 28) represents a direct
(56) variation.

22. (Tennis) A tennis court's dimensions can be represented by a width of $4x + 25$ feet and a
(58) length of $2x + 15$ feet. Write a polynomial for the area of the court.

Finding Special Products of Binomials

23. Determine the slope of the line graphed below.
(44)

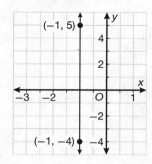

*24. Find the product of $(x + 2)(x + 9)$ using the FOIL method.
(58)

25. Translate the inequality $x - 2.5 > 4.7$ into words.
(45)

26. (Hobby) Eleanor has more than twice as many football trading cards as José. Eleanor has 79 trading cards. What is the greatest number of trading cards José might have?
(45)

27. Multi-Step The varsity football team started summer training with 85 players. 10 players dropped out after 1 week of training, and then 7 players dropped out after 2 weeks.
(47)
   a. What was the percent of decrease for the first week? What was the percent of decrease for the second week?
   b. What was the total percent of decrease for the 2 weeks of summer training?

28. (Law Enforcement) Arnold is using the graph to represent a speed limit of 45 miles per hour. What restrictions should also be placed on the graph?
(50)

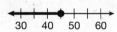

29. Before painting, the edges around a rectangular light fixture must be taped. The length of the fixture is $\frac{2a}{3a-2}$ yards, and the width is $\frac{a-2}{3a-2}$ yards. How much tape is needed?
(51)

30. Justify Identify the slope and y-intercept of the line $y = x - 4$. Explain.
(52)

Name _____ Date _____ Class _____

Simplifying Radical Expressions LESSON 61

Simplify.

*1. $\sqrt{12}$
(61)

*2. $\sqrt{200}$
(61)

Evaluate.

3. $x^{\frac{1}{4}}$ when $x = -16$
(46)

4. $x^{\frac{1}{3}}$ when $x = 343$
(46)

5. **Write** What does FOIL stand for in the term FOIL method?
(58)

6. (**Basketball**) A basketball court is $\frac{144x}{x+3}$ feet long and $\frac{432}{x+3}$ feet wide. A team runs laps
(51) around the court. How far have they run after one lap?

7. **Error Analysis** Students were asked to use the sum and difference pattern to find the
(60) product of $(x - 8)(x + 8)$. Which student is correct? Explain the error.

Student A	Student B
$(x - 8)(x + 8)$	$(x - 8)(x + 8)$
$= x^2 - 8^2$	$= x^2 - 8^2$
$= x^2 - 64$	$= x^2 + 64$

8. **Multiple Choice** A rectangular picture is twice as long as it is wide. The picture has a
(58) 4-inch-wide mat around the picture. Let x represent the picture's width. Which
product gives the area of the picture and mat?

A $(2x)(x)$
B $(2x + 4)(x + 2)$
C $(2x + 8)(x + 8)$
D $(-2x + 6)(6 + 2)$

Solve the systems by the method given.

*9. Use a graphing calculator to solve.
(55)
$$-5x + 8y = 7$$
$$3y = -2x - 9$$

10. Use substitution to solve.
(59)
$$2x - 3y = 3$$
$$x = 4y - 11$$

11. (**World Records**) In 2005, power-plant employees built the world's largest igloo in
(46) Quebec, Canada. The igloo is approximately in the shape of a half sphere and has a
volume of about 1728π cubic feet. What is the igloo's approximate diameter? (Hint:
The formula for the volume of a sphere is $V = \frac{4}{3}\pi r^3$.)

Simplifying Radical Expressions

LESSON 61

12. Multi-Step The following data shows the attendance at seven home games for a high school football team: 5846, 6023, 5921, 7244, 6832, 6496, 7012.
(48)
  a. Find the mean attendance value for the data set.
  b. Suppose the stadium was almost filled to capacity for the homecoming game and that the data value 11,994 is now added to the data set. How does this outlier affect the mean of the data?

Simplify.

13. $(4b - 3)^2$
(60)

14. $(-2x + 5)^2$
(60)

*15. (Fundraising) At a school fundraiser, students charge $10 to wash a car and $20 to wash an SUV. They make $1700 by washing 105 vehicles. How many of each kind do they wash?
(59)

16. Multi-Step Use these points (1, 3) and (2, 8) to answer the problems below.
(52)
  a. Find the slope of the line that passes through the points.
  b. Graph the line.
  c. Write the equation of the line in point-slope form.
  d. Write the equation of the line in slope-intercept form.
  e. Fill in the missing coordinates of the points $(x, -3)$ and $(-2, y)$.

17. (Physics) An object's weight on the moon varies directly with its weight on Earth. A 60-pound dog would weigh 9 pounds on the moon. How much would a 25-pound dog weigh on the moon?
(56)

18. Multiple Choice Which point is a solution of to the linear system?
(59)
$$y - 5x = 3$$
$$3x + 8y = 24$$

  A (1, 8)   **B** (0, 0)   **C** (0, 3)   **D** (4, 1.5)

*19. Write Explain how to simplify $\sqrt{18a^2}$, $a \geq 0$.
(61)

*20. Geometry A triangle has a length of $(x - 4)$ inches and a height of $(x + 4)$ inches. What is the area of the triangle?
(60)

21. Analyze What polynomial can be added to $x^2 + 5x + 1$ to get a sum of $4x^2 - 3$?
(53)

Find the LCM.

22. 21, 33, 13
(57)

23. 8, 32, 12
(57)

Simplifying Radical Expressions

LESSON 61

*24. **Multi-Step** Laura is building on to her new pool house. Write an expression for the area of the non-shaded region. Find the area.
  a. Find the area of the large square.
  b. Find the area of the smaller square.
  c. Write an expression for the area of the non-shaded region.

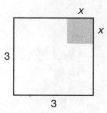

*25. **Measurement** Alan is building a square patio. He wants an 8-inch-wide border of flowers around his patio, and the total length of one side is $8x$. Write a polynomial that represents the area of the floor of the patio not including the border of flowers.

*26. **Multi-Step** The area of a circle is 20π cm². How long is the radius?
  a. Write an expression that can be used to find the length of the radius of a circle if given the area, A.
  b. Use the expression to find the length of the radius.

Translate each inequality into a sentence.

27. $\frac{n}{7} + 3 \geq 5$

28. $3g - 4 < -2$

*29. (**Circular Motion**) The tangential velocity of an object in circular motion can be found using the expression $\sqrt{ar}$, where a is the centripetal acceleration and r is the path radius of the circle. What is the tangential velocity of an object with a path radius of 15 cm and a centripetal acceleration of 60 cm/s²?

30. Make a box-and-whisker plot of the data below and title it "Shoe Sizes."

5, 10, 6, 7, 6.5, 7, 7, 8.5, 6.5, 8, 9, 7.5, 7

Name _____ Date _____ Class _____

Displaying Data in Stem-and-Leaf Plots and Histograms
LESSON 62

Find each of the following statistical measures using the data in the plot below.

Service Club Ages of Members

Stem	Leaves
2	5, 6
3	0, 2, 3, 4, 4, 4, 8, 9
4	1, 1, 2, 6, 7
5	2, 5, 7
6	0, 4

Key: 5| 6 means 56 years old

*1. median
(62)

*2. mode
(62)

*3. range
(62)

*4. relative frequency of 41
(62)

Simplify.

5. $\sqrt{88}$
(61)

*6. $\sqrt{720}$
(61)

7. $\sqrt{180}$
(60)

8. **Write** Explain in your own words how to graph the equation $y = -\frac{1}{3}x + 2$.
(49)

*9. **Error Analysis** Students were asked to find the product $(2n - 5)^2$ using the square of a binomial pattern. Which student is correct? Explain the error.
(60)

Student A	Student B
$(2n - 5)^2$ $= (2n)^2 - 2(2n)(5) + (5)^2$ $= 4n^2 - 20n + 25$	$(2n - 5)^2$ $= (2n)^2 - 5^2$ $= 4n^2 - 25$

10. **Analyze** The solution to a system of equations is (4, 6). One equation in the system is $2x + y = 14$. One term in the other equation is $3x$. What could the equation be?
(59)

11. **Multiple Choice** Which system of equations has a solution of $(2, -2)$?
(59)

A $x + y = 0$
   $2x - 3y = -2$

B $5x - 3y = 16$
   $4x + 9y = 10$

C $9x - 2y = 22$
   $3x + 6y = -6$

D $x + 2y = -2$
   $2x + y = -2$

Multiply.

12. $-4(x^2 + 4x - 1)$
(58)

13. $(b - 8)^2$
(60)

14. $\left(x + \frac{1}{2}\right)\left(x + \frac{1}{4}\right)$
(60)

© Saxon. All rights reserved. Saxon Algebra 1

Displaying Data in Stem-and-Leaf Plots and Histograms

LESSON 62

15. (Consumer Math) Consumers spend $95 million on Father's Day cards and $152 million on Mother's Day cards. What is the amount of increase or decrease from Mother's Day to Father's Day cards? What percentage of the combined total do consumers spend on Mother's Day cards?
(47)

*16. **Geometry** The stem-and-leaf plot shows the diameters of a shipment of bicycle tires to a bicycle shop. Find the circumference of the tire with the median diameter in the data set. Use 3.14 for pi.
(62)

Diameters of Bicycle Tires

Stem	Leaves
1	5, 5, 6, 7, 7, 8, 9, 9
2	0, 0, 0, 1, 2, 2, 4

Key: 1 | 5 means 15 inches

17. Solve the following system by graphing: $y = x + 6$
(55)
$y + 3x = 6$.

18. **Write** Explain what the whiskers in a box-and-whisker plot represent.
(54)

*19. (Fundraising) The amount of money raised for a charity event by the homerooms at Jefferson High School are shown below. Create a stem-and-leaf plot of the data.
(62)

$150, $125, $134, $129, $106, $157, $108, $135, $144, $149

20. (Commerce) The weekly profit, p, in dollars at Bill's TV repair shop can be estimated by the equation $p = 30n - 400$, where n is the number of TVs repaired in a week. Graph the equation and predict the profit for a week if 50 TVs are repaired.
(52)

Evaluate.

21. Evaluate $\sqrt[4]{x^2}$ when $x = 9$.
(46)

22. Evaluate $\sqrt[3]{x^6}$ when $x = 2$.
(46)

23. Write an equation for a direct variation that includes the point $(10, -90)$.
(56)

24. **Multi-Step** Levi deposits $500 into a savings account that earns interest compounded annually. Let r be the annual interest rate. After two years, the balance from the first deposit is given by the polynomial $500 + 1000r + 500r^2$. A year after making the first deposit, Levi makes another deposit. A year later, the balance from the second deposit is $600 + 600r$.
(53)

a. Find the balance of the accounts combined in terms of r after 2 years.

b. Find the balance after two years when $r = 0.03$.

Displaying Data in Stem-and-Leaf Plots and Histograms

LESSON 62

Find the missing number.

25. original price: $68
(47)
new price: _____
29% increase

26. original price: _____
(47)
new price: $98.60
15% decrease

27. (Triathlon Training) Training for a triathlon involves running, swimming, and biking.
(57) The athlete runs every $(2r - 2s)$ days, swims every $(4r^2 - 4rs)$ days, and bikes every $(8rs - 8s^2)$ days. When would the athlete have all three activities on the same day?

28. Multiple Choice Which of the following is not equivalent to $\sqrt{2800}$?
(61)
   A $10\sqrt{28}$ **B** $2\sqrt{700}$ **C** $20\sqrt{7}$ **D** $40\sqrt{7}$

*29. (Gardening) Isaac is planting a square flower garden. He wants a 10-inch-wide
(60) brick border around the outside edge of the garden. The total side length is $9x$. Write a special product that represents the area of the garden not including the brick border. Simplify.

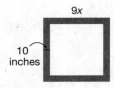

*30. Coordinate Geometry Lucia drew a square on grid paper that has an area of 25 units².
(61) What points could she have plotted?

Name _____ Date _____ Class _____

Solving Systems of Linear Equations by Elimination

LESSON 63

Simplify.

1. $\sqrt{256}$
(61)

2. $\sqrt{108}$
(61)

3. $\sqrt{294}$
(61)

4. $\left(\dfrac{r^{-3}t^{\frac{1}{2}}e}{rg^4 t^{\frac{3}{2}}}\right)^2$
(40)

5. (**Sports**) A sports court has area $\dfrac{t^3 n^{-2} s}{f^7 t b^5}\left(\dfrac{t^{-2}}{ns^3} - \dfrac{f^6 t^{-1}}{b}\right)$. Simplify.
(39)

6. Find the range for these 6 house sizes in a neighborhood (in square feet):
(48)
   1450, 1500, 2800, 1630, 1500, 1710

7. Tell whether the set of ordered pairs (5, 12), (3, 7.2), and (7, 16.8) represents a
(56) direct variation.

Find the LCM.

8. $2t^3 sv^5$, $6v^3 t^4$, and $10v^8 s^4$
(57)

9. $14dv^3$, $7s^2 v$, and $28s^7 v^5$
(57)

10. **Multiple Choice** Which of the following polynomials is equal to $(x + 5)(x - 5)$?
(60)
    A $x^2 - 25$ **B** $x^2 + 25$ **C** $x^2 - 5x + 25$ **D** $x^2 - 10x + 25$

*11. (**Landscaping**) Nasser wants to plant a 4-foot-high Yoshino Cherry tree near a
(63) 7-foot-high Snowdrift Crabapple tree. If the cherry tree grows at a rate of 16 inches a year and the crabapple tree grows at a rate of 13 inches per year, when will the two trees be the same height?

Solve each system of linear equations using the method indicated.

12. Solve by substitution: $6x - 2y = 12$
(59)
    $y = -5x + 10$

*13. Solve by elimination: $6x + 4y = 22$
(63)
    $-6x + 2y = -16$

*14. **Verify** Solve the system of linear equations by graphing: $\begin{array}{l} -x + y = 4 \\ 2x + y = 1 \end{array}$ Check your
(55) answer with a graphing calculator.

15. (**City Planning**) The city manager wants to build a rectangular walking
(53) track around the town's park. Below is a sketch of the new park. Write a polynomial expression for the perimeter of the rectangle. Simplify the polynomial.

   $2x$
   $3(x - 2)$

© Saxon. All rights reserved. 235 Saxon Algebra 1

Solving Systems of Linear Equations by Elimination

LESSON 63

*16. Multiple Choice Which ordered pair is a solution of the system of equations?
(63)

$$9x - 3y = 20$$
$$3x + 6y = 2$$

A $\left(-\frac{2}{3}, 2\right)$ **B** $\left(2, \frac{2}{3}\right)$ **C** $\left(2, -\frac{2}{3}\right)$ **D** $\left(-2, \frac{2}{3}\right)$

*17. Verify After solving a system of linear equations by elimination, how could you
(63) algebraically check your solution?

18. Multi-Step A group of 30 white-tailed deer all had 2 antler points at 1.5 years. At
(54) 4.5 years, their points were counted again.

Number of Points at 4.5 years	2	3	4	5	6	7	8	9
Number of Deer	1	0	4	3	7	4	10	1

a. Make a box-and-whisker plot for the data set at 4.5 years.

b. Can you predict how many points a deer would have at 4.5 years old if you know that it had 2 points at 1.5 years?

*19. Geometry Main Market Square in Krakow, Poland has an area of 40,000 m². Find
(61) the length of one side of the square.

20. (Football) A football field's dimensions can be represented by a width of $3x + 15$ feet
(58) and a length of $x + 10$ feet. Write a polynomial expression for the area of the field.

Find the missing number.

21. original price: $1527   **22.** original price: $25,720   **23.** original price: $10.25
(47) new price: _____   (47) new price: _____   (47) new price: _____
38% decrease   1.5% increase   215% increase

24. Write Write the two patterns for the square of a binomial.
(60)

25. (City Planning) A city is planning to build a new park in a shape of a square in the
(58) business district. It has a side length of $2x + 6$ feet. Find the area of this park.

*26. Multiple Choice Which of the following expressions is the simplest form of $\sqrt{76g^6}$?
(61)   **A** $2g^3\sqrt{19}$ **B** $2g^4\sqrt{19}$ **C** $6g^3\sqrt{2}$ **D** $6g^4\sqrt{2}$

Solving Systems of Linear Equations by Elimination

LESSON 63

***27. Error Analysis** Two students used the stem-and-leaf plot to find the mode height of the basketball players at their school. Which student is correct? Explain the error.
(62)

Heights of Players on Team

Stem	Leaves
16	5, 7
17	1, 5, 6, 8, 8, 9
18	4, 5, 5, 5
19	2, 7
20	1

Key: 16| 5 means 165 cm

Student A
The mode is 185 cm.

Student B
The mode is 170 cm.

***28. Write** Explain how to solve the system of linear equations.
(63)

$3x - 2y = 10$

$-\frac{9}{2}x + 3y = -15$

29. What must be true about the coefficient of a linear function if that function experiences a vertical stretch of the parent function?
(Inv 6)

***30. Probability** The histogram displays the number of hours students in Helene's homeroom average at their part-time jobs. What is the probability that a student randomly selected from the homeroom works 5–10 hours per week?
(62)

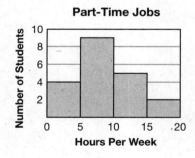

Part-Time Jobs

237

Saxon Algebra 1

Identifying, Writing, and Graphing Inverse Variation

LESSON 64

Find the LCM.

1. $8mn^4$ and $12m^5n^2$
(57)

2. $\frac{1}{2}wx^3$ and $\frac{1}{4}w^2x^6$
(57)

Tell whether each equation shows inverse variation. Write "yes" or "no".

*3. $y = \frac{x}{11}$
(64)

*4. $y = \frac{3}{x}$
(64)

*5. **Estimate** If y varies inversely as x and $y = 24$ when $x = 99$, estimate the value of y when $x = 50$.
(64)

6. What is the minimum and maximum number of books read over the summer by 10 students: 10, 9, 8, 10, 1, 8, 11, 20, 9, 10?
(48)

7. (**Budgeting**) The cost of a cell phone plan includes a monthly fee plus a charge per minute. The charge for 200 minutes is $50. The charge for 350 minutes is $57.50. How much is the monthly fee and how much is the charge per minute?
(59)

Find the product.

8. $(b + 2)^2$
(60)

9. $(x - 8)(x + 2)$ Use the FOIL method.
(58)

*10. **Multi-Step** Darius drew a square on grid paper that covers 121 cm².
(61)
   a. Write an expression to find the side length.
   b. Substitute known values into the equation.
   c. Find the side length.
   d. In which unit is the side length measured?

*11. **Multi-Step** Joseph lives 5 miles from the school and Maya lives 2 miles from the school. If Joseph walks from his house in an opposite direction from the school at a speed of 4 miles per hour, and Maya walks from her house in an opposite direction from the school at a speed of 6 miles per hour, after how long will they be the same distance from the school? What is that distance?
(63)

*12. **Multiple Choice** If y varies inversely as x, and $y = 9$ when $x = 12$, what is y when $x = 4$?
(64)
   A 3   B 5.3   C 27   D 108

Identifying, Writing, and Graphing Inverse Variation

LESSON 64

13. (**Recycling**) Use the table to compare the mean value of the amount of waste generated in the United States to the mean value of the amount of materials recovered for recycling from 1960 to 2005. Write a statement based on this comparison.

Generation and Materials Recovery of Municipal Solid Waste, 1960–2005 (in millions of tons)

Activity	1960	1970	1980	1990	2000	2003	2004	2005
Generation	88.1	121.1	151.6	205.2	237.6	240.4	247.3	245.7
Materials Recovery	5.6	8.0	14.5	33.2	69.1	74.9	77.7	79.0

14. Multi-Step A model sailboat calls for trim around the largest sail. The sail is triangular with side lengths $\frac{8}{2a^2 + 3a}$ centimeters, $\frac{5a + 1}{2a^2 + 3a}$ centimeters, and $\frac{3a + 3}{2a^2 + 3a}$ centimeters. How much trim will be used?

Find each of the statistical measures below using the data in the plot.

Annual Base Salary at Marketing Associates

Stem	Leaves
2	1, 1, 2, 2, 3, 3, 3, 6, 8
3	0, 1, 2, 6, 6, 9
4	2, 5, 8
5	3, 5, 7
6	4, 8

Key: 3|1 means $31,000

15. median

16. mode

17. range

18. relative frequency of $22,000

19. (**Health**) To calculate a personal body mass index, students first had to report their weight in pounds. Make a box-and-whisker plot of the weights below and determine if there is an outlier.

140, 145, 170, 157, 130, 155, 190, 180, 175, 120, 116, 118, 112, 103

20. Identify Statement 2 as the converse, inverse, or contrapositive of Statement 1. Then indicate the truth value of each statement.

Statement 1: If a figure is a rhombus, then it is not a rectangle.

Statement 2: If a figure is a rectangle, then it is not a rhombus.

*21. Statistics The standard deviation is the square root of the variance of a set of data. If the variance of a set of data is 24, what is the standard deviation in simplest form?

Identifying, Writing, and Graphing Inverse Variation

LESSON 64

22. Justify There are two different ways to write $\sqrt{\frac{75}{45}}$ in simplest form. Find both and then explain why each is correct.
(61)

*23. Error Analysis Two students solved the following systems of equations. Which student is correct? Explain the error.
(63)

Student A	Student B
$2x + 11y = 13$	$2x + 11y = 13$
$2x + 9y = 11$	$2x + 6y = 8$
$20y = 24$	$5y = 5$
$y = 1.2$	$y = 1$
$(-0.1, 1.2)$	$(1, 1)$

24. Geometry The areas of two similar rectangles add up to 39 square units. Twice the area of Rectangle A plus one-third the area of Rectangle B equals 33 square units. What are the areas of Rectangles A and B?
(63)

25. Graph $9x - 1.5y + 12 = 0$ using slope-intercept form.
(49)

26. Solve the following system of linear equations: $-3x + 2y = -6$.
(63)
$$-5x - 2y = 22$$

27. Multi-Step Use the system of linear equations to answer the problems below.
(55)
$$2x - y = 14$$
$$x + 4y = -2$$

    a. Graph the system.

    b. Determine the solution.

    c. Check your answer by using substitution.

28. (Boyle's Law) In chemistry, Boyle's Law states that the volume of a sample of gas is inversely related to its pressure if the temperature remains constant. Jameka recorded the pressure of a sample of gas inside a 450-cubic millimeter container to be 95 kPa. If the pressure increased to 475 kPa, what would the new volume of the gas be?
(64)

29. Write Describe a proportional situation that is represented by the equation $y = 6x$.
(56)

30. Generalize How would you describe the location of a graphed inverse variation based on the constant?
(64)

Name _____ Date _____ Class _____

Writing Equations of Parallel and Perpendicular Lines

LESSON 65

Simplify.

1. $\sqrt{360}$
(61)

2. $\sqrt{252}$
(61)

3. $\sqrt{384}$
(61)

Find the product using the FOIL method.

4. $(x^2 + 5)^2$
(58)

5. $(x - 2)(x - 9)$
(58)

6. Graph the inequality $c \geq -5$.
(50)

7. **Justify** Explain how you find the LCM of $2(x - 5)^3$ and $3(x - 5)^7$.
(57)

8. **Multiple Choice** The key of a stem-and-leaf plot shows $2|9 = 2.9$ cm. What is the value
(62) of a data point with a leaf of 4 and a stem of 7?

   A 4.7 cm B 7.4 cm C 47 cm D 74 cm

*9. **Sports** The results in seconds for the men's 50-meter freestyle swimming finals at
(62) the NCAA Division II Championship in 2007 are listed below. Create a stem-and-leaf plot to organize the data.

   20.43, 20.32, 20.36, 20.39, 20.67, 20.68, 20.68, 20.81, 20.62,
   20.97, 21.07, 21.24, 21.25, 21.31, 21.45, 21.56

10. **Multi-Step** The cost of parking at a concert is $22 for the first 2 hours and $4 for each
(52) additional hour. Write an equation that models the total cost y of parking a car in terms of the number of additional hours x. Using the linear equation, find the total cost of parking for 9 hours.

11. **Error Analysis** Two students attempted to write an inverse variation equation
(64) relating x and y when $x = 5$ and $y = 10$. Which student is correct? Explain the error.

Student A	Student B
$y = \dfrac{50}{x}$	$y = 2x$

*12. **Measurement** Giao wants to construct a picture frame made of wooden pieces that are
(63) 7 centimeters and 3 centimeters in length. He needs 20 pieces, and the total perimeter of the frame needs to be 108 centimeters. If Giao cuts eight 3-centimeter pieces first and is left with 0.8 meter of wood, will he have enough wood to cut the 7-centimeter pieces? Explain.

Solve each system of linear equations by the method indicated.

13. Use graphing.
(55)
$2x - y = 3$
$3x + y = 2$

14. Use elimination.
(63)
$5x + 7y = 41$
$3x + 7y = 47$

15. Use substitution.
(59)
$5x - 2y = 22$
$9x + y = 12$

© Saxon. All rights reserved.   243   Saxon Algebra 1

Writing Equations of Parallel and Perpendicular Lines

LESSON 65

16. (Banking) Ryan and Kathy both have savings accounts. Ryan has $12 in his account and plans to add $3 each week to it. Kathy does not have any money in her account, but plans to add $5 each week. How many weeks will it take until Ryan and Kathy have the same amount of money?
(55)

*17. Multi-Step The results for the women's 3-meter diving finals at the NCAA Division II Championship in 2007 are listed below.
(62)

499.15, 429.15, 409.75, 405.90, 395.65, 382.15, 353.20, 351.75,
342.30, 333.75, 328.20, 325.75, 315.20, 302.85, 292.90, 277.90

a. Write Explain how to determine the intervals to create a histogram for the data. Identify the intervals.

b. Create a histogram for this data using the intervals from part **a**.

c. Is a histogram or a stem-and-leaf plot a better display for this data? Explain.

18. Write the equation of the graphed line in slope-intercept form.
(49)

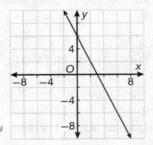

19. (Investing) Trey is making an investment of $(x - 11)$ dollars. The rate of interest on the investment is $(x - 11)$ percent. What is the interest gained after one year? (Hint: interest = principle × rate × time; $i = prt$)
(60)

20. (Construction) The school is constructing a new gym. They want a 12-foot-wide tile border around the outside edge of the square court, and the total side length is $7x$. Write a polynomial for the area of the court, not including the tile border.
(60)

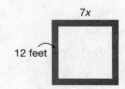

*21. Coordinate Geometry What kind of quadrilateral is the figure $WXYZ$? Justify your answer.
(65)

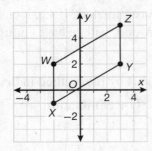

Writing Equations of Parallel and Perpendicular Lines

LESSON 65

22. State whether the equation $5xy = 40$ shows inverse variation. Write "yes" or "no".
(64)

*23. Write If line *a* is parallel to line *b*, could you find a line that makes lines *a* and *b* a reflection of each other? Explain.
(65)

*24. Analyze How many parallel lines can be found for the line $y = 2x + 1$?
(65)

*25. Find the slope of a line that is parallel to the line $6x + 3y = 36$.
(65)

*26. Multi-Step Draw a coordinate grid, and number it from -5 to $+5$ along the *x*- and *y*-axis. Will a line passing through the points $(-4, 0)$ and $(4, 2)$ be parallel to $y = -\frac{1}{4}x - 3$?
(65)

  a. Model Graph the points $(-4, 0)$ and $(4, 2)$.

  b. Find the equation for the line that passes through both points.

  c. Justify Is the line $y = -\frac{1}{4}x - 3$ parallel to the line you graphed? Explain.

*27. Error Analysis Two students solved the given system of equations. Which student is correct? Explain the error.
(63)

Student A	Student B
$3y = 2x + 8$	$3y = 2x + 8$
$8x = -2y + 24$	$8x = -2y + 24$
$11x = 32$	$14y = 56$
$x = 2\frac{10}{11}; \left(2\frac{10}{11}, 4\frac{20}{33}\right)$	$y = 4; (2, 4)$

28. (Paralympics) Suki is training in her wheelchair for the 100-meter race at the next Paralympics. In her initial acceleration phase, she averages a speed of 2.7 meters per second. At her maximum speed phase, she averages a speed of 6.6 meters per second. If she finishes a practice race in 18.1 seconds, how many meters long (to the nearest tenth of a meter) is her initial acceleration phase?
(63)

*29. Geometry The height *h* of a cylinder varies inversely with the area of the base *B* when the volume is constant. If the height of a cylinder is 6 centimeters when the area of the base is 12 square centimeters, what is the area of the base of the cylinder at right with the same volume?
(64)

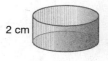
2 cm

30. Multi-Step Tyson often rides his bike 8 miles to visit a friend. His speed varies inversely with the time it takes to cover that distance. Today he rode at 16 mph to get to his friend's house, and then he rode at a rate of 12 mph to get home. How much longer did it take him to get home than to get to his friend's house?
(64)

Saxon Algebra 1

Name _____ Date _____ Class _____

Solving Inequalities by Adding or Subtracting

LESSON 66

Simplify the rational expression, if possible.

1. $\dfrac{11p}{6s^4} + \dfrac{p}{6s^4}$
(51)

2. $\dfrac{4x}{5w^4} - \dfrac{5x}{5w^4}$
(51)

3. $\dfrac{7y}{3x^4 + 1} + \dfrac{5y}{3x^4 + 1}$
(51)

Solve the inequality.

*4. $z + 10 \geq 3$
(66)

*5. $x - 4 \leq 9$
(66)

6. Graph the inequality $z \leq 1\tfrac{2}{3}$.
(50)

7. **Write** Explain the difference between $x > 5$ and $x \geq 5$.
(66)

8. Define a linear function that has shifted upward 3 units.
(Inv 6)

9. **Sports** In June 2000, the Russian Alexander Popov set a new world record of 21.64 seconds in men's freestyle swimming. Create the graph of an inequality that represents the times that could beat Popov's record. Explain any restrictions on the graph.
(50)

10. **Geometry** In two similar triangles, the lengths of the sides of the larger triangle are directly proportional to the lengths of the corresponding sides of the smaller triangle. Find the length of x.
(56)

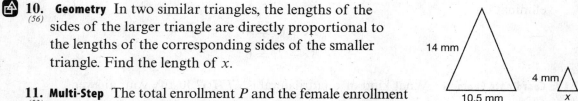

11. **Multi-Step** The total enrollment P and the female enrollment F of membership in a scholarship program (in thousands) can be modeled by the equations below, where t is the number of years since it was instituted.
(53)

$$P = 2387.74t + 155,211.46 \qquad F = 1223.58t + 79,589.03$$

a. Find a model that represents the male enrollment M in the scholarship program.

b. For the year 2010, the value of P is projected to be 298,475.86 and the value of F is projected to be 153,003.83. Use these figures to project the male enrollment in 2010.

*12. **Multiple Choice** What is the solution to $3z + 2 \leq z - 4 + 2 + z$?
(66)
    A $z \leq -4$ **B** $z \leq 0$ **C** $z \leq 3$ **D** $z \leq 4$

*13. **Coordinate Geometry** Find a line that is both a line of symmetry for the figure and also a perpendicular bisector of two of the sides.
(65)

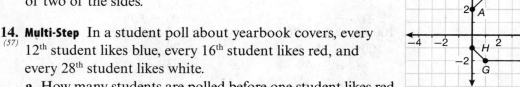

14. **Multi-Step** In a student poll about yearbook covers, every 12th student likes blue, every 16th student likes red, and every 28th student likes white.
(57)

a. How many students are polled before one student likes red, white, and blue?

b. If 1200 students are polled, how many like red, white, and blue?

© Saxon. All rights reserved.   247   Saxon Algebra 1

Solving Inequalities by Adding or Subtracting

LESSON 66

15. Solve the system of equations by substitution.
(59)

$$x = 10y - 2$$
$$2x - 18y = 8$$

*16. (Velocity) The tangential velocity of an object in circular motion can be found using
(61) the expression $\sqrt{\frac{Fr}{m}}$, where F is the centripetal force, r is the radius of the circle, and m is the mass of the object. What is the tangential velocity of an object with a path radius of 2 m, a centripetal force of 60 kgm/s², and a mass of 3 kg?

17. Multiple Choice Which system of equations has a solution of $y = 5$?
(63)
A $2x + 5y = 16$ **B** $2x + 5y = 19$
     $2x + y = -4$ $x - 5y = -13$
C $2x + 5y = -4$ **D** $2x + 5y = 11$
     $-2x + y = -8$ $x - 5y = -17$

18. Justify Explain the steps that are used to solve this system of linear equations by
(63) elimination.

$$2x + 6y = 4$$
$$3x - 7y = 6$$

*19. Coordinate Geometry What kind of quadrilateral is *EFHG*? Justify your answer.
(65)

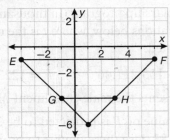

20. Error Analysis Two students are using the product rule below to find the missing value.
(64) Which student is correct? Explain the error.

If y varies inversely as x and $y = 0.2$ when $x = 100$, find y when $x = 4$.

Student A	Student B
$x_1 y_2 = x_2 y_1$	$x_1 y_1 = x_2 y_2$
$100 \cdot y_2 = 4 \cdot 0.2$	$100 \cdot 0.2 = 4 \cdot y_2$
$100 y_2 = 0.8$	$20 = 4y_2$
$y_2 = 0.008$	$5 = y_2$

21. Write a direct-variation equation relating x and y when $x = 5$ and $y = 30$.
(56)

248

Solving Inequalities by Adding or Subtracting

LESSON 66

22. Write an inverse-variation equation relating x and y when $x = 4$ and $y = 20$.
(64)

Multiply.

23. $(2x - 3)(2x - 3)$ **24.** $(t - 12)(t + 12)$ **25.** $(y^3 - 4)^2$
(60) (60) (60)

26. Justify Find the product of $(2y + 4)(3y + 5)$ using the Distributive Property.
(58) Then find the product using the FOIL method. Show that the answers are the same using either method.

27. Life Expectancy For most mammals, there is an inverse relationship between life
(64) span and heart rate. Use the table below to write an inverse-variation equation to represent this relationship. Then find the life span in years, rounded to the nearest tenth, of a hamster with an average heart rate of 450 beats per minute.

Animal	Heart Rate (beats per minute)	Life Span (in minutes)
Guinea Pig	280	3,571,429
Rabbit	205	4,878,049
Dog	115	8,695,952
Rat	328	3,048,780

*28. Coordinate Geometry Show that PQR is a right triangle.
(65)

*29. Justify True or False: The lines represented by $y = \frac{x}{3} - 1$ and $-4 = 12x + 4y$
(65) are parallel. Explain.

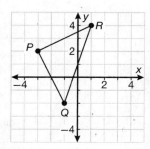

*30. Running John plans to run at least 5 miles more this week than he ran last week.
(66) He ran 25 miles last week. Write and solve an inequality describing the number of miles John plans to run this week.

Name _____ Date _____ Class _____

Solving and Classifying Special Systems of Linear Equations

LESSON 67

Find the product.

1. $(2b - 3)^2$
(60)

2. $(-b^3 + 5)^2$
(60)

Simplify.

3. $\sqrt{25x^4}$
(61)

4. $\sqrt{144x^6y}$
(61)

5. Simplify the rational expression $\frac{3x}{y^2} + xy^{-2}$, if possible.
(51)

6. In three hours, James read 18 pages of his history book. In four hours, he read 21 pages. Write the equation of the line that passes through the two points that represent the data in the problem. Use the equation to predict how many pages James will read in six hours.
(52)

Solve the inequality.

7. $z - 3 \geq 10$
(66)

8. $z - 5 < -2$
(66)

9. (**Hobbies**) The stem-and-leaf plot shows the number of cards each member of a baseball card enthusiasts' club has. What is the mode(s) of the data?
(62)

Cards in Collection

Stem	Leaves
30	9
31	1, 4, 5, 7, 7, 9
32	0, 1, 4, 4, 4, 7, 8
33	5, 6
34	2, 8, 9
35	0, 4, 6, 6, 6, 7,

Key: 31 | 5 means 315 cards

10. (**Transportation**) Two subway trains run through a station. One train goes through the station every 44 minutes and the other train goes through the station every 28 minutes. If they just went through the station at the same time, how many minutes will it be until the next time they are both at the station at the same time?
(57)

11. **Multi-Step** A bank teller receives several deposits in one day. She tallies how many of each amount she receives.
(54)
   a. Make a box-and-whisker plot for the set of data.
   b. List the lower extreme, lower quartile, median, upper quartile, and upper extreme of the data.

Amount	Tally
$25	4
$50	2
$60	6
$75	3
$80	7
$100	5

*12. Identify Statement 2 as the converse, inverse, or contrapositive of Statement 1. Then indicate the truth value of each statement.
(Inv 5)

Statement 1: If a figure is not a polygon, then it is not a square.

Statement 2: If a figure is a polygon, then it is a square.

Solving and Classifying Special Systems of Linear Equations

LESSON 67

13. Error Analysis Two students solve the inequality $x + 2 < 3$. Which student is correct? Explain the error.

Student A	Student B
$x + 2 < 3$	$x + 2 < 3$
$\underline{+2 \quad +2}$	$\underline{-2 \quad -2}$
$x < 5$	$x < 1$

*14. Sports The ages of the players of the Eastern Conference team during the 2007 All-Star Game are listed below. Create a histogram of the data.

25, 30, 24, 26, 30, 29, 21, 25, 33, 28, 34, 25

15. Multi-Step Find the volume of a toy building block that it is in the shape of a cube with a side length of $(2x + 2)$ inches.
a. Find the area of the base. Find the product of the length and the width.
b. Find the volume by multiplying the product found in part **a** by the height.

16. Verify Show that $(20, 15)$ is a solution to the system of equations.

$$5x - 2y = 70$$
$$3x + 4y = 120$$

*17. Coordinate Geometry For the figure below, find two lines that are both a line of symmetry and a perpendicular bisector of the sides.

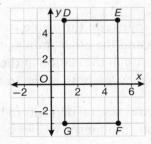

*18. Job Hunting Claudio is looking for a new job in sales. He interviews with Company A and they offer him a base salary of $32,000 plus a 1.5% bonus on his total sales for the year. When he interviews with Company B, they offer him a base salary of $26,000 plus a 3% bonus on his total sales for the year. How much would Claudio have to sell to make the same amount at each job?

*19. Analyze If an equation is multiplied by a number, a system of equations is formed that has infinitely many solutions. Classify the system formed by adding a number to a given equation.

Determine if the lines are parallel or perpendicular.

20. $y = 3x + 12$ and $y + 9 = 3x$

Solving and Classifying Special Systems of Linear Equations

LESSON 67

Solve each system of linear equations.

21. $2x - 3y = -17$
$2x - 9y = -47$

*22. $y = x - 5$
$y = -2x + 1$

23. Multiple Choice If y varies inversely as x, and $y = 7.5$ when $x = 5$, what is the value of k?
A 0 **B** 0.67 **C** 1.5 **D** 37.5

24. Write Will the graph of an inverse variation ever cross the x-axis? Explain.

25. Probability A teacher allows students to draw from a bag of various prizes, four of which are new graphing calculators. If there are 24 students and each student chooses 3 items, what is the probability that the first student will choose a graphing calculator on his or her first draw?

26. Geometry The triangle inequality states that the sum of the lengths of any two sides of a triangle must be greater than the length of the third side. Write an inequality for the length of the third side of the triangle.

*27. Multi-Step Mr. Sanchez is planning a business trip. He estimates that he will have to drive 55 miles on one highway, 48 miles on another, and then 72 more miles to arrive at his destination. He starts with 25,000 miles on his car's odometer and later finds that his estimate was high.

a. Write an inequality to represent the odometer reading at the end of the trip.

b. Solve the inequality.

c. The inequality only considers the greatest amount the odometer could read. Write and solve an inequality for the least it could read.

*28. Error Analysis Two students checked the graph of a system of equations to see if it was consistent and independent, consistent and dependent, or inconsistent. Which student is correct? Explain the error.

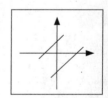

Student A	Student B
consistent and dependent	inconsistent

*29. Newspaper Delivery Mauricio and Aliyya are comparing their newspaper delivery speeds. Mauricio can deliver $y = 65x + 15$ papers per hour. Aliyya can deliver $13x - \frac{1}{5}y = -3$ papers per hour. Who delivers faster?

*30. Verify Wanda determined that the system of equations below has one solution in common. Find the common solution and check your answer.

$$2y = 4x + 1$$
$$y = 8x - 7$$

Name _____ Date _____ Class _____

Mutually Exclusive and Inclusive Events
LESSON 68

Determine the degree of each polynomial.

1. $14x^2y^3z^4$
(53)

2. $12q^2r + 4r - 10q^2r^6$
(53)

3. $5x^4z^3 + 4xz$
(53)

Simplify.

4. $\sqrt{\dfrac{1}{48}}$
(61)

5. $\sqrt{\dfrac{4}{25x^4}}$
(61)

Tell whether each of the following equations shows inverse variation. Write "yes" or "no".

6. $y = \dfrac{5}{x}$
(64)

7. $y = \dfrac{1}{2}x$
(64)

Solve each inequality.

8. $x + 1 > 1.1$
(66)

9. $x - 2.3 \le 7.6$
(66)

Find the probability of the following events.

*10. rolling a sum of 7 or a sum of 11 with two number cubes
(68)

*11. rolling a sum of 1 or a sum of 13 with two number cubes
(68)

 12. Probability A caterer is sorting silverware. There are 75 knives, 50 forks, and
(7) 75 spoons. Twenty of the forks and 30 of the spoons have red handles. If a piece of silverware is chosen at random, what is the probability that it will be a fork or have a red handle? Write and solve an expression to find the probability.

13. A car rental company charges a flat rate of $50 and an additional $.10 per mile
(52) to rent an automobile. Write an equation to model the total charge y in terms of x number of miles driven. Predict the cost after 100 miles driven.

14. (**Chemistry**) When Liquid A has a temperature of 0°F, Liquid B has a temperature of
(52) −8°F. When Liquid A has a temperature of 4°F, Liquid B has a temperature of −9°F. Write an equation of a line that passes through the two points that represent the data in this problem. Use the equation to predict the temperature of Liquid B if Liquid A has a temperature of −4°F.

*15. Use the following system to answer the problems below.
(55)
$$x = y + 2$$
$$2x = y$$

a. Graph the system.

 b. Determine the solution. Verify the solution using your graphing calculator.

16. (**Carpentry**) Jorge is adding a rectangular family room onto his house. The dimensions
(58) of the room can be represented by a width of $(x + 6)$ feet and a length of $(2x + 5)$ feet. Write a polynomial expression for the area A of the new room.

© Saxon. All rights reserved. 255 Saxon Algebra 1

Mutually Exclusive and Inclusive Events

LESSON 68

17. Multi-Step The sum of 6 times a boy's age and 5 times a girl's age is 150. The girl is 2 years less than twice the boy's age. Find how old each was 5 years ago.

18. Justify True or False: $(a + b)(a + b) = a^2 + b^2$. Justify by giving an example.

19. Craft Fair Celine sells her hand-painted chairs and tables at a local craft fair. She is selling her chairs for $30 a piece and her tables for $60 a piece. At the end of the day, she noted that she had sold 20 items for a total of $780. How many chairs did Celine sell? how many tables?

20. Error Analysis Two students want to find the equation of the line perpendicular to $y = -\frac{1}{5}x + 4$ that passes through $(1, -3)$. Which student is correct? Explain the error.

Student A	Student B
$y = 5x + b$	$y = 5x + b$
$1 = 5(-3) + b$	$-3 = 5(1) + b$
$1 = -15 + b$	$-3 = 5 + b$
$+15 \quad +15$	$-5 \quad -5$
$16 = b$	$-8 = b$
$y = 5x + 16$	$y = 5x - 8$

21. Multi-Step Draw a coordinate grid, numbering from -6 to $+6$ along the x- and y-axis. Will a line that passes through the points $(0, -4)$ and $(5, -2)$ be perpendicular to the line that passes through the points $(-2, 5)$ and $(0, 1)$?

   a. Model Line 1 passes through the points $(0, -4)$ and $(5, -2)$. Line 2 passes through the points $(-2, 5)$ and $(0, 1)$. Graph the lines.

   b. Write an equation for both lines.

   c. Justify Are the lines perpendicular? Explain.

22. Discounts A new pet store is selling dog food for $2 off. Another pet store is selling dog food for $14.99. Write and solve an inequality to represent the highest original price, before the discount, that will make the dog food at the new pet store cheaper.

23. Error Analysis Two students solve and graph the inequality $x - 5 \leq -8$. Which student is correct? Explain the error.

Student A	Student B
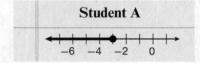	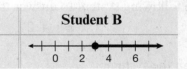

Find the common solution for each system of equations.

24. $y = \frac{3}{4}x + 3$
$y = x$

Mutually Exclusive and Inclusive Events

LESSON 68

25. Multi-Step Two satellite radio companies are offering different plans. One offers a flat fee of $15 per month for all stations. The second charges a rate of $y = \$10 + \$1.5x$, where x is the number of additional stations. Do the plans ever cost the same amount?

    a. How many stations can a consumer purchase from the second company before exceeding the flat fee of $15?

    b. Classify the system of equations for the plans offered by both companies.

    c. How is the classification of the equations misleading given the problem situation?

    d. Do the plans ever cost the same amount?

26. Geometry Two secants to a circle are shown at right. Classify the system formed by the equations for these two lines.

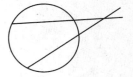

*27. Predict If two standard number cubes are rolled 100 times, predict the number of times that a sum of 7 or 11 will be rolled.

*28. Multiple Choice Which one of the following situations describes mutually exclusive events?

    A rolling doubles or a sum of 4 with two number cubes

    B choosing an odd number or a multiple of 3

    C tossing a coin heads-up or rolling a 2 with a number cube

    D rolling two 5s or a sum of 12 with two number cubes

*29. Weather Forecasting On a winter day in Maine, there was a 30% chance of only freezing rain, a 45% chance of only rain, and a 25% chance of only snow. What was the chance of getting only freezing rain or snow?

*30. Analyze Given two possible events, A and B, will the probability of either A or B occurring be higher if they are inclusive events or mutually exclusive events?

Adding and Subtracting Radical Expressions

LESSON 69

Add.

*1. $-6\sqrt{2} + 8\sqrt{2}$
(69)

*2. $-4\sqrt{7} - 5\sqrt{7}$
(69)

*3. $2\sqrt{3} + 5\sqrt{3}$
(69)

Determine the degree.

4. $9x$
(53)

5. $-3x^2 + 2x + 16$
(53)

6. $xy + 2$
(53)

Find the probability of the events described below.

*7. rolling a sum of 2 or a sum of 12 with two number cubes
(68)

8. rolling a sum greater than 1
(68)

9. **Volleyball** A volleyball is hit upward from a height of 1.4 meters. The expression
(53) $-4.3t^2 + 7.7t + 1.4$ can be used to find the height (in meters) of the volleyball
at time t. The ball was hit from a moving platform that has a height (in meters) of
$-3t^2 + 5t + 6$. What is the combined height?

10. Display the data using a box-and-whisker plot titled "Hours a Candle Burns".
(54) Identify any outliers.

$$4, 12, 3, 5, 7, 9, 4, 8, 3, 18, 5, 8, 4$$

11. **Multi-Step** The number of miles represented on a map varies directly with the number
(56) of inches on the map. In Texas, the distance from Austin to San Antonio is 79 miles;
on a map, it is 5 inches.
    a. Write the equation that represents this relationship.
    b. Graph this relationship.
    c. Estimate how far Waco is from Austin if they are $6\frac{1}{2}$ inches apart on a map.

12. **Produce Purchase** Bananas are $0.10 and apples are $0.25. The total cost for 35 pieces
(59) of fruit is $6.80. How many of each fruit were purchased?

13. **Multi-Step** Luke is building a new fence around his property.
(60) Write an expression for the area of the shaded region.
    a. Find the area of the smaller square.
    b. Write an expression for the area of the larger square using
       special products.
    c. Simplify part **b**.
    d. Write an expression for the area of the shaded region.

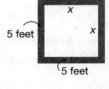

14. **Generalize** Use examples to explain why $\sqrt{x} \cdot \sqrt{x} = -x$ for all negative values of x.
(61)

15. Solve the system of linear equations given below.
(63)
$$6x + 15y = 15$$
$$7x - 3y = -3$$

Adding and Subtracting Radical Expressions

LESSON 69

16. (**Prom Night**) Jamal and his friends want to rent a limousine for prom night. The cost per person varies inversely with the number of people renting the limousine. If 4 people rent the limousine, it will cost them $180 each. How much would it cost per person if 12 people rented the limousine?

Determine if the lines are parallel or perpendicular.

17. $y = -\frac{3}{2}x + 8\frac{1}{2}$ and $y - \frac{2}{3}x = 0$

18. Data Analysis Rachel will receive an A in math for the semester if the mean of her test scores is at least 90. If her first two test scores are as shown in the table, write an inequality for the score Rachel needs on her third test to receive an A for the semester.

Test	Grade
Test 1	85
Test 2	95
Test 3	?

19. Analyze Graph the inequalities $x + 5 > 3$ and $x - 6 \leq -8$ on the same number line. What is true about their combined solutions?

20. Multiple Choice Which graph represents the solution to $x + 6 \geq 2x - 12$?

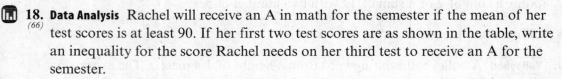

21. Find the common solution for the system of equations given below.

$$-\frac{1}{4}x + y = -2$$
$$-x + 4y = -8$$

22. Error Analysis Two students found a common solution for a system of equations. Which student is correct? Explain the error.

Student A	Student B
$y = -5x - 2$	$y = -5x - 2$
$6y = -30x - 12$	$6y = -30x - 12$
$-6y = +30x + 12$	$-6y = +30x + 12$
$6y = -30x - 12$	$6y = -30x - 12$
$0 = 0$	$0 = 0$
Solution:	Solution:
$y = -5x - 2$	$(0, 0)$

Adding and Subtracting Radical Expressions

LESSON 69

*23. **(Sanitation)** A garbage truck is on time with its collections if it maintains an average
(67) rate of $y = 75x + 5$. Currently, the truck is running at a rate of $\frac{1}{5}y - 15x = 1$. Is the garbage truck on schedule?

24. Error Analysis Two students found the probability of choosing either a black 2 or a
(68) king from a deck of cards. Which student is correct? Explain the error.

Student A	Student B
$P(A \text{ or } B) = \frac{2}{52} + \frac{4}{52}$ $= \frac{6}{52}$ $= \frac{3}{26}$	$P(A \text{ or } B) = \frac{2}{52} \cdot \frac{4}{52}$ $= \frac{8}{2704}$ $= \frac{1}{338}$

*25. Multi-Step Miranda baked several casseroles: 36 servings of chicken casserole,
(68) 24 servings of pasta casserole, 30 servings of beef casserole, and 28 servings of vegetarian casserole. She gives 1 serving to her brother. What is the probability that he will get a pasta casserole or a vegetarian casserole?
  a. What is the probability that the serving will be pasta?
  b. What is the probability that the serving will be vegetarian?
  c. What is the probability that the serving will be pasta or vegetarian?

26. If a circle is made into the spinner shown, what
(68) is the probability of landing on either a black space or a space worth 10 points?

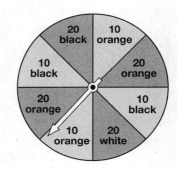

27. Analyze True or False: For the set of integers, $\sqrt{x^n}$ can be simplified when $x \geq 0$ and
(69) $n \geq 2$. Justify your answer.

*28. Error Analysis Ms. Nguyen asks her students if they can combine the radicals in the
(69) expression $\sqrt{13x} + \sqrt{23x} - \sqrt{33x}$. Student A says it is possible. Student B says that the radicals in the expressions $\sqrt{13x} + \sqrt{23x} - \sqrt{33x}$ do not combine. Which student is correct? Explain the error.

*29. Estimate Estimate the sum of $\sqrt{51} + \sqrt{63} + \sqrt{83} + \sqrt{104}$.
(69)

Adding and Subtracting Radical Expressions

LESSON 69

*30. (Great Pyramid) Each of the four sides of the base of the
(69) Great Pyramid measures 756 feet. If a scale model is
made with the measurement of $\sqrt{756}$ inches for each
side of the base, how many inches is the perimeter of the
base of the scale model of the pyramid rounded to the
nearest whole number?

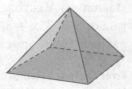

Solving Inequalities by Multiplying or Dividing

LESSON 70

Find the probability of the following events.

1. tossing a coin heads-up or rolling a 3 with a number cube

2. tossing a coin tails-up or rolling a number less than 4 on a number cube

3. tossing a coin tails-up and rolling a number greater than 6 on a number cube

Solve the inequality.

4. $y - 2 < \dfrac{1}{2}$

5. $y + \dfrac{3}{2} < \dfrac{1}{4}$

Add.

*6. $18\sqrt{3y} + 8\sqrt{3y}$

7. $\sqrt{3x} + 2\sqrt{3x}$

8. Write an inverse variation equation relating x and y when $x = 18$ and $y = 4.5$.

9. Determine if the inequality symbol needs to be reversed, and then solve.
$$-2a \geq -5$$

*10. A team scores 56, 42, 60, 43, 51, 22, 44, 55, and 49 points.
   a. Identify any outliers.
   b. On the same screen of the graphing calculator, make one box-and-whisker plot of these data without identifying any outliers, and make another box-and-whisker plot that does identify any outliers.
   c. How many points would you expect the team to score in the next game? Explain your answer.

11. **(Ages)** The ages at a family party are recorded.
   5, 30, 33, 42, 36, 40, 1, 44, 29, 61, 82, 63, 29, 38, 6, 11
   Make a box-and-whisker plot of these ages and determine if there is an outlier.

*12. Use the graphing calculator to determine the solution to the following system.
$$x + y = -2$$
$$y = 4x - 7$$

13. **Multi-Step** Balls are printed in a pattern. Every 18th ball has stripes, every 15th ball has polka dots, and every 30th ball has stars.
   a. How many balls are printed before there is a ball that has stripes, polka dots, and stars?
   b. How many balls have been printed at that point that have only two of the designs?

Solving Inequalities by Multiplying or Dividing

LESSON 70

14. **(Building)** Jason is building a square deck around his home. His home is also in the shape of a square. He wants the deck to be 9 feet wide around the outer edge of his house. The house has a total side length of $8x$. Write a special product and simplify to find the area of the house including the deck.
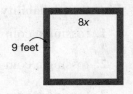

15. **(Golden Rectangle)** The length of a golden rectangle is equal to $x + x\sqrt{5}$. If $x = \sqrt{5}$, what is the length of the golden rectangle?

16. **Multi-Step** The data below show the number of customers served each day for a month at a diner.

    80, 86, 105, 109, 127, 148, 137, 148, 141, 140, 135, 146, 90, 95, 101, 83, 114, 148, 127, 86, 85, 91, 141, 136, 82, 148, 127, 149, 80, 86

    a. Create a stem-and-leaf plot of the data.

    b. Analyze Describe the distribution of the data. What conclusions can be drawn from this?

17. Solve the following system using the elimination method.

    $$-8x - 5y = -52$$
    $$4x + 3y = 28$$

18. **(Decibel Levels)** The relationship between the intensity of sound (W/m²) and the distance from the source of the sound is represented by the equation $I = \frac{k}{d^2}$. If you sit only 1 meter away from the stage at a rock concert, the intensity of sound is about 0.1 W/m². If Vanessa does not want the sound intensity to be any more than 0.0001 W/m², how close to the stage can she sit?

*19. **Coordinate Geometry** Prove that $ABCD$ is a rectangle.

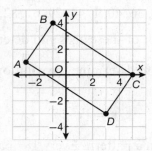

20. **Multiple Choice** Which one of the following systems has only one common solution?

    A $y = 21x + 6$
           $y = -7x$

    B $5x - 2y = 0$
           $\frac{5}{2}x - y = 0$

    C $-y = 13x - 6$
           $-2y = 26x + 9$

    D $x - 7y = 14$
           $\frac{1}{4}x - \frac{7}{4}y = \frac{7}{2}$

21. **Write** What kind of system is formed by two equations where one equation is a multiple of the other? Explain.

Solving Inequalities by Multiplying or Dividing

LESSON 70

22. Error Analysis Two students found the probability of choosing either a black card or a face card from a deck of cards. Which student is correct? Explain the error.

Student A	Student B
$P(A \text{ or } B) = \frac{26}{52} + \frac{12}{52}$ $= \frac{38}{52}$ $= \frac{19}{26}$	$P(A \text{ or } B) = \frac{26}{52} + \frac{12}{52} - \frac{6}{52}$ $= \frac{32}{52}$ $= \frac{8}{13}$

***23. (Fantasy Football)** In fantasy football, virtual teams are built by choosing from a pool of real players. In one pool of players, 72 play only defense, 65 play only offense, 8 play both, and 27 are on special teams. If at least one player must be chosen at random, what is the probability that the player will play either offense or special teams?

24. Geometry One side of a square measures $2\sqrt{9}$ meters. What is its perimeter?

***25.** What must be true about the coefficient of a linear function if that function experiences a vertical compression of the parent function?

26. Multi-Step A flag displayed in the school measures $6\sqrt{4}$ feet by $5\sqrt{4}$ feet. If the school decides to display 8 flags, what is the total measurement of the sides of all the flags?

***27. Multiple Choice** Which graph shows the solution set of $5f > -10$?

A 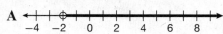   C

B   D

***28. Write** Explain in your own words how to solve the inequality $-\frac{2}{5}g \leq 6$.

***29. Analyze** Kyle wants to spend at most $100 for birthday presents for the 4 other members of his family this year. He plans to spend the same amount on each person. Write and solve an inequality to represent the situation. What restrictions exist on the solution set?

***30. (Cooking)** Marianna can afford to buy at most 20 pounds of ground turkey. Write and solve an inequality to determine the number of burgers she can make if each patty uses $\frac{1}{3}$ pound of ground turkey.

Making and Analyzing Scatter Plots

LESSON 71

1. Make a scatter plot from the data in the table.
(71)

x	1	2	3	4	5	6
y	26	24	21	14	9	5

Is the ordered pair (5, 2) a solution to the following systems?

2. $y = 7 - x$
(55) $y = \frac{1}{5}x + 1$

3. $5y - x = 5$
(55) $y = 2x - 8$

Subtract.

4. $31\sqrt{5} - 13\sqrt{5}$
(69)

*5. $\sqrt{27} - \sqrt{12}$
(69)

Find the common solution for each system of equations.

6. $-y = x + 8$
(67) $y = -x + 1$

*7. $6y - x = 12$
(67) $y = \frac{1}{6}x + 2$

Write an inverse variation equation relating x and y.

8. $x = \frac{2}{3}, y = 33$
(64)

9. $x = 6, y = 14$
(64)

*10. Error Analysis Two students solved the inequality $14 < -0.2k$ as shown below.
(70) Which student is correct? Explain the error.

Student A	Student B
$14 < -0.2k$	$14 < -0.2k$
$\frac{14}{0.2} > \frac{-0.2}{-0.2}k$	$\frac{14}{-0.2} > \frac{-0.2}{-0.2}k$
$70 > k$	$-70 > k$

*11. Multiple Choice When one set of data values increases as the other set of data values
(71) decreases, what type of correlation does this represent?
   A positive   **B** negative   **C** constant   **D** none

*12. (Elections) The table shows the voter turnout for National Federal Elections.
(71)

U.S. Voter Turnout (in millions)

Year	1990	1992	1994	1996	1998	2000	2002	2004
Number of Voters (in millions)	68	104	75	96	73	106	80	122

a. Use the data to make a scatter plot.

b. Does the scatter plot show any trend in the data?

Making and Analyzing Scatter Plots

LESSON 71

*13. Generalize If a scatter plot shows a negative correlation, what can you say about the relationship between the two sets of data values?
(71)

14. Error Analysis Student A and Student B combine the radicals. Which student is correct? Explain the error.
(69)

Student A	Student B
$8\sqrt{4y^2z^3} - 3yz\sqrt{49z}$	$8\sqrt{4y^2z^3} - 3yz\sqrt{49z}$
$= 8 \cdot 2yz\sqrt{z} - 3yz \cdot 7\sqrt{z}$	$= 8 \cdot 4yz\sqrt{z} - 3yz \cdot 7\sqrt{z}$
$= 16yz\sqrt{z} - 21yz\sqrt{z}$	$= 32yz\sqrt{z} - 21yz\sqrt{z}$
$= -5yz\sqrt{z}$	$= 11yz\sqrt{z}$

*15. (Biking) Randy biked $\sqrt{27}$ miles on the bike trail. He backtracked $\sqrt{3}$ miles, and then proceeded $\sqrt{12}$ miles to finish the trail. How far is Randy from his starting point?
(69)

16. Multiple Choice Which of the following situations describes inclusive events?
(68)

   A rolling a sum of 5 or a sum of 4 with two number cubes

   B rolling a sum of 3 or a factor of 6 with two number cubes

   C rolling two 5's or a sum of 8 with two number cubes

   D rolling two 4's or a sum of 9 with two number cubes

17. Write If a fair coin is used, what is the probability that one toss will result in either heads or tails? Explain why your answer makes sense.
(68)

*18. (Nutrition) Each day Paolo tries to consume at least 40 grams of protein. One day he has two soy shakes, each with 15 grams of protein, as well as a bowl of peanuts, containing 5 grams of protein. Write an inequality describing how much more protein Paolo should consume that day.
(66)

19. Line 1 passes through the points (2, −6) and (4, 6). Line 2 passes through the points (0, 1) and (6, 0). Are the lines parallel or perpendicular?
(65)

20. Data Analysis Use the table of values to draw a scatter plot.
(71)

x	72	60	65	50	56	69
y	32	31	30	28	28	38

21. Determine if the inequality symbol in the inequality $11b < 5$ needs to be reversed, and then solve.
(70)

*22. Geometry JoAnna is making a banner the shape of an equilateral triangle. She has 36 inches of cording to put around the banner. Write and solve an inequality to find the range of measures of one side of the banner.
(70)

Making and Analyzing Scatter Plots

LESSON 71

23. Multi-Step A company spends 2% of its sales on marketing. How much money does
(70) the company need to earn to spend at least $250,000 on marketing?
   a. Write an inequality to represent the situation.
   b. Solve the inequality.
   c. How much money is needed to spend $250,000 on marketing?

24. Generalize How would you choose which variable to eliminate when solving a system
(63) of linear equations?

25. Use the table to determine if there is a positive correlation, a negative correlation, or
(71) no correlation between the data sets.

x	3	6	10	13	15	17
y	100	88	73	62	51	38

26. Write Is it possible to determine the mode of a data set using a histogram? Explain.
(62)

27. (Pendulum) The period of a pendulum is equal to $2\pi\sqrt{\frac{l}{g}}$, where l is the length of
(61) the pendulum and g is the acceleration due to gravity. If the length of the pendulum
is 40 m and the acceleration due to gravity is 10 m/s^2, what is the period of
the pendulum?

*28. (Demography) The population of Fremont, California in 2005 was 200,770 and
(55) increased by 921 people by 2006. The population of Amarillo, Texas in 2005 was
183,106 and increased by 2419 people by 2006.
   a. Write a system of linear equations to represent the population of these cities
   assuming that they continued to grow at these yearly rates. Let x be the number of
   years after 2005 and y be the population.
   b. Use a graphing calculator to solve the system of equations. At what year would
   the populations be equal?

29. Write an equation for a direct variation that includes the point (6, 42).
(56)

30. Multi-Step A cereal box is in the shape of a rectangular prism. Find the volume of
(58) a cereal box that has a width of $(2x + 2)$ inches, a length of $(5x + 1)$ inches, and a
height of $(6x + 4)$ inches.
   a. Multiply the length times the width.
   b. Multiply the height times the product found in part **a**.

Name _____ Date _____ Class _____

Factoring Trinomials: $x^2 + bx + c$ — LESSON 72

1. (66) Solve the inequality $x + 2 + 3 > 6$. Then graph and check the solution.

Find the probability of the following events.

2. (68) choosing a vowel or a consonant from the alphabet

3. (68) rolling a sum that is a multiple of 4 or a set of doubles with two number cubes

Factor.

*4. (72) $x^2 + 11x + 24$

*5. (72) $k^2 - 3k - 40$

*6. (72) $m^2 + 9m + 20$

*7. (72) $x^2 + 33 + 14x$

*8. (72) **Error Analysis** Two students are factoring the following trinomial. Which student is correct? Explain the error.

$$x^2 - 5x - 6$$

Student A	Student B
$(-6)(1) = -6$	$(-6)(-1) = -6$
$(-6) + 1 = -5$	$(-6) - (-1) = -5$
$x^2 - 5x - 6 = (x + 1)(x - 6)$	$x^2 - 5x - 6 = (x - 1)(x - 6)$

*9. (72) (**Baking**) The area of the sheet-cake pans at a bakery are described by the trinomial $x^2 + 15x + 54$. What are the dimensions of a pan if $x = 11$?

*10. (72) **Analyze** How many possible pairs of number factors does c have in the following trinomial?

$$x^2 + bx + 36$$

*11. (72) **Model** These tiles represent the trinomial $x^2 + x - 6$.

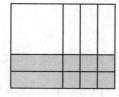

What does each of the shaded rectangles represent?

12. (71) Make a scatter plot from the data in the table.

x	5	10	15	20	25	30
y	11	13	16	20	21	25

Factoring Trinomials: $x^2 + bx + c$

LESSON 72

13. Error Analysis Two students are describing a trend line for a scatter plot but have
(71) different definitions. Which student is correct? Explain the error.

Student A	Student B
A trend line is a line on a scatter plot that goes through two data points and indicates a trend in the data.	A trend line is a line on a scatter plot that models the slope of the data points and indicates a trend in the data.

*14. Geometry Ten groups of students were given different circular objects. They measured
(71) the circumference and the diameter of the object. The table shows the results for
each group.

Diameter (in.)	4	3.75	8	6.25	5.5	5	7	1.5	3	9
Circumference (in.)	12.1	12	25	19	16	16	21	4.5	9.5	28

a. Make a scatter plot of the data and draw a trend line.

b. Write an equation that models the data.

c. How does your equation compare to the formula for the circumference of a circle, $C = \pi d$?

15. Multi-Step Use the scatter plot.
(71)
a. Use the trend line to estimate the corresponding y-value for an x-value of 18.

b. Use the trend line to estimate the x-value for a y-value of 50.

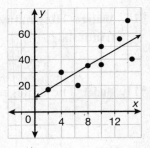

16. Write Can a perfect square have a negative square root? Explain.
(69)

17. (Air Traffic Control) An airplane approaching from the north is flying along a flight
(67) path of $y = 6x + 2$. The airport runway lies on the path $\frac{1}{2}y = 3x + 1$. Will the
airplane be able to land on the runway if it contiunes on its current path? Explain.

Factoring Trinomials: $x^2 + bx + c$

LESSON 72

18. Coordinate Geometry The triangle shown at right is formed
(67) by the intersection of three lines. The lines can be paired to
form three separate systems of two equations. Classify these
systems.

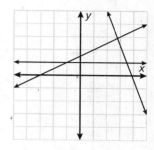

19. Determine if the lines described are parallel or
(65) perpendicular.

Line 1 passes through the points (−6, 3) and (6, 1).

Line 2 passes through the points (−6, −2) and (6, −4).

20. Model Make a table that relates the length and width of a rectangle with a constant
(64) area of 100 square feet.

*21. Multi-Step Yoon has a number of dimes and quarters she wants to put into rolls. She
(63) has 124 coins in all that add up to $20.50. If a roll of quarters is equal to ten dollars,
how many extra quarters does she have that will not fill up a roll?

22. Find the LCM of $(5x − 9)$ and $(3x + 8)$.
(57)

23. Multiple Choice Simplify $20\sqrt{7} − 12\sqrt{7} + 2\sqrt{7}$.
(69)
 A $10\sqrt{7}$ **B** $18\sqrt{7}$ **C** $34\sqrt{7}$ **D** $−12\sqrt{7}$

24. (Hobbies) Enrique recorded his personal best times in
(62) minutes for completing different levels of a computer
game in the stem-and-leaf plot shown. What is his
median time for completing a level?

**Personal Best Times
for Completing Levels
of Computer Game**

Stem	Leaves
0	8, 9
1	1, 2, 3, 3, 8, 9
2	2, 2, 5, 6
3	0, 3
4	1

25. Multi-Step The sum of two numbers is 36. Their
(59) difference is 8. Find each of the numbers. What is
their product?

26. Write an equation for a direct variation that includes the point
(56) (13, 13).

27. (Physics) The distance an object travels in a certain amount of time is directly
(56) proportional to its rate of travel. An object travels 105 meters at a rate of 3 meters
per second. How far will an object travel if it travels at 5 meters per second for the
same amount of time?

28. Solve $2 < −4a$.   **29.** Solve $\dfrac{-1}{3} < \dfrac{-1}{9}p$.
(70)   (70)

30. (Real Estate) Donna and James have to pay 20% of the cost of a house as a down
(70) payment. They have $35,000 saved for the down payment. Write and solve an
equation to determine the range for the sale price of a house they can make a down
payment on.

Saxon Algebra 1

Name _____ Date _____ Class _____

Solving Compound Inequalities

LESSON 73

Determine if the systems of equations are consistent and independent, consistent and dependent, or inconsistent.

1. $-\frac{1}{2}x + y = 5$
(67)
$x + y = 5$

2. $y = 5x - 3$
(67)
$2y - 10x = 8$

Solve the inequalities.

3. $x + 2 - 3 \le 6$
(66)

4. $z + 5 \ge 1.5$
(66)

*5. $-15 \le 2x + 7 \le -9$
(73)

*6. $x - 3 \ge 4$ OR $x + 2 < -5$
(73)

*7. Graph the compound inequality $-4 \le x \le 5$.
(73)

*8. Write a compound inequality that represents all real numbers that are no more than 1 or no less than 6.
(73)

*9. Multiple Choice Which inequality describes the graph?
(73)

$\leftarrow\!+\!+\!+\!\oplus\!+\!+\!+\!\oplus\!+\!+\!+\!\rightarrow$
$-4\ \ -2\ \ \ 0\ \ \ 2\ \ \ \ 4\ \ \ \ 6$

A $x < -1$ OR $x > 3$

B $x > -1$ OR $x < 3$

C $x < -1$ AND $x > 3$

D $x > -1$ AND $x < 3$

*10. (Noise) A human being cannot hear sound that is below 0 decibels and has a pain threshold of 120 decibels. Write a compound inequality that describes the decibel levels that human beings cannot hear or are too painful to hear.
(73)

*11. Write Explain the difference between the use of the words AND and OR in a compound inequality.
(73)

12. Factor $x^2 - 12x + 32$.
(72)

13. Add $x\sqrt[3]{xy} + x\sqrt[3]{xy}$.
(69)

*14. Geometry A rectangle is described by the trinomial $x^2 + 12x + 27$. What are the dimensions of the rectangle?
(72)

15. Multi-Step Factor the polynomial $8x + x^2 - 4 - 5x$.
(72)

16. Multiple Choice In which of the trinomials are the binomial factors equal?
(72)
A $(x^2 + 7x - 8)$ **B** $(x^2 + 6x + 9)$ **C** $(x^2 + 9x + 8)$ **D** $(x^2 + 7x + 12)$

Solving Compound Inequalities

LESSON 73

***17.** Use the scatter plot and the trend line to write an equation of the line.
(71)

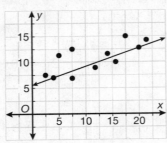

18. Error Analysis Two students are analyzing the data in the table and have drawn
(71) different conclusions. Which student is correct? Explain the error.

x	4	1	3	6	2	5
y	12	2	9	21	7	15

Student A	Student B
There is no correlation among the data because there is no constant increase or decrease in the data.	There is a positive correlation among the data because both sets of data values increase.

19. (Chemistry) The table shows atomic numbers and atomic mass for the first
(71) 10 elements of the periodic table of the elements. Make a scatter plot using the data values.

Atomic Symbol	H	He	Li	Be	B	C	N	O	F	Ne
Atomic Number	1	2	3	4	5	6	7	8	9	10
Atomic Weight	1.01	4.003	6.94	9.01	10.81	12.01	14.01	15.999	18.998	20.18

20. Which inequality is the solution of $-\frac{n}{12} < 36$?
(70)
    A $n < -432$ **B** $n > -432$ **C** $n < 432$ **D** $n > 432$

21. Analyze Consider the equation and inequality below. What parts of the solution steps
(70) are similar? What parts are different?

$$-2t = 4 \qquad -2t > 4$$

22. (Marketing) Four different collectible cards are being offered inside a cereal box. The
(68) probabilities of choosing a box with a given card are as follows: Card #1 (1 in 4), Card #2 (1 in 10), Card #3 (1 in 20), and Card #4 (6 in 10). If a box is chosen at random, what is the probability of getting either Card #3 or Card #4?

Solving Compound Inequalities

LESSON 73

23. Probability In some games, a single regular solid with 20 sides, labeled from 1 to 20, is used instead of a dot cube. What is the probability of rolling either a multiple of 2 or a multiple of 3 with a 20-sided solid?
(68)

24. Multi-Step Draw a coordinate grid, and number it from -6 to $+6$ along the x- and y-axis. Will a line that passes through the points $(-5, 1)$ and $(-2, 4)$ be perpendicular to the line that passes through the points $(-2, 2)$ and $(2, -2)$?
(65)
  a. **Model** Line 1 passes through the points $(-5, 1)$ and $(-2, 4)$ and Line 2 passes through the points $(-2, 2)$ and $(2, -2)$. Graph the lines.
  b. Write an equation for both lines.
  c. **Justify** Are the lines perpendicular? Explain.

25. Multi-Step There is an inverse square relationship between the electrostatic force between two charged objects and the distance between the two objects, meaning that $F = \frac{k}{d^2}$. If two objects are 10 centimeters apart and have an attractive force of 0.512 newtons, how far apart are the two objects when they have an attractive force of 0.0142 newtons? Round to the nearest centimeter.
(64)

*26. (Jewelry Appraisal) Maria has a white gold ring with a mass of 10.3 grams and a volume of 0.7 cubic centimeters. White gold is made of nickel and gold. Nickel's mass is 9 grams per cubic centimeter and gold's mass is 19 grams per cubic centimeter. What is the volume of each metal in the ring?
(63)

27. Multi-Step Jack is building a rectangular barn. Write an expression for the area of the region. Then find the area.
(60)

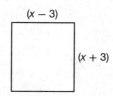

28. Find the product of $(x + 2)(x^2 + 2x + 2)$ using the Distributive Property.
(58)

29. Find the LCM of $(24r - 6d)$ and $(20r - 5d)$.
(57)

30. (Books) Every $14bn$ books donated to the library are nonfiction. Every $38b^9n$ books donated are new. How many books have been donated when a new, nonfiction book is donated?
(57)

Name _____ Date _____ Class _____

Solving Absolute-Value Equations — LESSON 74

Factor.

1. (72) $x^2 + 12x - 28$
2. (72) $15x + 50 + x^2$
3. (72) $18 + x^2 + 11x$
*4. (72) $3x - 18 + x^2$

Solve.

*5. (74) $|n| = 13$

*6. (74) Solve $|x + 7| = 3$.

*7. (67) Find the common solution for the system of equations: $-2y + 3x = 4$.
$$-y + \frac{3}{2}x = 2$$

*8. (74) **(Carpentry)** The shelf in a cabinet is set into a notch at a height of 12 inches. So that the shelf can be adjusted to different heights, the cabinetmaker cuts notches at one-inch intervals above and below the shelf. Write and solve an absolute-value equation for the maximum and minimum heights of a shelf set two notches above or below the 12-inch height.

*9. (74) **Write** Why does the equation $|x - 6| = 2x - 3$ have no solution for $x = -3$?

*10. (74) **Model** Use a number line to diagram the solution of the absolute-value equation $|x - 3| = 5$. Then write the solution set to the absolute-value equation.

*11. (74) **Measurement** The diameter of a circular clamp is 10 centimeters. Its diameter can be changed by changing its circumference.
   a. If the circumference of the clamp is changed by ± 5cm, write an equation to find the clamp's minimum and maximum diameters.
   b. Solve the equation.

12. (73) Graph the compound inequality $x \geq 5$ OR $x < 0$.

13. (73) Solve the inequality $6x < 12$ OR $3x > 15$.

*14. (74) **Error Analysis** Two students are solving $|z + 3| = 5$. Which student is correct? Explain the error.

Student A	Student B
$\|z + 3\| = 5$	$\|z + 3\| = 5$
$\|z\| + 3 = 5$	$z + 3 = 5 \quad\quad z + 3 = -5$
$\|z\| = 5 - 3 = 2$	$z = 5 - 3 \quad\quad z = -5 - 3$
$z = 2 \quad\quad z = -2$	$z = 2 \quad\quad z = -8$
The solution set is $\{2, -2\}$.	The solution set is $\{2, -8\}$.

Solving Absolute-Value Equations

LESSON 74

*15. Geometry The third side of a triangle must be greater than the difference of the other two sides and less than the sum of the other two sides. Suppose that two sides of a triangle have lengths 6 inches and 11 inches. Write a compound inequality that describes the possible values for the third side of the triangle.
(73)

16. Multi-Step Use the compound inequality $3x > 45$ OR $-2x \geq 24$.
(73)
    a. Solve the compound inequality.
    b. Graph the solution.

17. Factor the trinomial $x^2 - 3x - 40$.
(72)

18. (Landscaping) To beautify a city park, landscapers use two sets of plans to lay out rectangular flower beds. One set of plans for the area is described by the trinomial $x^2 + 9x + 20$. The other set is described by the trinomial $x^2 + 21x + 20$.
(72)
    a. Determine the binomial factors of each set.
    b. Which set can be used to lay out longer, narrower beds? Explain.

19. Multiple Choice A _____ is a graph made up of separate, disconnected points.
(71)
    A discrete graph   **B** continuous graph   **C** trend graph   **D** linear graph

20. Determine if the inequality symbol in the expression $-\frac{2}{3}c \leq 6$ needs to be reversed, and then solve.
(70)

21. If a square measures $\sqrt{144}$ inches on one side, what is the perimeter of the square?
(69)

22. Determine the probability of rolling a sum of 9 or an odd number with two number cubes.
(68)

23. Multi-Step The library charges a late fee based on the number of days a book is overdue. The equation for the fee is $y = \$0.25d + \$0.05$.
(67)
    a. What is the fee for a book that is 10 days overdue?
    b. Thirty days is the maximum number of days for which the library charges. What is the maximum amount the library charges?
    c. Classify the system of equations for the library fee and $y = \$7.55$.

24. Justify If a, b, x, and y are all greater than 0 and $x > a$ and $y < b$, how does $\frac{x}{y}$ compare to $\frac{a}{b}$? Justify your answer.
(66)

Solving Absolute-Value Equations

LESSON 74

***25. Predict** Use the trend line to predict the corresponding y-value for an x-value of 40.
(71)

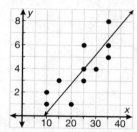

26. Justify True or False: The lines represented by $y = \frac{x}{4} - 2$ and $4y = x + 8$ are parallel. Explain.
(65)

27. (Construction) To build Ariane's house will take a constant number of individual work days. If a construction crew of 15 people can build the house in 20 days, how many people does it take to finish the house in 5 days?
(64)

28. Multi-Step A container has a square base. The area is 800 cm². The size of a book is 25 cm by 30 cm. Can the book lie flat on the base of the container?
(61)
  a. Find the side length of the container.
  b. Compare the side length with the dimensions of the book.
  c. Use the comparison to determine if the book will fit.

29. Solve the system by substitution: $3x + y = 13$.
(59)
$$2x - 4y = 4$$

30. Sue purchases a rectangular billboard sign for her new business. The length can be represented by $6x^2 + 6x + 6$ and the width is $(x + 8)$. What is the area of Sue's new billboard?
(58)

Factoring Trinomials: $ax^2 + bx + c$

LESSON 75

Factor completely.

*1. $6x^2 + 13x + 6$

*2. $3x^2 - 14x - 5$

*3. $18 - 15x + 2x^2$

*4. $-15 + 7x + 2x^2$

Simplify.

5. $22c\sqrt{de} - 9\sqrt{de}$

6. $8\sqrt{7} - 4\sqrt{11} - 3\sqrt{7} + 7\sqrt{11}$

*7. **Write** Explain why b in $ax^2 + bx + c$ is not the sum of the factors of c.

*8. **Justify** Show that $7x^2 - 12x + 10$ is prime (cannot be factored).

*9. **Multiple Choice** Evaluate $2x^2 - 10x + 14$ if $x = -2$.
   A 2 **B** 10 **C** 42 **D** 50

*10. For a situation that represents a direct variation, what does the graph of that situation look like?

11. Solve $|z| = 5$.

12. **Geometry** Sliding-glass tubes form a square with a perimeter of 36 inches. Both the length and width of the square can be changed by ±1.5 inches.
   a. Write an absolute-value equation to find the greatest and least perimeter of the square.
   b. What are the greatest and least perimeters?

*13. **Multi-Step** Given the equation $|x + 2| + 6 = 17$, answer the questions below.
   a. Isolate the absolute value expression.
   b. Use the definition of absolute value to rewrite the absolute-value equation as two equations.
   c. What is the solution set?

*14. **Physics** A steel bar at a temperature of 300°C has the length L. If the temperature of the bar is raised or lowered 100°C, its length will increase or decrease by 0.12%, respectively. Write an absolute-value equation for the maximum and minimum lengths of the bar when it is heated to 400°C or cooled to 200°C.

15. Solve the compound inequality $-14 \leq -3x + 10 \leq -5$.

16. **Error Analysis** Two students graph the compound inequality $-2 < x < 1$ as shown. Which student is correct? Explain the error.

Student A

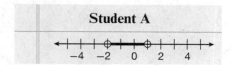

Student B

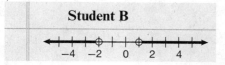

Factoring Trinomials: $ax^2 + bx + c$

LESSON 75

17. **Temperature** The predicted high temperature for the day is 88°F and the predicted low temperature for the day is 65°F. Write a compound inequality that shows the range of temperatures predicted for the day.
(73)

18. **Multiple Choice** Which binomial is a factor of $x^2 + 5x - 6$?
(72)
   A $x + 1$ **B** $x + 2$ **C** $x + 3$ **D** $x + 6$

19. **Verify** Show that 2 is a solution of the compound inequality $x < 3$ OR $x > 6$.
(73)

20. Use the scatter plot and the trend line to write an equation of the line.
(71)

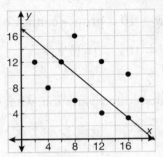

21. **Measurement** Tyrone bought 24 quarts of juice for a party. He plans to serve the juice in $\frac{1}{4}$-quart cups. How many servings can he plan to make with the juice he has available?
(70)

22. **Fitness** To make the swim team, Maria needs to swim 10 laps in under 8 minutes. Write and solve an inequality to find the maximum average time per lap she could swim to still make the team.
(70)

23. Suppose Carrie walks home from school at a rate of 4 miles per hour. Write an equation that relates Carrie's time to the distance traveled home. Is this a direct or indirect variation? What represents the constant of variation?
(Inv 7)

24. Find the probability of choosing an odd number card from a pack of cards numbered 1 through 9 or the number 1.
(68)

25. **Web Authoring** A company designs web sites and charges a fee of $y = \$250 + \$50w$. The charge includes a base fee of $250, plus a weekly fee based on the number of weeks the client requires full support. If the equation $y = 800$ represents the initial amount a company pays, for how many weeks do they receive full support?
(67)

26. **Analyze** The equation $y = 4$ describes a horizontal line with zero slope. The equation $x = 4$ describes a vertical line with infinite slope. Classify the system formed by these two equations.
(67)

Factoring Trinomials: $ax^2 + bx + c$

LESSON 75

27. Multi-Step Charlene is shopping for used DVDs. She buys some DVDs from Bin A
(66) that cost $7 each and some from Bin B that cost $5 each. Charlene has $45.
  a. Write an inequality representing the situation. (Hint: You must use two different variables to represent DVDs from each bin.)
  b. Can Charlene buy 4 DVDs from Bin A and 4 from Bin B?
  c. Charlene decides to buy 3 DVDs from Bin A. How many can she choose from Bin B?

28. Multi-Step The amounts of measurable rain (in inches) per day recorded
(62) for Seattle in July 2007 are listed below.

$$0.06, 0.04, 0.01, 0.28, 0.01, 0.15, 0.21, 0.38, 0.30, 0.10$$

  a. Create a histogram of the data.
  b. Is this the best representation of the data for the entire month? Why or why not?

29. Find the product: $(4y - 4)(4y + 4)$.
(60)

30. Solve the system of equations by substitution.
(59)
$$5x + 3y = 1$$
$$8x + 4y = 4$$

Saxon Algebra 1

Name _____ Date _____ Class _____

Multiplying Radical Expressions LESSON 76

Solve.

1. $\dfrac{5}{6} \leq -2p$
   (70)

2. $|x + 4| = 5$
   (74)

Factor.

3. $12x^2 - 25x + 7$
   (75)

4. $x^2 + 10x - 39$
   (72)

5. $5z^2 + 2z - 7$
   (75)

6. $3x^2 + 25x - 18$
   (75)

Simplify.

*7. $4\sqrt{3} \cdot 6\sqrt{6} \cdot 3\sqrt{3} \cdot 2\sqrt{2}$
   (76)

8. $-17\sqrt{7s} - 4\sqrt{7s}$
   (69)

*9. $(4\sqrt{5})^2$
   (76)

*10. $3\sqrt{2} \cdot 4\sqrt{12} - 6\sqrt{54}$
   (76)

11. $\sqrt{\dfrac{x^3}{60}}$
    (61)

12. Use mental math to find the product of $17 \cdot 23$.
    (60)

*13. **(City Parks)** The City Works Department wants to build a new fence around the town's park. Write an expression for the area of the figure at right. Find the area.
    (76)

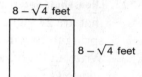

$8 - \sqrt{4}$ feet

$8 - \sqrt{4}$ feet

*14. **Justify** Explain how to find the product of $\sqrt{2}(\sqrt{3} - \sqrt{8})$.
    (76)

*15. **Write** Find two radical expressions that when multiplied together equal a perfect square.
    (76)

*16. **Multiple Choice** Which expression is not equivalent to $\sqrt{48}$?
    (76)
       A $\sqrt{3 \cdot 16}$ **B** $16\sqrt{3}$ **C** $\sqrt{4^2 \cdot 3}$ **D** $4\sqrt{3}$

17. **(Vacation)** The cost of a vacation, in dollars, is represented by the expression $46x^2 - 9x + 95$, where x is the number of nights. What is the cost of a 3-night vacation?
    (9)

18. **Error Analysis** Two students factor $5x^2 - 6x - 8$. Which student is correct? Explain the error.
    (75)

Student A	Student B
$(5x + 4)(x - 2)$	$(5x - 4)(x + 2)$

*19. **Geometry** The sum of the squares of the two legs of a right triangle is $16x^2 - 40x + 25$. Factor the expression, and then take the square root to find the expression that represents the length of the hypotenuse.
    (75)

Multiplying Radical Expressions

LESSON 76

20. Multi-Step The area of a rectangular ottoman is represented by $2x^2 + 3x - 27$ square inches.
(75)
  a. Evaluate the expression for $x = 12$.
  b. Factor the expression completely.

21. Error Analysis Two students are solving $|3x| = 6$. Which student is correct? Explain the error.
(74)

Student A
$\|3x\| = 6$
$3x = 6 \qquad 3x = -6$
$3x - 3 = 6 - 3 \qquad 3x - 3 = -6 - 3$
$x = 3 \qquad x = -9$
The solution set of $\|3x\| = 6$ is $\{6, -6\}$.

Student B
$\|3x\| = 6$
$3x = 6 \qquad 3x = -6$
$\dfrac{3x}{3} = \dfrac{6}{3} \qquad \dfrac{3x}{3} = \dfrac{-6}{3}$
$x = 2 \qquad x = -2$
The solution set of $\|3x\| = 6$ is $\{2, -2\}$.

*22. (Household Security) A window guard is designed to be placed directly into windows that are 27 inches wide. The guard can be compressed or expanded with a tension spring to fit windows that are three inches narrower or wider than the standard width. Write and solve an absolute-value equation for the maximum and minimum width of windows in which the guard is designed to fit.
(74)

23. Multiple Choice Which inequality describes the graph?
(73)

```
  -6  -4  -2   0   2   4   6
```

  A $-3 < x < 1$ **B** $-3 \leq x < 1$ **C** $-3 < x \leq 1$ **D** $-3 \leq x \leq 1$

24. Analyze What are the solutions to the compound inequality $x > 3$ OR $x < 5$? Explain your answer.
(73)

*25. (Climate) The heat index displays apparent air temperatures in relation to the actual air temperature. The table shows the apparent air temperature for different humidity levels given that the actual air temperature is 85°.
(71)

Humidity Level (percent)	10	20	30	40	50	60	70	80	90	100
Apparent Air Temperature (°F)	80	82	84	86	88	90	93	97	102	108

  a. Use the data to make a scatter plot.
  b. Do the data values show a positive correlation, a negative correlation, or no correlation?

26. Data Analysis Use the values from the table to draw a scatter plot.
(71)

x	1	2	3	4	5	6
y	6	11	19	24	28	34

288   Saxon Algebra 1

Multiplying Radical Expressions

LESSON 76

27. Analyze A book has 12 chapters about flying birds, 5 chapters about flying insects, 4 chapters about land insects, and 4 chapters about land mammals. If the probability of picking a chapter about certain types of animals is $\frac{8}{25}$, what type of animal is the chapter about?
(68)

28. Multi-Step A new exercise facility recently opened up. Its initial promotional rate for new members, represented by $y = \$20 + \$6x$, includes a monthly fee plus an additional amount for x services.
(67)

   a. How many services do new members receive before paying the same amount as a regular fee of $32?

   b. Write an equation describing the regular rate for the facility.

   c. Classify the system of equations for the promotional and regular facility rates.

29. (Snowfall) In 1947, Mount Locke, Texas, had 23.5 inches of snow—its greatest yearly snowfall. The greatest monthly snowfall recorded was 20.5 inches, which occurred in January, 1958. Since the record wasn't broken in 1958, how many more inches of snow could Mount Locke, Texas, have had that year?
(66)

30. Multi-Step The sum of Kaleigh and Dwayne's ages is 30. Three times Kaleigh's age minus twice Dwayne's age equals 5. How old is César if he is five years younger than Kaleigh?
(63)

Solving Two-Step and Multi-Step Inequalities

LESSON 77

Use the stem-and-leaf plot to find the following statistical measures.

1. Find the median.
(62)

2. Find the mode.
(62)

Average Milk Production

Stem	Leaves
3	5, 7
4	1, 6, 7, 7, 7, 8, 8, 9, 9
5	0, 2, 3, 6
6	1, 1

Key: 6 | 1 means 61 pounds

3. Factor the expression $6x^2 - 10x - 4$ completely.
(75)

4. Solve $9 > 0.3r$.
(70)

*5. Solve $5 + 4x > 37$, and then graph the solution.
(77)

Solve.

*6. $\dfrac{x}{-3} - 2 \leq 1$
(77)

*7. $-3x + 2 \leq 1$
(77)

*8. $\dfrac{x}{5} - 4 > 9$
(77)

9. $-5 < r - 6 < -2$
(73)

Simplify.

10. $4\sqrt{3}x \sqrt{4}x$
(76)

11. $\sqrt{400g^6}$
(61)

*12. **Generalize** Describe how solving inequalities is different from solving equations.
(77)

*13. **Multiple Choice** What is the solution to $6 - 7y < 48$?
(77)
 A $y < -6$ **B** $y > -6$ **C** $y > -48$ **D** $y < -48$

*14. **Hobbies** Building ships in a bottle is Jeff's favorite hobby. He has $42 to purchase
(77) two ship kits and two bottles. Each ship kit is $18. To find how much he can spend on each bottle, solve the inequality $2(18) + 2b \leq 42$.

15. **Body Surface Area** Physicians can estimate the body surface area of an adult
(76) (in square meters) using the formula $x = \sqrt{\dfrac{HW}{3125}}$ called BSA, where H is height in inches and W is weight is pounds. Find the BSA of a person who is $\sqrt{5184}$ inches tall and weighs $\sqrt{32,400}$ pounds. Round to the nearest whole number.

16. The brightness of a photograph is represented by the expression $12x^2 - 2x - 4$.
(75) Factor this expression completely.

17. **Error Analysis** Students were asked to find the product of $\sqrt{75} \cdot \sqrt{2}$. Which student is
(76) correct? Explain the error.

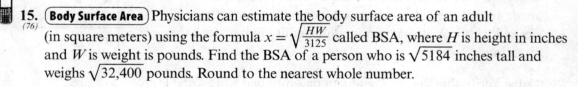

	Student A	Student B
	$\sqrt{75} \cdot \sqrt{2}$	$\sqrt{75} \cdot \sqrt{2}$
	$\sqrt{150}$	$\sqrt{150}$
	$5\sqrt{6}$	$5\sqrt{30}$

Saxon Algebra 1

Solving Two-Step and Multi-Step Inequalities

LESSON 77

***18. Multi-Step** A tsunami is a big ocean wave that can be caused by underwater volcanic eruptions, earthquakes, or hurricanes. The equation $S = \sqrt{g \cdot d}$ models the speed of a tsunami, where g is the acceleration due to gravity, which is 9.8 meters/second^2, and where d is the depth of the ocean in meters.
  a. Suppose a tsunami begins in the ocean at a depth of 1000 meters. What is the speed of the tsunami?
  b. Suppose a tsunami begins in the ocean at a depth of 2000 meters. What is the speed of the tsunami?

*19. Geometry A triangle has a base of $3 + \sqrt{15}$ inches and a height of $5 - \sqrt{20}$ inches. Find the area.

20. Error Analysis Two students factor $15x^2 + 16x - 15$. Which student is correct? Explain the error.

Student A	Student B
$(3x + 5)(5x - 3)$	$(3x - 5)(5x + 3)$

21. Multiple Choice Solve $|x - 3| + 2 = 0$.
  A $\{5, 1\}$ **B** $\{4, -1\}$ **C** $\{1\}$ **D** $\emptyset$

22. Write Why is there no solution for the absolute-value equation $|x + 11| + 3 = 1$?

23. (Weaving) The area of rugs designed by a weaver are described by the trinomial $x^2 + 30x - 400$. What are the dimensions of a rug if x is 40?

24. State whether the data show a positive correlation, a negative correlation, or no correlation.

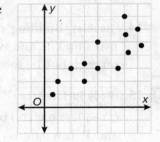

*25. Measurement A manufacturing company is making screws that have a width of 4 millimeters. Each screw must be made so that the actual width is ±0.03 millimeters of the desired width. Write a compound inequality to represent this situation.

26. Verify Verify that $3gh\sqrt{275g^7h^9}$ can be simplified to $15g^4h^5\sqrt{11gh}$.

27. Multi-Step Gwynedd's piggy bank contains 57 quarters, 24 dimes, 35 nickels, and 60 pennies. Gwynedd turns her piggy bank over and one coin falls out. What is the probability that the coin will be worth more than 5 cents?
  a. What is the probability that the coin is a quarter or dime?
  b. What is the probability that the coin is a quarter or worth more than 5 cents?
  c. Explain how the previous two answers are related.

Solving Two-Step and Multi-Step Inequalities

28. (Cell Phones) Two different cell phone service plans are represented by the given equations. Classify this system of equations.
(67)

$$p = 6x + 24$$
$$p = 8x + 20$$

29. Multi-Step Anya has a number of square bricks with a side length of 1 foot that she wants to use to create a rectangular patio in her backyard. At first, she arranges them so the rectangle is 12 bricks long by 3 bricks wide, but then she decides she does not like that arrangement. How can Anya arrange the bricks if she would rather have a square-shaped patio?
(64)

30. Error Analysis Two students simplified the expression as shown. Which student is correct? Explain the error.
(61)

Student A	Student B
$\sqrt{80y^2z^2} = 16yz\sqrt{5}$	$\sqrt{80y^2z^2} = 4yz\sqrt{5}$

Name _____ Date _____ Class _____

Graphing Rational Functions — LESSON 78

Solve each system of linear equations.

1. $3x - 2y = 17$
(63) $-4x - 3y = 17$

2. $y = 2x + 4$
(63) $-x - 3y = 9$

Factor.

3. $x^2 + 10xy + 21y^2$
(72)

4. $-30 - 13x + x^2$
(72)

*5. Find the excluded values for $y = \frac{4}{7m}$.
(78)

*6. Find the excluded values for $y = \frac{m-2}{3m+9}$.
(78)

*7. Find the vertical asymptote for $y = \frac{5}{x-3} + \frac{2}{5}$.
(78)

8. Solve $3 - 9m < 30$ and graph the solution.
(77)

9. Use the Distributive Property to find the product of $5(\sqrt{4} + \sqrt{36})$.
(76)

*10. (Awards) Jason plans to buy 20 prizes for the next school carnival and to divide
(78) them equally between the winners of the school trivia contest. There will also be an additional 6 prizes given to each winner at the awards assembly. The number of prizes for each winner, y, is given by $y = \frac{20}{x} + 6$, where x is the number of winners. Find the asymptotes and graph the function.

*11. Verify Show that the value $y = 5$ will not satisfy the function $y = \frac{1}{x} + 5$.
(78)

*12. Justify What are the equations of the asymptotes in the graph of
(78) $y = \frac{2.3}{x+1.9} + 0.3$?

*13. Multiple Choice What is the vertical asymptote for the rational function $y = \frac{6.1}{x+1.5} + 3.1$?
(78)
 A $x = 3.1$ **B** $x = 1.5$
 C $x = -1.5$ **D** $x = 6.1$

*14. (Budgeting) You have $30 to spend at the mall and your mother gives you an
(77) additional $15. You must buy a new shirt that is on sale for $12. You also would like to purchase 3 CDs. To find the maximum amount you can spend on each CD, solve the inequality $30 + 15 \geq 12 + 3c$.

Graphing Rational Functions

LESSON 78

15. Error Analysis Two students solve $8p - 7 \leq 3p + 18$. Which student is correct? Explain the error.

Student A	Student B
$8p - 7 \leq 3p + 18$	$8p - 7 \leq 3p + 18$
$5p - 7 \leq 18$	$-7 \leq -5p + 18$
$5p \leq 25$	$-25 \leq -5p$
$p \leq 5$	$5 \leq p$

16. Geometry The triangle inequality theorem states that the sum of any two sides of a triangle must be greater than the third side. A triangle has sides $4g + 10$, $3g - 13$, and 46 as the longest side. Therefore, $(4g + 10) + (3g - 13) > 46$. Solve for g.

17. Multi-Step You can hike the trails at 2.5 miles per hour. You break for 1 hour to rest and eat lunch. You want to hike at least 15 miles. To find the number of hours you will hike, solve the inequality $2.5(h - 1) \geq 15$.

18. Error Analysis Students were asked to find the product of $\sqrt{2}(7 + \sqrt{14})$. Which student is correct? Explain the error.

Student A	Student B
$\sqrt{2}(7 + \sqrt{14})$	$\sqrt{2}(7 + \sqrt{14})$
$\sqrt{14} + \sqrt{28}$	$7\sqrt{2} + \sqrt{28}$
$\sqrt{14} + 2\sqrt{7}$	$7\sqrt{2} + 2\sqrt{7}$

*19. Art Michael is submitting his first painting to the county fair art show. His painting has a side length of $7 + \sqrt{32}$ inches and a width of $9 + \sqrt{50}$ inches. What is the area of Michael's painting?

20. Verify Show that $10x^2 - 11xy - 6y^2 = (2x - 3y)(5x + 2y)$.

21. Multiple Choice Which expression is the factored form of $20x^2 + 49x + 9$?

A $(10x + 3)(2x + 3)$

B $(20x + 3)(x + 3)$

C $(5x + 1)(4x + 9)$

D $(5x + 9)(4x + 1)$

22. Measurement The area of a rectangular tray is represented by $4x^2 - 16x + 16$ square inches. Evaluate the expression for $x = -3$.

23. Solve $|n| = 12$.

Graphing Rational Functions

LESSON 78

24. **Dining** A restaurant offers discounted meals to children 12 years old and under or to senior citizens who are at least 65 years old. Write a compound inequality that represents this situation.
(73)

25. State whether the data show a positive correlation, a negative correlation, or no correlation.
(71)

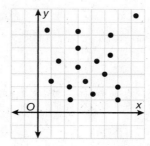

26. **Verify** Solve the inequality $\frac{n}{5} \leq -3$. Then select several values of n to substitute into the inequality to verify the answer.
(70)

27. **Multi-Step** The larger square has an area of $216x^2$ and the smaller square has an area of $125x^2$. What is the combined perimeter of the two squares?
(69)

28. **Game Play** Derek is playing a board game with three number cubes. What is the probability of Derek rolling three of a kind or a sum that is an odd number?
(68)

29. **Multiple Choice** What slope would the line parallel to $3x - y = -3$ have?
(65)

   A $-\frac{1}{3}$ **B** $\frac{1}{3}$

   C 3   **D** -3

30. Create a histogram to display the following data representing average daily milk production (in gallons) of seventeen dairies: 49, 41, 50, 47, 35, 47, 61, 49, 60, 46, 53, 48, 37, 47, 56, 52, 48. Use 35–39 as the starting interval.
(62)

Name _____ Date _____ Class _____

Factoring Trinomials by Using the GCF   LESSON 79

Assuming that y varies inversely as x, find the missing value.

1. If $y = 6$ and $x = 9$, what is y when $x = 12$?
(64)

Solve each system of linear equations.

2. $5x = 2y + 10$
(63) $-3y = -2x + 4$

3. $x - y = 2$
(63) $y + 2x = 1$

4. Solve the compound inequality $3b - 2 < -8$ OR $4b + 3 > 11$.
(73)

Factor.

5. $x^2 - 4x - 45$
(72)

*6. $k^4 + 6k^3 + 8k^2$
(79)

7. $5x^2 + 3x - 2$
(75)

*8. $2x^3 + 16x^2 + 30x$
(79)

*9. $abx^2 - 5abx - 24ab$
(79)

10. $15mx^2 + 9mx - 6m$
(79)

11. Find the excluded values for $y = \frac{9m}{m+3}$.
(78)

12. Solve $16 + (-6) \geq 2(d + 4)$ and graph the solution.
(77)

*13. Write Can you factor out the GCF after factoring a trinomial?
(79)

*14. Justify When asked to factor $3x^2 + 45x + 132$, one student writes $3(x + 4)(x + 11)$.
(79) Another student writes $3(x + 11)(x + 4)$. Show that both answers are correct. Explain your answer.

*15. Multiple Choice Which expression is factored completely?
(79)
  A $2(5x - 10)(x - 5)$ **B** $10(x^2 - 7x + 10)$
  C $5(2x - 4)(x - 5)$ **D** $10(x - 5)(x - 2)$

*16. (**Physics**) An object is thrown from inside a hole. Its height after x seconds is
(79) represented by the expression $-16x^2 + 32x - 16$. Factor the expression completely.

17. Error Analysis Students were asked to find the vertical asymptote of the expression
(78) $y = \frac{1}{x+2} + 8$. Which student is correct? Explain the error.

Student A	Student B
$x = -2$	$y = 8$

Factoring Trinomials by Using the GCF

LESSON 79

*18. Multi-Step A band teacher has a budget of $50,000 to buy new instruments. He will receive 1 free instrument when he places his order. The number of instruments, y, that he can get is given by $y = \frac{50,000}{x} + 1$, where x is the price per instrument.
 a. What is the horizontal asymptote of this rational function?
 b. What is the vertical asymptote?
 c. If the price per instrument is $1000, how many instruments will he receive?

19. (Party Planning) A party planner has a budget of $1350.00 to buy a steak dinner for the entire guest list for a company retirement party. The planner will receive 15 free dinners when she places the order. The number of steak dinners, y, that the planner can buy is given by $y = \frac{1350}{x} + 15$, where x is the price per steak dinner.
 a. What is the horizontal asymptote of this rational function?
 b. What is the vertical asymptote?
 c. If the price per dinner is $25, how many dinners will the planner receive?

*20. Geometry Write a rational function that shows the relationship between the side lengths of the similar triangles shown.

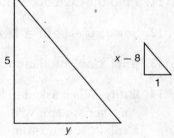

21. (Automotive Care) Steve's car is really dirty and is almost on empty. He has $20. A car wash is $7 and gas is $3 per gallon. To find how many gallons of gas he can buy with a car wash, solve the inequality $7 + 3g \leq 20$.

22. Error Analysis Two students solve $3h + 15 > 6$. Which student is correct? Explain the error.

Student A	Student B
$3h + 15 > 6$	$3h + 15 > 6$
$3h > -9$	$3h > -9$
$h < -3$	$h > -3$

23. Write Explain the process for multiplying binomials with radicals.

24. Multiple Choice Between which two numbers does $\sqrt{124}$ fall on a number line?
 A 9 and 11   **B** 8 and 11   **C** 12 and 14   **D** 11 and 13

25. Measurement Lou is building a square sandbox. The sandbox has a side length of $6\sqrt{36}$ inches. What is the area of the sandbox?

Factoring Trinomials by Using the GCF   79

26. (**Fine Arts**) In an antique shop, a painting is inclined on a display easel at an angle
(74) of 65°. The angle of inclination can be increased or decreased by a maximum of 8°
from the present setting by an adjustment screw. Write and solve an absolute-value
equation for the maximum and minimum angles of inclination for the easel.

27. Write Describe the characteristics of the trend line for scatter plots with a positive
(71) correlation, a negative correlation, and no correlation.

28. Multi-Step The video store charges $4.80 per rental of a DVD. The store spends $5100
(70) on expenses each month. How many videos does the store have to rent to make more
money than it spends on expenses?
    a. What are the total expenses of the store?

    b. Write an inequality to show how many DVDs need to be rented to make a profit.

    c. Solve the inequality.

    d. How many DVDs need to be rented a month to make a profit?

29. (**Sewing**) Deandra wants to put trimming around a rectangular tablecloth that
(69) measures $2\sqrt{49}$ feet by $\sqrt{81}$ feet. How many feet of trimming will Deandra need?

*30. Multi-Step A fitness center has a small elevator for members to use and to move
(66) fitness equipment. The elevator has a maximum capacity of 750 pounds.
    a. Troy weighs 185 pounds. Write an inequality representing how much more weight
the elevator can carry if Troy uses the elevator.

    b. Alicia needs to move weight equipment to the second floor. She has four
75-pound weights, four 50-pound weights, four 25-pound weights, and six
20-pound weights. Will she be able to bring all of the weights up in one elevator
trip? Explain.

Calculating Frequency Distributions

LESSON 80

1. Write an equation for a line that passes through $(-1, 4)$ and is parallel to $2y + 10x = -36$.

Assuming that y varies inversely as x, find the missing value for each problem.

2. If $y = 108$ and $x = 3$, what is x when $y = 3$?

3. If $y = 56$ and $x = 7$, what is x when $y = 4$?

4. Find the horizontal asymptote: $y = \frac{7}{x+5}$.

5. Find the horizontal asymptote: $y = \frac{3}{2x+3} - 5$.

Solve.

6. $11|x| = 55$

7. $x - 5 \geq 0$ OR $x + 1 < -2$

Factor.

*8. $c^{12} + 11c^{11} + 24c^{10}$

9. $42x^4 + 77x^3 - 70x^2$

*10. $-3m^2 - 30m - 48$

*11. $(x-1)x^2 + 7x(x-1) + 10(x-1)$

*12. A student rolls a number cube, numbered 1–6, and spins the spinner. Make a table of the possible outcomes.

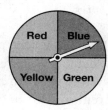

*13. **Generalize** Explain when you would choose a graph or a table to show a frequency distribution.

14. **Multiple Choice** Which expression represents the probability of getting a 4 or 5 on three number cubes, numbered 1–6, and heads on four coins?

   A $P = \frac{4}{6} \cdot \frac{5}{6} \cdot \frac{4}{2}$

   B $P = \frac{1}{6} \cdot \frac{1}{6} \cdot \frac{1}{2}$

   C $P = \left(\frac{1}{6}\right)^2 \cdot \left(\frac{1}{2}\right)^4$

   D $P = \left(\frac{1}{3}\right)^3 \cdot \left(\frac{1}{2}\right)^4$

*15. **(Chemistry)** Experiments in a chemistry lab either succeed or fail. They also can be characterized as chemical or physical reactions. Make a table to show all the possible outcomes.

*16. **Probability** A bag contains 3 red marbles and 7 blue marbles. A marble is drawn and replaced. Use an equation to show the probability of drawing 3 red marbles in a row.

Calculating Frequency Distributions

LESSON 80

17. (Physics) An object is thrown from a tower. The expression $-16x^2 + 32x + 48$ represents its height after x minutes. Factor this expression completely.
(79)

18. Error Analysis Two students factor $2x^2 - 2x - 12$. Which student is correct? Explain the error.
(79)

Student A	Student B
$(2x + 4)(x - 3)$	$2(x + 2)(x - 3)$

*19. Geometry A square has area $9m^4 - 54m^3 + 81m^2$ square inches. What is its side length?
(79)

20. Multi-Step The volume of a rectangular cereal box is represented by the expression $3x^3 + 3x^2 - 18x$.
(79)
a. Factor the expression completely.
b. What are the dimensions of the box?

21. Error Analysis Students were asked to find the excluded values for $\frac{5m}{m+3}$. Which student is correct? Explain the error.
(78)

Student A	Student B
$m + 3 \neq 0$	$m + 3 \neq 0$
$m \neq 3$	$m \neq -3$
$m \neq 3$	$m \neq -3$

*22. (Bakery) Erica plans to buy 100 cookies for the parent-teacher conference and to divide them equally among the number of parents attending. There will also be an additional dozen cookies given to each parent at the end of school party. The number of cookies for each parent, y, is given by $y = \frac{100}{x} + 12$, where x is the number of parents. Find the asymptotes and graph the function.
(78)

23. Analyze Explain why you would not reverse the inequality sign for $6x < -42$ and why you would for $-6x < 42$.
(77)

24. Multiple Choice What is the solution to $5 - 3(2 - m) \geq 29$?
(77)
A $m \leq -10$ **B** $m \geq -10$ **C** $m \leq 10$ **D** $m \geq 10$

25. Use the Distributive Property to find the product of $\sqrt{4}(3 + \sqrt{6})$.
(76)

26. (Softball) The speed of a pitched ball is represented by the expression $9x^2 - 36x - 13$. Factor the expression completely.
(75)

27. Justify Why would you eliminate the factor pair $(5)(9z^2)$ in determining the binomial factors of $x^2 + 18xz + 45z^2$?
(72)

Calculating Frequency Distributions

28. Multi-Step Use the table of values.
(71)

x	40	100	120	20	80	60
y	80	161	196	34	141	105

a. Enter the data into a graphing calculator and find an equation for the line of best fit.

b. How can you tell if there is a positive or negative correlation for the data from the equation?

29. (**Real Estate**) A realtor gets 6.5% commission for each house sold. Write and solve
(70) an inequality to find the house prices that will give the realtor a commission of at least $20,000.

30. Multi-Step Two trains leave the rail yard at the same time. One travels east and the
(67) other west. The distance in miles traveled by the westbound train is $d = 40t + 12$. The distance in miles traveled by the eastbound train is $\frac{1}{4}d = 10t + 3$.
a. Classify these two equations.
b. After a time of 10 hours, which train will have traveled farther?

Solving Inequalities with Variables on Both Sides
LESSON 81

Factor completely.

1. $w^2 - 13w + 36$
2. $-q^2 + q + 42$
3. $30x^2 - 7xy - 2y^2$
4. $x^2 - 11 + 6x - 44$

Solve.

5. $|x - 3| = 14$
6. $|x + 4| = 7.5$
7. $-5 - \dfrac{n}{8} \geq -6$
8. $12 - 3d \leq -3$

Solve and graph. Then check the solution.

*9. $6v + 5 > -2v - 3$
10. $y + 4.5 < 10$

Write an equation for each of the lines described.

11. a line that passes through $(1, -2)$ and is perpendicular to $y = 2x + 6$

12. a line that passes through $(6, 5)$ and is parallel to $y = -x + 4$

13. **Multi-Step** To win a game, Alvaro needs to spin a black section or the number 10.

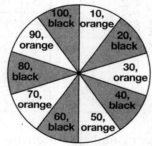

   a. What is the probability of spinning a black section?
   b. What is the probability of spinning a 10?
   c. Are the events inclusive or mutually exclusive?
   d. What is the probability of Alvaro hitting a black section or the number 10?

14. **Employment** The table below shows the number of people employed in the United States. Make a scatter plot of the data.

U.S. Employment (in millions)

Year	1970	1975	1980	1985	1990	1995	2000
Employment (in millions)	79	86	99	107	119	125	137

15. **Verify** Show that 2 is a solution of the compound inequality $x < 3$ OR $x > 6$.

Solving Inequalities with Variables on Both Sides

LESSON 81

16. (Gardening) Debra is planting a square garden. The side length is $8 + \sqrt{8}$ inches. Write an expression to find the area of the garden, and then find the area.

17. Measurement To find the values of x for which the triangle would have a perimeter of more than 81 units, solve the inequality $x + (x + 13) + (2x + 12) > 81$.

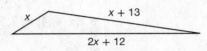

*18. Write How do changes to the value of b affect the graph of $y = \dfrac{a}{x - b} + c$?

*19. Two number cubes are rolled and their values are added. One number cube is labeled 1–6. The other has two of each of the numbers 1, 2, and 3. The possible outcomes are displayed in the table.

What is the theoretical probability of each possible sum?

		Cube 1					
		1	2	3	4	5	6
Cube 2	1	2	3	4	5	6	7
	2	3	4	5	6	7	8
	3	4	5	6	7	8	9
	1	2	3	4	5	6	7
	2	3	4	5	6	7	8
	3	4	5	6	7	8	9

20. Multiple Choice What is the horizontal asymptote for the rational expression $y = \dfrac{7}{x - 2} + 4$?

A $y = -2$ **B** $y = 4$ **C** $y = -4$ **D** $y = 7$

21. (Football) A football is kicked into the air. In the expression $-5t^2 + 25t - 30$, t represents the time when the ball is 30 feet in the air. Factor the expression completely.

22. Error Analysis Two students factor $9m^2x^3 + 81mx^3 + 126x^3$. Which student is correct? Explain the error.

Student A	Student B
$9m^2x^3 + 81mx^3 + 126x^3$ $= 9x^3(m^2 + 9m + 14)$ $= 9x^3(m + 2)(m + 7)$	$9m^2x^3 + 81mx^3 + 126x^3$ The GCF is $9x^3$. Factor out $9x^3$. $m^2 + 9m + 14$ $= (m + 2)(m + 7)$

Solving Inequalities with Variables on Both Sides

LESSON 81

***23.** **(Games)** Two number cubes are rolled and their values are added. Find the probability that the sum is less than or equal to 7.
(80)

***24.** **Geometry** Students made five tetrahedrons, a three-dimensional figure with four triangular faces, and labeled the faces 1–4. Use an equation to find the probability that all five tetrahedrons land on 3.
(80)

***25.** **Error Analysis** Two students find the probability of rolling two number cubes and getting a sum less than 6. Which student is correct? Explain the error.
(80)

Student A	Student B
$\dfrac{10}{36} = \dfrac{5}{18}$	$\dfrac{15}{36} = \dfrac{5}{12}$

26. **Multi-Step** A game has two spinners. After spinning both spinners, the sum of the spins is found.
(80)

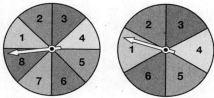

a. Make a table of all possible outcomes.

b. What is the probability that the sum is greater than 8?

***27.** **Multiple Choice** What is the first step in solving the inequality $2(x + 5) > x + 12$?
(81)

A Combine the variables.

B Use the Addition Property of Inequality.

C Apply the Distributive Property.

D Use the Multiplication Property of Inequality.

***28.** **(Travel)** A car rental company charges $40 a day with no additional mileage fees. Another company charges $24 each day plus $0.16 per mile. How many miles would have to be driven in one day for the first company to offer the better deal?
(81)

***29.** **Write** Explain how to solve the inequality $2x + 5 > -3(x - 15)$. Identify the solution.
(81)

***30.** **Estimate** During their freshman year, Malcolm averaged 17.3 points per game and Frederico averaged 15.2 points per game. In their sophomore year, Malcolm averaged 19.1 points per game and Frederico averaged 18.4 points per game. If the trend continues, in which years will Frederico have a better average than Malcolm?
(81)

Name _____ Date _____ Class _____

Solving Multi-Step Compound Inequalities

LESSON 82

Factor completely.

1. (75) $2x^2 + 9xy + 7y^2$

2. (75) $-4m^2 + 8mn + 5n^2$

Find the product.

3. (76) $(\sqrt{3} - 12)^2$

4. (76) $(2x + \sqrt{3})(2x - \sqrt{3})$

Find the excluded values.

5. (78) $\dfrac{m - 6}{2m - 10}$

6. (78) $\dfrac{y + 4}{-2y - 6}$

Solve the inequality. Then graph and check the solution.

7. (66) $2z - 6 \leq z$

8. (81) Solve and graph $2x + 9 > -x + 18$.

Determine if the following systems of equations are consistent and independent, consistent and dependent, or inconsistent.

9. (67) $y = 10x - 2$
$y = 10x + 8$

10. (67) $y = 3x$
$2y = 6x$

11. **Multi-Step** (69) The perimeter of a square area rug is 48 feet. What is the length of each side? Express your answer as a radical number.

12. **Quilting** (72) A quilter uses a series of rectangular patterns to design quilt blocks. The area of the quilt block can be represented by the trinomial $(x^2 + 7x + 12)$ cm². If he plans a quilt block with x having a value of 20 centimeters, what is the dimension of the longer side of the block?

13. **Multi-Step** (73) A real number is less than 12 or is greater than 15.
   a. Write a compound inequality that represents the situation.
   b. Graph the solution.

14. **Estimate** (74) What is the lesser value of q in the solution set of $|q - 24.9| = 5.1$?

15. **Cell Phones** (77) You pay $10 a month plus $0.30 per minute for your cell phone. You budget $20 each month for your bill. To find the maximum minutes you can use your phone, solve the inequality $10 + 0.3m \leq 20$.

16. **Probability** (78) Write a rational function that expresses the following probability. Find the probability y of randomly choosing a red marble out of a bag full of x number of marbles that contains only one red marble.

17. **Verify** (79) Show that $-8u^5y + 56u^4y - 80u^3y = -8u^3y(u - 5)(u - 2)$.

Solving Multi-Step Compound Inequalities

LESSON 82

18. Multiple Choice Which expression is the complete factored form of
$3x^6 + 6x^5 - 45x^4$?

A $3x^4(x-3)(x+5)$
B $x^4(3x-9)(x+5)$
C $x^4(x-3)(3x-15)$
D $(3x-9)(x^5+5x^4)$

*19. Use the graph to find the theoretical probability of receiving each grade.

Frequency of Grades

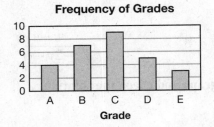

20. (Biology) A Punnett square shows the probability distribution for genes. A short pea plant contributes two short genes, labeled "t." A tall pea plant contributes two tall genes, labeled "T." The plant will be short if it inherits the combination "tt." "TT" means the plant will be tall. The combination "Tt" also results in a tall plant. What is the probability that this plant will be short? Explain your answer.

	T	T
t	Tt	Tt
t	Tt	Tt

*21. Error Analysis Two students use an equation to find the probability of getting heads on four coins and rolling a 2 or 3 on a two number cubes labeled 1–6. Which student is correct? Explain the error.

Student A	Student B
$P(\text{4 heads and two 2 or 3}) =$	$P(\text{4 heads and two 2 or 3}) =$
$\left(\frac{1}{2}\right)^4 \cdot \left(\frac{1}{6} \cdot \frac{1}{6}\right)^2 = \frac{1}{20,736}$	$\left(\frac{1}{2}\right)^4 \cdot \left(\frac{1}{3}\right)^2 = \frac{1}{144}$

*22. Multi-Step Veejay is throwing a party. It costs $75 to rent a skating arena plus $3 per person to rent skates. It costs $100 to rent a bowling alley plus $2 per person to rent bowling shoes. How many people would Veejay have to invite to his party for the bowling alley to cost more than the skating arena?
a. Write an inequality to answer.
b. Solve the inequality.
c. Explain the correct domain of the solution set.

23. (Music) Amber normally listens to 1 new CD and 7 old CDs every day. She starts to listen to 2 more new CDs each day and 1 less old CD each day. How many days will it take her to listen to more new CDs than old CDs?

Solving Multi-Step Compound Inequalities

LESSON 82

*24. **Geometry** The length of a rectangle is greater than its width. The length is $4x + 7$ and the width is $5x - 2$. What does the value of x have to be for this statement to be true?
(81)

*25. **Error Analysis** Two students are told to write an inequality that is a contradiction. Which student is correct? Explain the error.
(81)

Student A	Student B
$-2 + x > x + 3$	$2x + 24 < 3x + 24$

*26. Solve the inequality. Justify your steps.
(82) $-17 > -2x - 7$ OR $27 > 3(x + 6)$

*27. **Multiple Choice** What is the solution to $32 < 7x + 11 < 39$?
(82)
A $21 > x > 28$ **B** $3 < x > 4$
C $3 < x < 4$ **D** $3 > x > 4$

*28. (**Cholesterol Levels**) An average level of HDL, a type of good cholesterol, for a person is usually no more than 60, and an unhealthy level is lower than 40. A doctor sees 4 patients and tests their HDL levels. The first 3 levels are 45, 52, and 60. What can the fourth patient's HDL level be if the average of all four patients' levels fall in the average, but not unhealthy, range of levels?
(82)

*29. **Formulate** Half of Mr. Rubenstein's math class studied for the test and the other half did not. Everyone who studied for the test got a score no lower than 90; everyone who did not study got a score lower than 70. Write the scores of the class as an inequality.
(82)

*30. **Estimate** Felipe wants to earn a grade between 90 and 100 in math. There are 4 major tests over the year, which are averaged to determine his final grade. Felipe scored 94, 88, and 91 on the first 3 tests. What must he score on the last test for his average grade to fall between 90 and 100? Round his scores to solve.
(82)

Name _____ Date _____ Class _____

Factoring Special Products
LESSON 83

Find the product.

1. $(7 + \sqrt{6})(4 - \sqrt{9})$
   (76)

2. $(x + \sqrt{12})(x - \sqrt{3})$
   (76)

Solve.

*3. $\dfrac{-b}{4} + \dfrac{3}{8} \geq \dfrac{3b}{4} - \dfrac{5}{8}$
 (81)

4. $11h + 9 \leq 5h - 21$
   (81)

Factor completely.

5. $3x^5 - 3x^4 - 216x^3$
   (79)

6. $-12x^3 - 48x$
   (79)

Determine whether the polynomial is a perfect-square trinomial or a difference of two squares. Then factor the polynomial.

*7. $x^2 + 10x + 25$
 (83)

*8. $x^2 + 12x + 36$
 (83)

9. **Geometry** Show that TUV is a right triangle.
   (65)

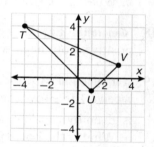

*10. **Analyze** The expression $-7x + 2y$ is one factor of a difference of two squares. What
  (83) is the expanded polynomial?

11. (**Hiking**) Raul and his friends are hiking a 4-mile trail. After 2 hours of hiking,
    (66) they turn off of the path to find a spot for lunch, and then hike back to the trail
    and continue to the end. Write an inequality to represent x, the total distance
    they hiked.

Determine if the following systems are consistent and independent, consistent and dependent, or inconsistent.

12. $y = 2x + 5$
    (67) $y - 2x = 1$

13. $3y = 2x + 4$
    (67) $3x = 4.5y - 6$

14. Find the probability of rolling a sum of 10 or a set of doubles with two
    (68) number cubes.

© Saxon. All rights reserved. Saxon Algebra 1

Factoring Special Products

LESSON 83

15. Multi-Step The local bank offers a savings account with a 3% annual interest rate.
(70) Alfred wants to earn at least $60 in interest. How much should he deposit?
  a. Write an inequality to represent the situation.
  b. Solve the inequality.
  c. How much should he deposit?
  d. Graph the solution.

16. (Food Packaging) The label of a certain cheese states that it weighs 8 ounces. The actual
(73) weight of the product sold is allowed to be 0.2 ounces above or below that. Write a compound inequality that represents this situation.

17. Multi-Step Solve $\frac{|x+3|}{4} = 6$.
(74)
  a. Isolate the absolute-value expression.
  b. Use the definition of absolute value to rewrite the absolute-value equation as two equations.
  c. What is the solution set?

18. Generalize Explain how you know if the second terms of binomial factors are both
(75) positive, both negative, or have opposite signs.

19. (Consumerism) A preschool has a budget of $1000 to buy new outside toys. They will
(78) receive 1 free toy when they place the order. The number of toys, y, that they can get is given by $y = \frac{1000}{x} + 1$, where x is the price per toy.
  a. What is the horizontal asymptote of this rational function?
  b. What is the vertical asymptote?
  c. If the price per toy is $5, how many toys will they receive?

20. Measurement The length of the hypotenuse of a right triangle is found by adding the
(79) squares of the two legs and then taking the square root. The sum of the squares of the legs is $16m^6 + 320m^5 + 1600m^4$. Find the length of the hypotenuse by factoring.

21. Verify The table shows that the theoretical probability of landing on heads two times
(80) when flipping two coins is $\frac{1}{4}$. Use an equation to show that the probability is correct.

22. Multiple Choice Using the table, what is the probability of rolling an even number and
(80) spinning a yellow or a green?

  A $\frac{1}{3}$ **B** $\frac{1}{6}$

  C $\frac{1}{18}$ **D** $\frac{1}{1944}$

Factoring Special Products

LESSON 83

*23. Write Describe how factoring can help find $45^2 - 15^2$.
(83)

*24. Error Analysis Beth claims that the inequality $12x + 67 \geq 52 + 5x + 15$ is always true for any value of x. Is Beth correct? Explain the error.
(81)

25. (Finances) Chad works 20 hours each week. Juan works 10 hours and makes an additional $50 in tips each week. They both get paid the same amount per hour. How much money do they each have to earn per hour for Chad to make more money than Juan?
(81)

26. Error Analysis Two students solve the following compound inequality. Which student is correct? Explain the error.
(82)

Student A	Student B
$10 < -2x + 2 < 16$	$10 < -2x + 2 < 16$
$-4 < x < -7$	$-4 > x > -7$

*27. Multi-Step Yvonne learns that a refrigerator should be kept at a temperature of no more than 40°F but warmer than 32°F.
(82)
   a. Write an expression to show the possible range of proper refrigerator temperatures.

   b. Yvonne tests the temperatures of some refrigerators at an appliance store. The first 4 temperatures are 35°, 40°, 20°, and 45°. What should the last temperature be if the average of the temperatures is within the proper temperature range?

*28. Justify Solve the inequality $28 < 2(x + 3) < 42$ and justify each step.
(82)

*29. Multiple Choice Which expression is a perfect-square trinomial?
(83)
   A $9x^2 + 49$ **B** $64x^2 - 100$
   C $6x^2 + 48x - 96$ **D** $49x^2 - 28x + 4$

*30. (Home Improvement) A square storage shed sits in the corner of a square deck that has a side length of s feet. The shed has a side length of 8 feet. Harper wants to apply a coat of paint to the deck. Write and factor an expression to find the area of the deck Harper will paint, not including the storage shed.
(83)

Name _____ Date _____ Class _____

Identifying Quadratic Functions
LESSON 84

Determine whether the polynomial is a perfect-square trinomial or a difference of two squares. Then factor the polynomial.

1. $q^2 + 18q + 81$
(83)

*2. $36x^2 - 144$
(83)

Simplify.

3. $\sqrt{12} + \sqrt{48} - \sqrt{27}$
(69)

4. $\sqrt{18} + \sqrt{32} + \sqrt{50}$
(69)

Solve. Graph the solution.

5. $2p + 7 > p - 10$
(77)

6. $16 < 2x + 8$ OR $15 > 7x + 1$
(82)

*7. Rewrite $x + 15x^2 - y = 4$ in the standard form of a quadratic function, if possible.
(84)

Find the probability of the following events.

8. spinning a blue section or a letter B
(68)

9. spinning a gray section or a letter D
(68)

10. spinning a white section or a letter C
(68)

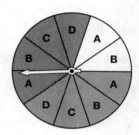

11. (**Computer Electronics**) The failure rate for desktop computers is 5% for the first year of
(68) use. For notebooks it is 15%. If you purchase both a new computer and a notebook, what is the probability that either one will fail in the first year?

12. **Multi-Step** Use the scatter plot.
(71)
  a. Using two points on the line, find an equation for the trend line.
  b. Does the graph show a positive correlation, a negative correlation, or no correlation?

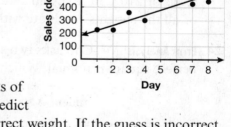

13. (**Entertainment**) People are often employed by
(74) amusement parks to predict the ages and weights of patrons. For a fee, one guesser claims she can predict a patron's weight within three pounds of the correct weight. If the guess is incorrect, the patron receives a prize. Write and solve an absolute-value equation for the maximum and minimum values of a correct guess for a person weighing 162 pounds.

14. **Multi-Step** The success rate on an exam is represented by $72x^2 - 156x + 72$.
(79)
  a. Evaluate the expression for $x = 2$.
  b. Factor the expression completely.

15. **Verify** Show that $\sqrt{14} \cdot \sqrt{21} = 7\sqrt{6}$.
(76)

Identifying Quadratic Functions

LESSON 84

16. Find the vertical asymptote: $y = \dfrac{1.6}{x + 2.5 + 7.8}$.
(78)

17. **Baseball** A pop fly is hit into the infield. In the expression $-5t^2 + 40t - 35$, t represents the time that the ball was 37 meters high. Factor the expression completely.
(79)

18. One student is selected from a school committee that has 12 seniors, 8 juniors, 10 sophomores, and 4 freshmen. Make a graph showing the frequency distribution.
(80)

19. **Data Analysis** Use the frequency distribution from the table to make a bar graph.
(80)

	Pasta Salad	Cucumber Salad	Caesar Salad	Carrot Salad
Number	12	16	22	10

*20. The value of a varies jointly with b and c. What is the constant of variation if $a = 18$, $b = 2$, and $c = 3$? Write an equation expressing the given relationship.
(Inv 8)

21. **Model** Graph the solution for $2(x + 9) - 14 > 3x + 7 + 2x$.
(81)

22. **Multiple Choice** What is the justification for subtracting 11 from all parts of the inequality $32 < 7x + 11 < 39$?
(82)
   A Combine the variable.
   B Addition Property of Inequality
   C Distributive Property of Inequality
   D Multiplication Property of Inequality

23. **Health Checks** A borderline unhealthy cholesterol level is between 200 and 240. Five patients come to the doctor with borderline cholesterol levels. The first 4 have levels of 210, 230, 225, and 235. What could the fifth patient's level be if the average of all the patients' levels are within the borderline unhealthy range?
(82)

*24. **Error Analysis** Ms. Cho asks two students to factor the polynomial only if it is a perfect-square trinomial. Which student is correct? Explain the error.
(83)

Student A	Student B
$x^2 + 8x - 16 = (x - 4)^2$	$x^2 + 8x - 16$ is not a perfect-square trinomial.

320   Saxon Algebra 1

Identifying Quadratic Functions

LESSON 84

*25. Multi-Step A cylindrical thermos has a radius of r.
(83) Beneath the outer surface is an insulating layer. The volume, in cubic centimeters, that the thermos can hold is given by the expression $30\pi r^2 - 60\pi r + 30\pi$.
  a. Factor the polynomial representing the volume of the thermos.
  b. How thick is the insulating layer?
  c. What is the height of the thermos?

*26. Geometry The surface area of a cube is given by the expression
(83) $6x^2 + 36x + 54$. What is the length of one side in terms of x?

*27. Multiple Choice Which function does the
(84) table of values represent?

x	-2	-1	0	1	2
y	6	6	4	0	-6

  A $y = x + 8$
  B $y = -x^2 + 12$
  C $y = x^2 + 2$
  D $y = -x^2 - 3x + 4$

*28. Formulate Write the equation of a function with degree 1 and of a function
(84) with degree 2.

*29. Generalize What is the relationship of the graph of $y = x^2$ to the graph of $y = -x^2$?
(84)

*30. (Water Fountains) A circular fountain sits in front of a city library. A pool of water
(84) surrounds a sculpture that sits on a circular platform in the middle. The radius of the sculpture is half the radius of the entire fountain. Write an equation to represent the area of the pool. Is the equation a quadratic function?

Name _____ Date _____ Class _____

Solving Problems Using the Pythagorean Theorem
LESSON 85

Simplify.

1. (69) $3\sqrt{45} - \sqrt{5}$

2. (39) $\dfrac{p^{-1}}{w}\left(\dfrac{wx}{cp^{-2}q^{-4}} + 5pq^{-3}\right)$

Factor completely.

3. (79) $-3t^3 - 27t^2 - 24t$

4. (79) $4x^4 - 16x^2$

5. (72) $2x^2 + 14 - 9x - x^2$

Determine whether the polynomial is a perfect-square trinomial or a difference of two squares. Then factor the polynomial.

6. (83) $3g^2 - 12$

7. (83) $9x^2 - 24x + 16$

Write the equations in the standard form of a quadratic function, if possible.

8. (84) $4 + y = -8 + 16x$

9. (84) $y + x^2 = 3x^2 - 10x + 12$

Solve.

10. (24) $0.7 + 0.05y = 0.715$

11. (28) $\dfrac{1}{2} + \dfrac{3}{4}x = \dfrac{1}{6}x + 2$

12. (70) Find the solution of $-1.2x \geq -4.8$. Then graph the solution set.

13. (65) Write an equation for a line that passes through $(1, 5)$ and is parallel to $y = -3\dfrac{1}{2}x - 9$.

14. (69) **Gardening** Niko wants to build a fence around his garden. If his garden measures $2\sqrt{4}$ feet by $\sqrt{25}$ feet, how many feet of fencing does Niko need?

15. (77) **Verify** Show that the solution to $\dfrac{x}{-5} + 6 \leq 10$ is $x \geq -20$.

*16. (85) Use the Pythagorean Theorem to find the missing side length. Give the answer in simplest radical form.

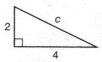

17. (78) Find the vertical asymptote: $y = \dfrac{2.4}{x + 4.5} + 6.9$.

18. (Inv 7) Given that y varies inversely with x, identify the constant of variation when $x = 4$ and $y = 2$.

Solving Problems Using the Pythagorean Theorem

LESSON 85

19. Statistics In 2005 the population of North Dakota was about 635,000—a decrease of about 7000 from five years earlier. The population of Wyoming in 2005 was 509,000—an increase of about 15,000 from five years earlier. If this trend continues, around what year will Wyoming's population exceed North Dakota's?
(81)

*20. **Sail Dimensions** A main sail can be modeled by a right triangle whose sides are called the leach edge, the luff edge, and the foot. If the luff edge measures 27.5 feet and the foot measures 10 feet, use the Pythagorean Theorem to estimate the length of the leach edge to the nearest foot.
(85)

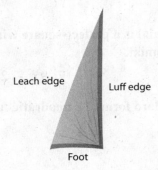

*21. **Multi-Step** In cooking school, Larissa learns that some foods should never be kept in the "danger zone": the temperature at which bacteria grow the fastest, potentially causing food poisoning. She learns that the danger zone is 5°C to 60°C.
(82)

   a. Write an inequality that shows the temperatures that are in the danger zone in degrees Celsius.

   b. Find an inequality that shows the temperatures that are in the danger zone in degrees Fahrenheit by substituting the expression $\frac{5}{9}(f - 32)$ for the variable used in the inequality from part a, and then solve for f.

   c. Write an inequality to show at what temperature food should be kept in degrees Fahrenheit.

22. Verify Solve the inequality $24 < 2x + 6 < 36$. Check to make sure that the equation really is an AND inequality.
(82)

*23. **Error Analysis** Two students factor the polynomial. Which student is correct? Explain the error.
(83)

Student A	Student B
$25x^2 - 36 = (5x - 6)^2$	$25x^2 - 36 = (5x - 6)(5x + 6)$

Solving Problems Using the Pythagorean Theorem

LESSON 85

***24. (Tires)** A truck's tire has an outside radius of r inches. The area of the side of the tire, not including the inside rim, is $\pi r^2 - 81\pi$ inches. What is the diameter of the rim?

25. Geometry Graph the quadratic function representing the total surface area of a cube with side length x.

***26. Write** Explain why the Pythagorean Theorem cannot be used to find the missing side length c.

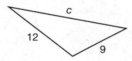

***27. Formulate** One leg of a right triangle is twice the length of the other leg. The length of the hypotenuse is $\sqrt{45}$ centimeters. Let x represent the length of the shorter leg. Use the Pythagorean Theorem to write and solve an equation to find the length of the legs.

***28. Multiple Choice** What is the perimeter of the triangle to the nearest inch?

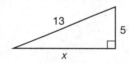

A 212 inches   **B** 30 inches   **C** 32 inches   **D** 12 inches

***29. (Grades)** The grade a student earns on a project is represented by $6x^2 - 11x + 35$ where x is the number of hours spent working on the project.

a. Factor the polynomial.

b. What grade is earned when a student works 4 hours on the project?

***30. Multi-Step** Write an expression for the area of the square. Then find the area.

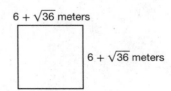

Name _____ Date _____ Class _____

Calculating the Midpoint and Length of a Segment
LESSON 86

Solve and graph the solution set.

1. $15y < 60$
   (70)

2. $16 < 6x + 10$ OR $-16 > 6x - 10$
   (82)

Factor completely.

3. $-2g^2 - 8g + 90$
   (79)

4. $20b^2 + 21b - 5$
   (75)

5. $-13w^2 + 38w - 25$
   (79)

Write each equation in the standard form of a quadratic function, if possible.

6. $y - 14x = -20x^2$
   (84)

7. $x - 5x = -2x^2 + 7$
   (84)

*8. Find the distance between $(4, -1)$ and $(7, 3)$ using the distance formula.
(86)

9. **(Personal Finance)** George's credit card company offers 4% cash back on all purchases.
(70) Write and solve an inequality to determine how many charges he needs to make in one year to earn at least $100 cash back.

10. Use the table to determine if there is a positive correlation, a negative correlation, or
(71) no correlation between the data sets.

x	2	4	6	8	10	12
y	12	25	40	51	61	75

11. **Multi-Step** A real number is at most 13 and at least 5.
(73)
   a. Write two separate inequalities to describe the problem.
   b. Write the two inequalities as one compound inequality.
   c. Graph the compound inequality.

12. **(Carpentry)** Louis is building a new rectangular room onto his house. The room has a
(76) side length of $3 + \sqrt{15}$ feet and a width of $4 + \sqrt{36}$ feet. What is the area of Louis's new room?

13. **Multi-Step** The temperature in Texas has never been above 120 degrees Fahrenheit.
(77) Describe this using Celsius temperature by solving the inequality $120 \geq \frac{9}{5}C + 32$.

14. **Write** How do changes to the value of c affect the graph of $y = \frac{a}{x-b} + c$?
(78)

15. A sandwich maker chooses a meat and a vegetable at random to put on a sandwich.
(80) There are three meats: turkey, ham, and chicken. There are 5 vegetables: lettuce, tomato, cucumber, onion, and peppers. Make a table of the possible outcomes.

16. **(Hobbies)** Kelly goes to a local store that has a monthly fee of $5 and rents games
(81) there for $1.75 a week. An online company has no monthly fee but rents games for $2.25 a week. How many games would Kelly have to rent per month for the local store to be the better deal?

Calculating the Midpoint and Length of a Segment

LESSON 86

17. **Measurement** One square has an area of 16 square units and another square has an area of 36 square units. A third square has an area greater than that of the smaller square and less than that of the larger square. What are the possible lengths of the sides of the third square?
(82)

18. **Multiple Choice** What is the factored form of $32x^2 - 50y^2$?
(83)
   A $2(4x + 5y)^2$ **B** $2(4x - 5y)^2$
   C $2(4x + 5y)(4x - 5y)$ **D** $2(16x + 25y)(16x - 25y)$

*19. **Verify** Rewrite the expression $y^2 - x^2 - 8x - 41$ as a difference of two squares minus a perfect-square polynomial to show that $y^2 - x^2 - 8x - 41 = (y + 5)(y - 5) - (x + 4)^2$.
(83)

20. **Error Analysis** Two students are asked if the equation $7x^2 + 24 = y - 6x(2 - 3x^2)$ is a quadratic function. Which student is correct? Explain the error.
(84)

Student A	Student B
$y = -18x^3 + 7x^2 + 12x + 24$	$y = 10x^2 + 12x + 24$
no	yes

21. **Economics** A company has developed a new product. To determine the selling price, the company uses the function $y = -55x^2 + 1500x$ to predict the profit for selling the product for x dollars. Does the graph of this function open upward or downward? Explain why this might be given the context of the situation.
(84)

22. Use the Pythagorean Theorem to find the missing side length. Give the answer in simplest radical form.
(85)

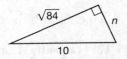

*23. **Geometry** A right isosceles triangle is a right triangle whose legs are equal in length.
(85)
   a. Find the length of the hypotenuse of a right isosceles triangle with leg lengths of 3.
   b. Find the length of the hypotenuse of a right isosceles triangle with leg lengths of 5.
   c. **Formulate** Use the results of parts **a** and **b** to suggest a formula for the hypotenuse of a right isosceles triangle with legs of length a.

*24. **Error Analysis** Two students use the Pythagorean Theorem to find length p. Which student is correct? Explain the error.
(85)

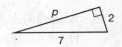

Student A	Student B
$2^2 + 7^2 = p^2$	$2^2 + p^2 = 7^2$
$4 + 49 = p^2$	$4 + p^2 = 49$
$53 = p^2$	$p^2 = 45$
$\sqrt{53} = p$	$p = \sqrt{45}$
	$p = 3\sqrt{5}$

Calculating the Midpoint and Length of a Segment

LESSON 86

*25. Use the diagram of city streets from Example 1 on page 563. What is the
(86) direct distance (in city blocks) from the corner of A St. and 4th Ave. to the corner
of E St. and 2nd Ave.? Give your answer in simplest radical form and to the nearest
tenth of a city block.

*26. Multi-Step Marisol is flying a kite as shown in the diagram.
(85)
a. Use the Pythagorean Theorem to find the length h.

b. How high is the kite off of the ground?

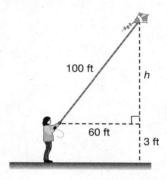

*27. Write To find the distance between (5, 3) and (−2, 9), Dan lets $(x_1, y_1) = (5, 3)$ in the
(86) distance formula. Dawn lets $(x_1, y_1) = (−2, 9)$. Explain why Dan and Dawn will get
the same result.

*28. Justify Is the triangle with vertices at (−3, 3), (1, 0), and (4, 4) a right triangle?
(86) Justify your answer.

*29. Multiple Choice Which points are not endpoints of a line segment with midpoint
(86) (7, −3)?

A (1, 2) and (13, −8) **B** (5, 0) and (9, −5)

C (2, −9) and (12, 3) **D** (4, −2) and (10, −4)

*30. Baseball A baseball diamond is a square that is 90 feet long on each side. Use a
(86) coordinate grid to model positions of players on the field; place home plate at (0, 0),
first base at (90, 0), second base at (90, 90), and third base at (0, 90). An outfielder
located at (50, 300) throws to the third-baseman. How long is the throw? Round you
answer to the nearest foot.

Name _____ Date _____ Class _____

Factoring Polynomials by Grouping

LESSON 87

Factor.

1. (72) $x^2 + 3xy - 54y^2$

*2. (87) $64a^2b - 16a^3 + 18a^2b - 9$

Solve.

3. (81) $2g + 9 - 4g < 5 + 6g - 2$

4. (81) $6(k - 5) > 3k - 26$

Use an equation to find the probability of the event.

5. (80) rolling 4 on two number cubes and a coin landing on heads

6. (80) rolling a number less than 4 on two number cubes and a coin landing on heads

Determine whether the polynomial is a perfect-square trinomial or a difference of two squares. Then factor the polynomial.

7. (83) $100 - c^6$

8. (83) $4x^2 + 20x + 25$

Find the distance between the given points. Give the answer in simplest radical form.

9. (86) $(1, 3)$ and $(4, 7)$

*10. (86) $(2, -1)$ and $(6, 3)$

11. (71) Which situation would most likely be represented by a negative correlation: hours of practice and your golf score, hours of practice and the number of baskets you make in basketball, or hours of practice and the cost of a computer? Hint: In golf, the lower the score, the better.

12. (71) **Entertainment** The table shows the average ticket price for a movie. Make a scatter plot from the data.

Year	1990	1995	2000	2001	2002	2003	2004	2005
Ticket Price (dollars)	4.23	4.35	5.40	5.65	5.80	6.03	6.21	6.41

13. (74) **Multi-Step** A grower ships fruit to a processing plant in 50-pound cases. The plant will not accept cases that differ from this weight by more than ±0.5 pound.
    a. Write an absolute-value equation to find the minimum and maximum weights of the cases that the processing plant will accept.
    b. What are the maximum and minimum acceptable weights?

14. (77) **Hiking** Heidi can hike the mountains at 2 miles per hour. She has already hiked 5 miles and wants to be sure to turn around before she hikes more than 9 miles. To find the number of hours she can hike before she needs to turn around, solve the inequality $2m + 5 \leq 9$.

Factoring Polynomials by Grouping

LESSON 87

15. Multi-Step The local public library has a budget of $3000 to buy new children's books.
(78) The library will receive 100 free books when it places its order. The number of books, y, that they can get is given by $y = \frac{3000}{x} + 100$, where x is the price per book.
a. What is the horizontal asymptote of this rational function?
b. What is the vertical asymptote?
c. If the price per book is $20, how many books will the library receive?

16. Generalize How do you know when a trinomial is factored completely?
(79)

17. (Painting Job) Marco paints walls and charges a $20 set-up fee. He charges at least $60
(82) per big wall, and for small walls he charges no more than $40 per wall. He has just completed a job painting either all big walls or all small walls. His invoice states that he received $2420 in payment. How many walls could he have painted?

18. Measurement A map has a scale 1 cm:500 m. A circular pond on the map has an area
(83) of $9x^2\pi - 6x\pi + \pi$ square centimeters. What is the actual diameter of the pond?

19. Multiple Choice The graph of which function opens downward?
(84)
A $-8y + 3x^2 = 4 + 7x$ **B** $-12x^2 + 15y = 18$
C $-y + 36x = x^2 + 40$ **D** $-15 + 9y = 45x^2 - 3x$

*20. Write Explain how to graph a quadratic function such as $y + 28x - 3 = 50x^2 + 7$.
(84)

21. Use the Pythagorean Theorem to find the missing side length.
(85) Give the answer in simplest radical form.

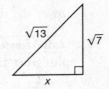

*22. Do the lengths 10, $5\sqrt{5}$, and 15 form a right triangle?
(85)

23. (Art) Theresa is painting on a triangular canvas. The lengths of the sides of the
(85) canvas are 24 inches, 32 inches, and 42 inches. Is her canvas a right triangle?

*24. Error Analysis Two students use the distance formula to find the length of the line
(86) segment with endpoints $(-1, 6)$ and $(4, -2)$. Which student is correct? Explain the error.

Student A	Student B
$d = \sqrt{(4+1)^2 + (-2-6)^2}$	$d = \sqrt{(4-1)^2 + (-2-6)^2}$
$d = \sqrt{25 + 64}$	$d = \sqrt{9 + 64}$
$d = \sqrt{89}$	$d = \sqrt{73}$

© Saxon. All rights reserved.   332   Saxon Algebra 1

Factoring Polynomials by Grouping

LESSON 87

*25. Geometry A midsegment of a triangle is a line segment joining the midpoints of two sides of the triangle. A triangle has vertices at $P = (3, 2)$, $Q = (3, 8)$, and $R = (7, 6)$.
(86)
a. Find the endpoints M and N of the midsegment that joins sides $\overline{PQ}$ and $\overline{QR}$.

b. Verify Show that the length of the midsegment $\overline{MN}$ is half the length of $\overline{PR}$.

26. Multi-Step Use a coordinate plane to model the following situation in a football game:
(86)
A quarterback is on the 25 yard line at (25, 10); he has two receivers, one at (30, 40) and the other at (50, 20).

a. Find the quarterback's distance to the receiver at (30, 40). Give your answer to the nearest yard.

b. Find the quarterback's distance to the receiver at (50, 20). Give your answer to the nearest yard.

c. Which receiver is closer to the quarterback?

*27. Multi-Step The area of a right triangle is $x^2 + 2x$. The base, b, equals $x + 2$. What is the length of the height, h?
(87)
a. Write the formula for the area of a triangle given the base and height.

b. Set the area equal to the product of the base and the height.

c. Factor.

d. Divide by the length of the base to find the length of the height.

*28. Write Is a binomial also a polynomial? Explain.
(87)

*29. Formulate Write an expression to show the price of 2 discounted books when the first one is $20 + n$ and each one after that is $(20 + n)(n - 5)$.
(87)

*30. (Cost) The baseball team gets new uniforms. The first group of 5 costs $y^2 + 5$ dollars, and after that, each group of 5 costs $y + 1$ dollars. How much do the uniforms cost if the team buys 15?
(87)

Name _____ Date _____ Class _____

Multiplying and Dividing Rational Expressions

LESSON 88

1. (81) Solve $6(x + 2) - 4x > 2 + x + 2$.

2. (82) Solve and graph the inequality $4 \geq 2(x + 3)$ OR $23 < 8x + 7$.

Rewrite the equation in the standard form of a quadratic function, if possible.

3. (84) $x(y - 2x) = 18x^2$

4. (84) $2(y - 2x) = 6x^2$

Find the distance between the given points. Give the answer in simplest radical form.

5. (86) $(-4, -5)$ and $(2, -3)$

6. (86) $(3, -2)$ and $(1, 0)$

Find the midpoint of the line segment with the given endpoints.

7. (86) $(-4, -5)$ and $(2, -3)$

8. (86) $(3, -2)$ and $(1, 0)$

Find the product or quotient.

*9. (88) $\dfrac{2y^2 + 10y}{y + 5} \cdot \dfrac{2x}{2x^3}$

*10. (88) $\dfrac{6y}{x} \div \dfrac{x + y}{3y}$

*11. (88) $\dfrac{7x}{y} \div \dfrac{4}{y}$

Factor.

12. (72) $7x - 60 + x^2$

13. (87) $64a^3b - 32a^3 + 16a^2b - 8a^2$

14. (72) **(Woodworking)** To inlay designs on the top of chest, a crafter uses a series of patterns. The patterns are represented by the trinomial $x^2 + 4x - 21$. What are the dimensions of the patterns?

15. (73) Write a compound inequality that describes the graph.

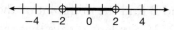

16. (78) **(Athletic Directing)** An athletic director has a budget of $700 to buy uniforms. He will receive 5 free uniforms when he places his order. The number of uniforms, y, that he can get is given by $y = \dfrac{700}{x} + 5$, where x is the price per uniform.
   a. What is the horizontal asymptote of this rational function?
   b. What is the vertical asymptote?
   c. If the price per uniform is $50, how many uniforms will he receive?
   d. Can the athletic director receive only 5 uniforms? Why or why not?

Multiplying and Dividing Rational Expressions

LESSON 88

17. Multi-Step The area of a triangular sail in square feet is represented by the equation $\frac{1}{2}x^2 + \frac{7}{2}x + 5$.
(79)
   a. Factor this expression completely in terms of the area of a triangle.
   b. Let $x = 3$. What is the area of the sail in square feet?

18. Write Why is making tables and graphs a good way to show probability distribution?
(80)

19. (Paving) A new restaurant is opening in a square building with a side length of s feet. A parking lot surrounds the building in the form of a larger square. The restaurant sits in the middle of the lot. The plot of land on which the restaurant and parking lot lie has an area of $9s^2 + 54s + 81$ square feet. How far does the parking lot extend from the building?
(83)

*20. Analyze The figure shows the first four right triangles in the *Wheel of Theodorus* (named after a fifth-century Greek philosopher).
(85)

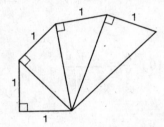

   a. Find the length of the hypotenuse of each of the four triangles.
   b. Predict Suppose the pattern of the triangles is continued until the figure contains 10 triangles. Predict what the length of the hypotenuse of the tenth triangle will be.

21. Multiple Choice Which lengths represent the lengths of the sides of a right triangle?
(85)
   A 3, 4, 6   **B** $\sqrt{13}$, 5, 12
   C $\sqrt{15}$, 7, 8 **D** 6, 8, 12

22. Error Analysis Students A and B find the midpoint of the line segment with endpoints $P = (12, -5)$ and $Q = (-2, 3)$. Student A says that the x-coordinate of the midpoint is $\frac{12 - (-2)}{2} = \frac{14}{2} = 7$. Student B says that the x-coordinate of the midpoint is $\frac{12 + (-2)}{2} = \frac{10}{2} = 5$. Which student is correct? Explain the error.
(86)

Multiplying and Dividing Rational Expressions

LESSON 88

23. (**Software**) A game designer is working on a computer basketball game. On the computer screen, the coordinates of the corners of the court are (20, 20), (20, 70), (114, 70), and (114, 20). Find the length of a diagonal of the court to the nearest tenth of a unit.
(86)

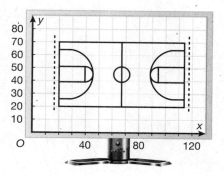

*24. Measurement Malik is adding 5 feet to the length of a screened-in porch. If the porch is originally a square room with area x^2 and the new area measures 50 square feet, what are the dimensions of the new porch?
(87)

*25. Geometry A rectangle has a length of $12x^2 + 3y$ and a width of $6x^2 + y$. What is its area? What is the simplest way to express the area?
(87)

26. Multiple Choice What is the result of a complete factoring of $25x^2 - 81$?
(87)
   A $25(x + 9)(x - 9)$

   B $5x \cdot 5x - 81$

   C $5^2 x^2 - 9^2$

   D $(5x + 9)(5x - 9)$

*27. Generalize What is the method for multiplying and dividing exponents when working with rational expressions?
(88)

*28. Multiple Choice Multiply $\frac{y^2 + 6y + 5}{y^2} \cdot \frac{y}{y + 1}$.
(88)

   A $\frac{y + 5}{y^2}$ **B** $\frac{y + 5}{y}$

   C $\frac{(y + 1)(y + 5)}{y^2}$ **D** $y(y + 5)$

*29. (**Murals**) Lucy paints murals. She charges $10.00 per square foot. If she paints a mural that is $c^2 + 7c + 10$ ft by $\frac{1}{c + 5}$ ft, how much does Lucy charge for painting the mural?
(88)

30. Multi-Step The area of a square patio is represented by $81x^2 - 36x + 4$ square feet.
(75)
   a. Factor the expression completely.

   b. Why do you think this expression is called a perfect square?

Saxon Algebra 1

Name _____ Date _____ Class _____

Identifying Characteristics of Quadratic Functions

LESSON 89

Solve each equation and check your answer.

1. $-2|r+2| = -30$ ₍₇₄₎
2. $3|r+6| = 15$ ₍₇₄₎
3. Solve and graph the inequality $35 < 3x + 8$ OR $72 \geq 9(x+1)$. ₍₈₂₎

Determine whether the polynomial is a perfect-square trinomial or a difference of two squares. Then factor the polynomial.

4. $100y^2 - 80y + 16$. ₍₈₃₎
5. $81x^2 - 1$ ₍₈₃₎
6. Factor $9c^2 - 42c + 49$. ₍₈₇₎
7. Find the distance between $(2, 2)$ and the $(4, 4)$. Give your answer in simplest radical form. ₍₈₆₎

Find the quotient or product.

8. $\dfrac{15a + 5b}{5ab} \div 3a + b$ ₍₈₈₎
9. $4x + 2 \div \dfrac{2x+1}{3y}$ ₍₈₈₎

Find the axis of symmetry for the graph of each equation.

*10. $y = x^2 + 4x + 6$ ₍₈₉₎
*11. $y = x^2 - x - 6$ ₍₈₉₎

*12. **Analyze** Describe how you could use symmetry to graph a quadratic equation. ₍₈₉₎

13. **Traffic** You are driving on a highway that has a speed limit of 65 miles per hour and a minimum speed of 45 miles per hour. Write a compound inequality that describes this situation. ₍₇₃₎

14. Use the graph to write a compound inequality as two separate inequalities. Then write a compound inequality without using the word AND. ₍₇₃₎

15. **Multi-Step** Jordan is making a model of a trapezoid for her math class. Write an expression for the area of the trapezoid shown at right. Find the area. ₍₇₆₎

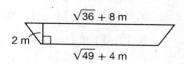

16. **Tennis** A tennis ball is served at 87 miles per hour. In the expression $-16x^2 + 128x - 240$, x represents the time the ball is 246 feet high. Factor this expression completely. ₍₇₉₎

Identifying Characteristics of Quadratic Functions

LESSON 89

17. Multi-Step A spinner is spun 100 times. The spinner lands on brown 50 times,
(80) red 25 times, and yellow 25 times.
  a. Find the experimental probability for each color.
  b. Draw a spinner that would also give this theoretical probability distribution.

18. (Communication) A 400-anytime-minutes plan costs $35 plus $0.09 per minute overage.
(81) Another 400-anytime-minutes plan costs $45 plus $0.06 per minute overage. After how many overage minutes is the first plan more expensive?

19. Verify Solve the inequality $2x + 8 > 2 + 5x + 6$. Then select some values of x and
(81) substitute them into the inequality to verify the solution set.

20. (Bridges) A footbridge follows the graph of the equation $y = -\frac{1}{9}x^2 + 1$. If the
(84) x-axis represents the ground and each unit on the graph represents 1 foot, what is the horizontal distance across the bridge?

21. Geometry A rectangular prism and a cylinder are both 10 inches tall. The base
(84) of the prism is a square with a side length x. The cylinder has a radius x. Graph the functions of the volumes of each solid in a coordinate plane. Then compare the graphs.

22. Use the Pythagorean Theorem to find the missing side length.
(85) Give the answer in simplest radical form.

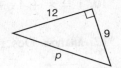

23. Multiple Choice Which point is the midpoint of the line segment
(86) joining $(3, 9)$ and $(-1, 2)$?

  A $\left(2, \frac{11}{2}\right)$ **B** $\left(2, \frac{7}{2}\right)$ **C** $\left(-2, -\frac{7}{2}\right)$ **D** $\left(1, \frac{11}{2}\right)$

24. Describe the transformation of $f(x) = -4x$ from the linear parent function.
(Inv 6)

*25. Multi-Step A company is making various boxes to ship different-sized globes.
(88) The boxes have a volume of $\frac{6r^3h^2}{4r^2h} \cdot \frac{8rh^2}{3r^2h} \cdot \frac{2r^2h}{rh^2}$ and the globes have a volume of $\frac{4\pi rh}{3}$.
What fraction of the box do the globes take up?
  a. Solve for the volume of the box.
  b. Find the fraction of the box that the globes take up.

*26. Geometry A rectangle has length $\frac{3x^2 + x}{y}$ and width $x + 2y$. What is its area?
(88)

*27. Error Analysis Two students are asked to solve the equation $\frac{6s^2 - 3s}{15} \div 2s - 1$. Which
(88) student is correct? Explain the error.

Student A	Student B
$\frac{6s^2 - 3s}{15} \div 2s - 1$	$\frac{6s^2 - 3s}{15} \div 2s - 1$
$\frac{6s^2 - 3s}{15} \cdot \frac{2s - 1}{1} = \frac{s(2s-1)^2}{5}$	$\frac{6s^2 - 3s}{15} \cdot \frac{1}{2s - 1} = \frac{s}{5}$

Identifying Characteristics of Quadratic Functions
LESSON 89

*28. **Multiple Choice** What is the x-coordinate of the vertex of the graph of a quadratic
(89) function whose zeros are 0 and −8?

    A −4   **B** 0   **C** 4   **D** 8

*29. (Space) If it were possible to play ball on Saturn, the function $y = -5.5x^2 + 44x$ would
(89) approximate the height of a ball kicked straight up at a velocity of 44 meters per
second, where x is time in seconds. Find the maximum height of the ball and the time
it takes the ball to reach that height.

*30. **Analyze** How does the value of a in a quadratic equation indicate if the graph of the
(89) equation has a minimum or a maximum?

Name _____ Date _____ Class _____

Adding and Subtracting Rational Expressions

LESSON 90

Find the distance between the given points. Give the answer in simplest radical form.

1. $(-3, -1)$ and $(4, 2)$
2. $(1, 1)$ and $(9, 1)$

Factor completely.

3. $12 + 17x + 6x^2$
4. $21t + 4t^2 - 49$

Determine whether the polynomial is a perfect-square trinomial or a difference of two squares. Then factor the polynomial.

5. $9x^4 + 42x^2y + 49y^2$
6. $x^6 + 16x^3 + 64$

7. Graph the function $y = \frac{-x^2}{2}$.

8. Multiply $\frac{2x - 16}{6x^2} \cdot \frac{3y^2 + 3x}{3x - 24}$.

9. Solve $|n + 3| + 4 = 4$.

10. Find the axis of symmetry for the graph of the equation $y = -3x^2 + 8x + 1$.

Add or subtract.

*11. $\frac{7x}{y} - \frac{2}{y}$

*12. $\frac{4y}{x} - \frac{5y}{2x}$

13. **Marine Biology** A marine biologist netted a sample of wild sardines from the northern Pacific. In his analysis of the sample, the average length of an adult male fish was 210 millimeters. He noted that the greatest variation in length from the average length in the sample was ±33 millimeters. Write and solve an absolute-value equation to find the lengths of the longest and shortest sardines in the sample.

14. **Multi-Step** A student has grades 75 and 90 on his first two math tests. He wants to have an average of at least 80 after his third test. To find what score he needs, solve the inequality $\frac{75 + 90 + x}{3} \geq 80$.

15. **Chess** All the chess pieces are put in a bag and one is drawn randomly. There are 2 kings, 2 queens, 4 rooks, 4 bishops, 4 knights, and 16 pawns. Make a graph to show the probability distribution of the pieces.

16. **Multi-Step** Marisa is changing her dog's food from a puppy formula to an adult formula. The vet recommended that she exchange 2 ounces of adult formula for 3 ounces of puppy formula each day. Her dog usually eats 20 ounces of food per day. After how many days will the dog's daily diet include more adult formula than puppy formula?
    a. Write and solve an inequality.
    b. Explain the correct domain of the solution set.

Adding and Subtracting Rational Expressions

LESSON 90

17. Write What does an AND inequality mean?
(82)

18. Coordinate Geometry Use the points (2, 2), (5, 2), and (2, −3).
(85)
   a. **Write** Explain why the triangle is a right triangle.
   b. What are the lengths of the legs of the triangle?
   c. Use the Pythagorean Theorem to find the length of the hypotenuse.

19. (Astronomy) Eli is looking at three stars on a star map. He measures and records
(85) the distances between each pair of stars: 9 centimeters, 12 centimeters, and 15 centimeters. Do the stars on the map determine a right triangle?

20. Multi-Step The area of a square is equal to $x^2 + 6x + 9$. What is the side length of
(87) the square?
   a. Write the formula for the area of a square.
   b. Factor the area.
   c. Set the expressions equal to each other.
   d. Find the square root of each side.

21. The area of a triangle varies jointly with the base and height of the triangle. Given that
(Inv 8) the area of a triangle is 6 square feet, the base is 3 feet, and the height is 4 feet, find the constant of variation. Express the area of a triangle.

22. Multiple Choice Divide $\dfrac{3mn^2}{4m^2n} \div \dfrac{9mn}{8m^3n^2}$.
(88)

A $\dfrac{2mn^2}{3m}$ **B** $\dfrac{2mn}{3n^2}$ **C** $\dfrac{2mn^2}{3}$ **D** $\dfrac{27}{32m^3}$

23. (Travel Times) Use the formula $\dfrac{d}{r} = t$ to find how long it takes to travel a distance of
(88) $x^2 - 25$ miles at the rate of $x + 5$ miles per hour.

24. Error Analysis Two students tried to find the axis of symmetry for the equation
(89) $y = x^2 - 4x - 3$. Which student is correct? Explain the error.

Student A	Student B
$x = \dfrac{-b}{2a} = \dfrac{-4}{2(1)} = \dfrac{-4}{2} = -2$	$x = \dfrac{-(-4)}{2(1)} = \dfrac{4}{2} = 2$

***25. Multi-Step** The population of Ireland between the years 1901 and 2006 can be
(89) approximated by the function $y = 0.000285x^2 - 0.0232x + 3.336$, where x is the number of years after 1900 and y is the population for that year in millions of people.
   a. **Justify** Tell how you know the function has a minimum rather than a maximum.
   b. **Estimate** What was the minimum population and when did it occur? Round to the nearest year.
   c. **Predict** Use the function to predict the population of Ireland in 2020.

Adding and Subtracting Rational Expressions

LESSON 90

*26. Geometry The equation $y = -x^2 + 35x$ gives the area of a rectangle with a fixed
(89) perimeter of 70 units, where x is the width of the rectangle. Explain how to use
the equation to find the greatest possible area of the rectangle.

*27. Multiple Choice Simplify $\dfrac{6}{9q^2} - \dfrac{3q}{9q^2}$.
(90)

  A $\dfrac{1}{3q}$ **B** $\dfrac{2}{3q}$ **C** $\dfrac{2-q}{3q^2}$ **D** $\dfrac{3(2-q)}{q^2}$

*28. Business Ms. Suarez owns a manufacturing company. Her total profits for one year
(90) are $x^2 + 44x + 420$, where x is the number of units manufactured. Her profits for
the spring are $x^2 + 7x + 100$, while her profits for the winter are $17x + 40$. Write a
simplified expression for the fraction of the total profits that come from the spring
and winter seasons.

*29. Justify When adding $\dfrac{5n}{n^4 p^3} + \dfrac{6p}{n^2 p^5}$, why multiply the first expression by $\dfrac{p^2}{p^2}$ and the
(90) second by $\dfrac{n^2}{n^2}$?

*30. Write Describe the steps in adding and simplifying $\dfrac{2}{x^2 + 6x + 9} + \dfrac{x}{x+3}$.
(90)

Name _____ Date _____ Class _____

Solving Absolute-Value Inequalities

LESSON 91

1. Find the axis of symmetry for the graph of the equation $y = -\frac{1}{2}x^2 + x - 3$.
   (89)

Add or subtract.

2. $\dfrac{6rs}{r^2s^2} + \dfrac{18r}{r^2s^2}$
   (90)

3. $\dfrac{b}{2b+1} - \dfrac{6}{b-4}$
   (90)

Factor.

4. $-4y^4 + 8y^3 + 5y^2 - 10y$
   (87)

5. $3a^2 - 27$
   (87)

Evaluate.

6. $4x^2 + 6x - 15$ for $x = 6$
   (75)

7. $9x^2 - 11x + 32$ for $x = -5$
   (75)

*8. Solve and graph the inequality $|x| < 96$.
   (91)

*9. **Write** Explain what $|x| \geq 54$ means on a number line.
   (91)

*10. **Justify** When solving an absolute-value inequality, a student gets $|x| \geq -5$. Justify
   (91) that any value for x makes this inequality true.

*11. **Multiple Choice** Which inequality is represented by the graph?
   (91)

   ← ++⊙++++++++++++⊙++ →
   −12 −8 −4 0 4 8 12

   A $|x| < 9$ **B** $|x| > 9$ **C** $|x| \leq 9$ **D** $|x| < -9$

*12. (Track) A runner finishes a sprint in 8.54 seconds. The timer's accuracy is plus or
   (91) minus 0.3 seconds. Solve and graph the inequality $|t - 8.54| \leq 0.3$.

*13. **Error Analysis** Two students simplified $\dfrac{2c}{c-6} + \dfrac{12}{6-c}$. Which student is correct?
   (90) Explain the error.

Student A	Student B
$\dfrac{2c - 12}{c - 6}$	$\dfrac{2c + 12}{c - 6}$
$\dfrac{2(c - 6)}{c - 6} = 2$	$\dfrac{2(c + 6)}{c - 6} = -2$

*14. **Multi-Step** A farmer has a rectangular plot of land with an area of $x^2 + 22x + 72$
   (90) square meters. He sets aside x^2 square meters for grazing and $2x - 8$ square meters
   for a chicken coop.
   a. Write a simplified expression for the total fraction of the field the farmer has
      set aside.
   b. **Estimate** About what percent of the field has the farmer set aside if $x = 30$?

© Saxon. All rights reserved.   347   Saxon Algebra 1

Solving Absolute-Value Inequalities

LESSON 91

15. Geometry Write a simplified expression for the total fraction of the larger rectangle that the triangle and smaller rectangle cover.

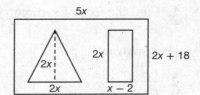

16. Find the product of $(\sqrt{4} - 6)^2$.

17. Analyze Why is it necessary to understand factoring when dealing with rational expressions?

18. Multi-Step The base of triangle ABC is $x^2 + y$. The height is $\frac{4x + 2xy}{x^3 + xy}$. What is the area of triangle ABC?
   a. Multiply the base of the triangle by its height
   b. Multiply the product from part **a** by $\frac{1}{2}$.

*19. Error Analysis Two students tried to find the axis of symmetry for the equation $y = 8x + 2x^2$. Which student is correct? Explain the error.

Student A	Student B
$x = \frac{-b}{2a} = \frac{-2}{2(8)} = \frac{-2}{16} = -\frac{1}{8}$	$x = \frac{-8}{2(2)} = \frac{-8}{4} = -2$

*20. Space If it were possible to play ball on Jupiter, the function $y = -13x^2 + 39x$ would approximate the height of a ball kicked straight up at a velocity of 39 meters per second, where x is time in seconds. Find the maximum height the ball reaches and the time it takes the ball to reach that height. (Hint: Find the time the ball reaches its maximum height first.)

21. Measurement The coordinates of two landmarks on a city map are $A(5, 3)$ and $B(7, 10)$. Each grid line represents 0.05 miles. Find the distance between landmarks A and B.

22. Archeology Archeologists use coordinate grids to record locations of artifacts. Jonah recorded that he found one old coin at $(41, 37)$, and a second old coin at $(5, 2)$. Each unit on his grid represents 0.25 feet. How far apart were the coins? Round your answer to the nearest tenth of a foot.

23. Find the length t to the nearest tenth.

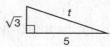

Solving Absolute-Value Inequalities

LESSON 91

24. (House Painting) A house painter leans a 34-foot ladder against a house with the bottom of the ladder 7 feet from the base of the house. Will the top of the ladder touch the house above or below a windowsill that is 33 feet off the ground?
   ₍₈₅₎

25. Graph the function $y = 4x^2$.
   ₍₈₄₎

26. **Generalize** What is the factored form of $a^{2m} + 2a^m b^n + b^{2n}$?
   ₍₈₃₎

27. (Shopping) Roger has $40 to buy CDs. The CDs cost $5 each. He will definitely buy at least 3 CDs. How many CDs can Roger buy? Use inequalities to solve the problems.
   ₍₈₂₎

28. **Multi-Step** A summer school program has a budget of $1000 to buy T-shirts. Twenty free T-shirts will be received when they place their order. The number of T-shirts y that the program can get is given by $y = \frac{1000}{x} + 20$, where x is the price per T-shirt.
   ₍₇₈₎
   a. What is the horizontal asymptote of this rational function?
   b. What is the vertical asymptote?
   c. If the price per T-shirt is $10, how many T-shirts can the program receive?

*29. Suppose the area of a rectangle is represented by the expression $4x^2 + 9x + 2$. Find possible expressions for the length and width of the rectangle.
   _(Inv 9)

30. Simplify the expression $6\sqrt{8} \cdot \sqrt{5}$.
   ₍₇₆₎

Simplifying Complex Fractions

LESSON 92

Find the product or quotient.

1. (88) $\dfrac{15x^4}{x-4} \cdot \dfrac{x^2 - 10x + 24}{3x^3 + 12x^2}$

2. (88) $\dfrac{x^2 + 12x + 36}{x^2 - 36} \div \dfrac{1}{x - 6}$

Solve.

3. (77) $-3(r - 2) > -2(-6)$

4. (77) $\dfrac{y}{4} + \dfrac{1}{2} < \dfrac{2}{3}$

Simplify.

*5. (92) $\dfrac{\dfrac{5x}{10x+20}}{\dfrac{15}{x+2}}$

6. (76) $8\sqrt{9} \cdot 2\sqrt{5}$

*7. **Write** Under what conditions is a rational expression undefined?
(92)

*8. **Justify** Give an example to show $\dfrac{\dfrac{a}{b}}{\dfrac{c}{d}} = \dfrac{a}{b} \cdot \dfrac{d}{c} = \dfrac{ad}{bc}$.
(92)

*9. **(Skating)** It took Jim $\dfrac{15}{x^2 + 2x - 3}$ minutes to skate to the park that was $\dfrac{2x}{8x - 8} + \dfrac{x}{4x + 12}$ miles away. Find his rate in miles per minute.
(92)

*10. **Multiple Choice** Two fractions have a denominator of $x^2 + 6x + 9$ and $x^2 - 9$. What is the least common denominator?
(92)
   A $x^2 + 9$ **B** $x^2 + 6x + 9$
   C $2x^2 + 9$ **D** $(x + 3)^2(x - 3)$

11. Solve and graph the inequality $|x| < 65$.
(91)

*12. **Error Analysis** Two students solve the inequality $|x - 15| < -4$. Which student is correct? Explain the error.
(91)

Student A	Student B
$\|x - 15\| < -4$	$\|x - 15\| < -4$
$-4 < x - 15 < 4$	no solution
$11 < x < 19$	

351

Simplifying Complex Fractions

LESSON 92

*13. Geometry A triangle has sides measuring 16 inches and 23 inches. The triangle
(91) inequality states that the length of the third side must be greater than 7 inches and
less than 39 inches. Write this inequality and graph it.

*14. Multi-Step The grades on a math test were all within the range of 80 points plus or
(91) minus 15 points.
  a. Write an absolute-value inequality to show the range of the grades.
  b. Solve the inequality to find the actual range of the grades.

*15. Error Analysis Two students simplified $\frac{x+2}{2x-5} - \frac{3-x}{2x-5}$. Which student is correct?
(90) Explain the error.

Student A	Student B
$\frac{x+2-3+x}{2x-5} = \frac{2x-1}{2x-5}$	$\frac{x+2-3-x}{2x-5} = \frac{-1}{2x-5}$

*16. Canoeing Vanya paddled a canoe upstream for 4 miles. He then turned the canoe
(90) around and paddled downstream for 3 miles. The current flowed at a rate of
c miles per hour. Write a simplified expression to represent his total canoeing time if
he kept a constant paddling rate of 6 miles
per hour.

17. Multiple Choice Which equation's graph has a maximum?
(89)
  A $y = -5 + x^2$ **B** $y = -x^2 + 5x$
  C $y = x^2 + 5$ **D** $y = 5x^2 - 1$

18. Write Describe 2 ways to find the axis of symmetry for a parabola.
(89)

19. Justify Give an example of a quadratic function and an example of a function that is
(84) not quadratic. Explain why each function is or is not quadratic.

20. Find the length a to the nearest tenth.
(85)

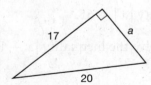

21. Subtract $\frac{5f+6}{f^2+7f-8} - \frac{f+10}{f^2+7f-8}$.
(90)

22. Let $A = (-5, 3)$, $B = (0, 7)$, $C = (12, 7)$, and $D = (7, 3)$. Use the distance formula
(86) to determine whether $ABCD$ is a parallelogram.

Simplifying Complex Fractions

LESSON 92

23. (Geography) Ithaca, New York is almost directly west of Oneonta, New York and directly north of Athens, Pennsylvania. The three cities form a triangle that is nearly a right triangle. Use the distance formula to estimate the distance from Athens to Oneonta. Each unit on the grid represents 5 miles.
(86)

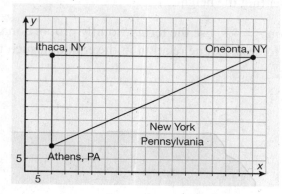

24. Error Analysis Two students factor the polynomial $(3x^2 + 6) - (4x^3 + 8x)$. Which student is correct? Explain the error.
(87)

Student A	Student B
$3(x^2 + 2) - 4x(x^2 - 2)$	$3(x^2 + 2) - 4x(x^2 + 2)$
$(3 - 4x)(x^2 - 4)$	$(3 - 4x)(x^2 + 2)$

25. Probability A bag has $2x + 1$ red marbles, $3x$ blue marbles, and $x + 2$ green marbles. What is the probability of picking a red marble?
(87)

26. Multi-Step A toolbox is 2 feet high. Its volume is represented by the expression $2x^2 - 8x + 6$.
(79)
  a. Factor the expression completely.
  b. Identify the expressions that represent the length and width of the toolbox.

27. (Hardiness Zones) The state of Kansas falls into USDA hardiness zones 5b and 6a. This means that plants in these zones must be able to tolerate an average minimum temperature range greater than or equal to $-15°F$, and less than $-5°F$. Write an inequality to represent the temperature range of the hardiness zones in Kansas.
(82)

28. Multi-Step A rectangular playing field has a perimeter of $8x$ feet. Its length is 2 feet greater than $2x$.
(83)
  a. Write expressions for the length and width of the rectangle.
  b. Write an expression for the area of the field.
  c. The playing field is in a city park. The park is a large square with a side length of x^2. Write and factor an expression for the area of the park that does not include the playing field.

Simplifying Complex Fractions

LESSON 92

29. **(Vertical Motion)** The height h of an object t seconds after it begins to fall is given by
$_{(83)}$ the equation $h = -16t^2 + vt + s$, where v is the initial velocity and s is the initial height. When an object falls, its initial velocity is zero. Write an equation for the height of an object t seconds after it begins to fall from 14,400 feet. Then factor the expression representing the height.

30. Given the equation $r = \frac{kst}{p}$, use the terms "jointly proportional to" and "inversely
$_{(Inv\ 8)}$ proportional to" to describe the relation in the equation such that k is the constant of variation.

Dividing Polynomials

LESSON 93

1. Find the distance between $(-3, 2)$ and $(9, -3)$. Give the answer in simplest radical form.
   (86)

2. Solve $\frac{5}{16}y + \frac{3}{8} \geq \frac{1}{2}$, and graph the solution.
   (77)

Factor.

3. $2x^2 + 12x + 16$
   (79)

4. $3x^3 - 5x^2 - 9x + 15$
   (87)

*5. Find the quotient: $\frac{4x^3 + 42x^2 - 2x}{2x}$.
   (93)

6. Find the axis of symmetry for the graph of the equation $y = x^2 - 2x$.
   (89)

Simplify.

7. $\dfrac{\frac{7x^4}{4x+18}}{\frac{3x^2}{6x+27}}$
   (92)

*8. $\dfrac{\frac{1}{x^3}}{\frac{1}{x^3} + \frac{1}{x^3}}$
   (92)

*9. **Write** Explain how to check that $(5x + 6)$ is the correct quotient of $(15x^2 + 13x - 6) \div (3x - 1)$.
   (93)

*10. **Justify** Show that the quotient of $(x^2 - 4) \div (x + 2)$ can be found using two different methods.
   (93)

*11. (**Swimming**) The city has decided to open a new public pool. The area of the new rectangular pool is $(x^2 - 16x + 63)$ square feet and the width is $(x - 7)$ feet. What is the length?
   (93)

*12. **Multiple Choice** Simplify $\dfrac{x^3 - 7x + 3x^2 - 21}{x + 3}$.
   (93)

   A $x^2 - 7$

   B $x^3 - 7$

   C -3

   D $\dfrac{x^2 - 7}{2x}$

Dividing Polynomials

LESSON 93

*13. Error Analysis Students were asked to simplify $\dfrac{\frac{6x^2 - 6x}{8x^2 + 8x}}{\frac{3x - 3}{4x^2 + 4x}}$. Which student is correct? Explain the error.
(92)

Student A	Student B
$\dfrac{6x^2 - 6x}{8x^2 + 8x} \cdot \dfrac{4x^2 + 4x}{3x - 3}$	$\dfrac{6x^2 - 6}{8x^2 + 8x} \cdot \dfrac{4x^2 + 4x}{3x - 3}$
$= \dfrac{6x(x - 1)}{8x(x + 1)} \cdot \dfrac{4x(x + 1)}{3(x - 1)}$	$= \dfrac{6x(x - 1)}{8x(x + 1)} \cdot \dfrac{4x(x + 1)}{3(x - 1)}$
$= x$	$= \dfrac{24x^2}{24x}$

14. Multi-Step Brent rode his scooter $\dfrac{8x^2 - 48x}{24x^5}$ minutes to get to baseball practice that
(92) was $\dfrac{7x - 42}{4x^2}$ miles away.

   a. Find his rate in miles per minute.

   b. If the rate is divided by $\frac{1}{x}$, what is the new rate?

*15. Geometry The area of a parallelogram is $\dfrac{m + n}{5}$ square inches and the height is
(92) $\dfrac{m^2 + n^2}{15}$ inches. What is the length of the base?

16. Solve and graph the inequality $|x| > 84$.
(91)

*17. (Census) The census of England and Wales has a margin of error of ±104,000 people.
(91) The 2001 census found the population to be 52,041,916. Write an absolute-value inequality to express the possible range of the population. Then solve the inequality to find the actual range for the population.

18. Error Analysis Two students solve the inequality $|x + 11| > -15$. Which student is
(91) correct? Explain the error.

Student A	Student B
$|x + 11| > -15$	$|x + 11| > -15$
x can be any real number.	$x = \emptyset$

19. Multiple Choice When simplifying $\dfrac{1}{x^2 - 5x - 50} + \dfrac{1}{2x - 20}$, what is the numerator?
(90)

   A 1 **B** 2

   C $x + 5$ **D** $x + 7$

*20. Analyze Explain what can be done so that $\dfrac{1}{3 - r}$ and $\dfrac{5r}{r - 3}$ have like denominators.
(90)

356 Saxon Algebra 1

Dividing Polynomials

LESSON 93

21. Probability Mr. Brunetti writes quadratic equations on pieces of paper and puts them in a hat, and then tells his students each to choose two at random. After each student picks, his or her two papers go back in the hat. The functions are $y = 4x^2 - 3x + 7$, $y = -7 + x^2$, $y = -2x + 6x^2$, $y = 0.5x^2 + 1.1$, and $y = -\frac{1}{2}x^2 + 7x + 5$. What is the probability of a student choosing two functions that have a minimum?
(89)

22. Error Analysis Two students divide the following rational expression. Which student is correct? Explain the error.
(88)

Student A	Student B
$\frac{m^2}{6m^2} \div (m^2 + 2)$	$\frac{m^2}{6m^2} \div (m^2 + 2)$
$\frac{m^2}{6m^2} \cdot \frac{1}{m^2 + 2} = \frac{1}{6(m^2 + 2)}$	$\frac{m^2}{6m^2} \cdot \frac{m^2 + 2}{1} = \frac{m^2 + 2}{6}$

23. (Cost) A bakery sells specialty rolls by the dozen. The first dozen costs $6b + b^2$ dollars. Each dozen thereafter costs $4b + b^2$ dollars. If Marcello buys 4 dozen rolls, how much does he pay?
(87)

24. Find the vertical and horizontal asymptotes and graph $y = \frac{4}{x+2}$.
(78)

25. (Flooring) Theo is installing new kitchen tiles. The design on the tile includes a square within a square. The smaller square has a side length of s centimeters. The expression $4s^2 + 12s + 9$ describes the area of the entire tile. What is the difference between the length of the tile and the length of the square within the tile?
(83)

26. Multi-Step Tony is sketching the view from the top of a 256-foot-tall observation tower and accidentally drops his pencil.
(84)
  a. Use the formula $h = -16t^2 + 256$ to make a table of values showing the height h of the pencil, 1, 2, and 3 seconds after it is dropped.
  b. Graph the function.
  c. About how long does it take for the pencil to hit the ground?

27. (Pendulums) The time it takes a pendulum to swing back and forth depends on its length. The formula $l = 2.45\frac{t^2}{\pi^2}$ approximates this relationship. Graph the function using 3.14 for π. Use the graph to estimate the time it takes a pendulum that is 1 meter long to swing back and forth.
(84)

28. Write Explain how to determine whether a triangle with side lengths 5, 7, and 10 is a right triangle.
(85)

Dividing Polynomials

LESSON 93

29. Multi-Step A bag of marbles contains 3 red, 5 blue, 2 purple, and 4 clear marbles.
(80)
  a. Make a graph that represents the frequency distribution.
  b. What is the probability of drawing a red or a clear marble?

30. Given the following table, find the value of the constant of variation and
(Inv 8) complete the missing values in the table given that y varies directly with x and inversely with z.

y	x	z
1	1	3
3	2	
	4	2
9		2
2		12

358 Saxon Algebra 1

Solving Multi-Step Absolute-Value Equations

LESSON 94

Add or subtract.

1. (90) $\dfrac{m^2}{m-4} - \dfrac{16}{m-4}$

2. (90) $\dfrac{-66}{w^2 - w - 30} + \dfrac{w}{w-6}$

*3. (94) **Write** Explain why some absolute-value equations have no solutions.

*4. (94) **Multiple Choice** The solution set $\{-12, 60\}$ correctly solves which absolute-value equation?

    A $6\left|\dfrac{x}{4} - 1\right| = 42$ **B** $-2\left|\dfrac{x}{4} - 1\right| = 16$

    C $8\left|\dfrac{x}{3} - 2\right| = 48$ **D** $-5\left|\dfrac{x}{6} - 4\right| = -30$

5. (88) **(Refurbishing)** Rudy has $x + y$ junk cars in his lot. He fixes them up and sells each car for $\dfrac{\$400 + \$100x}{y}$. If he sells 30% of them, how much profit will he make?

6. (89) **(Physics)** The function $y = -16x^2 + 80x$ models the height of a droplet of water from an in-ground sprinkler x seconds after it shoots straight up from ground level. Explain how you know when the droplet will hit the ground.

Factor.

7. (87) $2a^2 + 8ab + 6a + 24b$

8. (79) $zx^{10} - 4zx^9 - 21zx^8$

9. (88) Find the product of $\dfrac{b-4}{b+9} \cdot (b^2 + 11b + 18)$.

*10. (93) **Error Analysis** Students were asked to simplify $(15x^4 + 4x^2 + 3x^3) \div (x - 6)$. Which student set their problem up correctly? Explain the error.

Student A	Student B
$(x-6)\overline{)15x^4 + 4x^2 + 3x^3}$	$(x-6)\overline{)15x^4 + 3x^3 + 4x^2 + 0x + 0}$

Simplify.

*11. (92) $\dfrac{\dfrac{1}{10x-10}}{\dfrac{x^5}{10x^2 - 10}}$

12. (92) $\dfrac{\dfrac{2x}{3x+12}}{\dfrac{6x^2}{x^2 + 8x + 16}}$

Solving Multi-Step Absolute-Value Equations

LESSON 94

*13. **(Canoeing)** A canoe rental company charges $10 for the canoe and an additional charge per person. There are 4 people going on the trip and they have planned on spending a total of $50. They hope that the total cost is within $20 of the planned spending total. What is the minimum and maximum they can be charged per person?
(94)

Find the quotient.

14. $(1 + 4x^4 - 10x^2) \div (x + 2)$
(93)

15. $\dfrac{25x^3 + 20x^2 - 5x}{5x}$
(93)

*16. Solve the equation $\dfrac{|x|}{11} + 9 = 15$ and graph the solution.
(94)

*17. **Justify** Show that the solution set to the equation $\dfrac{|x|}{-3} + 1 = 5$ is Ø.
(94)

*18. **Multi-Step** Marty measured the area of his rectangular classroom. He determined that the area is $(-2x^2 + x^3 - 98 - 49x)$ square feet. The length is $(x + 6)$ feet.
(93)
a. What is the width?
b. If the area is $(x^2 - 36)$ square feet, what is the width?

19. **Geometry** The area of a triangle is $(10y^2 + 6y)$ square centimeters. The base is $(5y + 3)$ centimeters. What is the height?
(93)

*20. **Error Analysis** Students were asked to simplify $\dfrac{\frac{4x}{8x+16}}{\frac{12}{x+2}}$. Which student is correct? Explain the error.
(92)

Student A	Student B
$\dfrac{4x}{8x+16} \cdot \dfrac{12}{x+2}$	$\dfrac{4x}{8x+16} \cdot \dfrac{x+2}{12}$
$= \dfrac{4x}{8(x+2)} \cdot \dfrac{12}{x+2}$	$= \dfrac{4x}{8(x+2)} \cdot \dfrac{x+2}{12}$
$= \dfrac{6x}{(x+2)(x+2)}$	$= \dfrac{x}{24}$

*21. **(Commuting)** It took Taylor $\dfrac{1}{x^2 + 3x - 40}$ minutes to get to work, which was $\dfrac{x^2}{6x + 48}$ miles away. Find his rate in miles per minute.
(92)

22. **Verify** Show that 0 is a solution to the inequality $|x - 14| < 30$.
(91)

Solving Multi-Step Absolute-Value Equations

23. Multiple Choice Which inequality is represented by the graph?
(91)

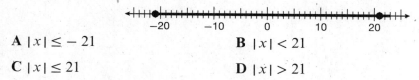

A $|x| \leq -21$ **B** $|x| < 21$

C $|x| \leq 21$ **D** $|x| > 21$

24. Analyze When a line segment is horizontal, which expression under the radical in the
(86) distance formula is 0: $(x_2 - x_1)^2$ or $(y_2 - y_1)^2$?

25. Multi-Step Ornella hiked 8 miles on easy trails and 3 miles on difficult trails. Her
(90) hiking rate on the easy trails was 2.5 times faster than her rate on the difficult trails.
  a. Write a simplified rational expression for Ornella's total hiking time.
  b. Find Ornella's hiking time if her hiking rate on the difficult trails was
  2 miles per hour.

26. Find the vertical and horizontal asymptotes and graph $y = \frac{1}{x-2} - 4$.
(78)

27. Multi-Step Phone Plan A costs $12 per month for local calls and $0.06 per minute for
(81) long-distance calls. Phone Plan B costs $15 per month for local calls plus $0.04 per
minute for long-distance calls. How many minutes of long-distance calls would Jenna
have to make for Plan B to cost less than Plan A?
  a. Formulate Write an inequality to answer.
  b. Solve the inequality and answer the question.
  c. Graph the solution.

28. (United States Flag) An official American flag should have a length that is 1.9 times its
(84) width. The area of an official American flag can be found by the function
$y = 1.9x^2$. Graph the function. Then use the graph to approximate the width of a flag
that has an area of 47.5 square feet.

29. Multi-Step A rectangular garden that is 25 feet wide has a diagonal length that is
(85) 50 feet long.
  a. Find the length of the garden in simplest radical form.
  b. Find the perimeter of the garden to the nearest tenth of a foot.

30. The volume of a sphere is $V = \frac{4}{3}\pi r^3$. Describe in words the relationship between the
(Inv 8) volume of a sphere and its radius. Identify the constant of variation.

Combining Rational Expressions with Unlike Denominators

LESSON 95

Factor completely.

1. $3x^3 - 9x^2 - 30x$ (Inv 9)

2. $8x^3y^2 + 4x^2y - 12xy^3$ (Inv 9)

3. $32x^3 - 24x^4 + 4x^5$ (79)

4. $mn^3 - 10mn^2 + 24mn$ (79)

Find the quotient.

5. $\dfrac{4m}{17r} \div \dfrac{12m^2}{5r}$ (88)

6. $(x^2 - 16x + 64) \div (x - 8)$ (93)

7. Find the zeros of the function shown. (89)

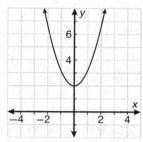

*8. Find the LCD of $\dfrac{4}{x+4} - \dfrac{8}{x^2 + 6x + 8}$. (95)

*9. Add $\dfrac{2x}{2x^2 - 128} + \dfrac{5}{x^2 - 7x - 8}$. (95)

*10. Solve the equation $9|x| - 22 = 14$ and graph the solution. (94)

11. **Error Analysis** Students were asked to subtract $\dfrac{6x^2}{x^2 + 6x - 16} - \dfrac{3}{8x - 16}$. Which student is correct? Explain the error. (95)

Student A	Student B
$\dfrac{6x^2}{x^2 + 6x - 16} - \dfrac{3}{8x - 16}$	$\dfrac{6x^2}{x^2 + 6x - 16} - \dfrac{3}{8x - 16}$
$\dfrac{6x^2}{(x+8)(x-2)} - \dfrac{3}{8(x-2)}$	$\dfrac{6x^2}{(x+8)(x-2)} - \dfrac{3}{8(x-2)}$
$\dfrac{6x^2(8)}{8(x+8)(x-2)} - \dfrac{3(x+8)}{8(x-2)(x+8)}$	$\dfrac{6x^2(8)}{8(x+8)(x-2)} - \dfrac{3(x+8)}{8(x-2)(x+8)}$
$\dfrac{48x^2 - 3x - 24}{8(x+8)(x-2)}$	$\dfrac{48x^2 - 3x + 24}{8(x+8)(x-2)}$

*12. **Generalize** Explain how to find the LCD of two algebraic rational expressions. (95)

*13. **Running** Michele is training for a marathon. She ran $\dfrac{3x^2}{x^2 - 100}$ miles on Monday and $\dfrac{x-1}{2x-20}$ miles on Tuesday. How many miles did she run in all? (95)

Combining Rational Expressions with Unlike Denominators

LESSON 95

*14. Multi-Step The girls' track team sprinted $\frac{2x}{4x^2 - 196}$ meters Thursday and $\frac{12x}{x^2 + x - 56}$ meters Friday.

a. What was the total distance that the track team sprinted?

b. If their rate was $\frac{2x}{x+8}$ meters per minute, how much time did it take them to sprint on Thursday and Friday?

*15. Error Analysis Two students solve the equation $6|x + 3| - 8 = -2$. Which student is correct? Explain the error.

Student A	Student B
$6\|x + 3\| - 8 = -2$ no solution Absolute values cannot equal a negative number.	$6\|x + 3\| - 8 = -2$ $6\|x + 3\| = 6$ $\|x + 3\| = 1$ $x + 3 = 1 \quad$ and $x + 3 = -1$ $x = -2 \quad$ and $\quad x = -4$

16. Geometry The perimeter of a square must be 34 inches plus or minus 2 inches. What is the longest and the shortest length each side can be?

*17. Multi-Step A student budgets $35 for lunch and rides each week. He gives his friend $5 for gas and then pays for 5 lunches a week. He has a $2 cushion in his budget, meaning that he can spend $2 more or less than he budgeted.

a. Write an absolute-value equation for the minimum and maximum he can spend on each lunch.

b. What is the maximum and the minimum he can spend on each lunch?

18. Error Analysis Students were asked to simplify $\frac{x^2 + 10x + 24}{x}$. Which student is correct? Explain the error.

Student A	Student B
$\frac{(x + 6)(x + 4)}{\cancel{x}} = (x + 6)(4)$	$\frac{(x + 6)(x + 4)}{x}$

*19. (Carpeting) The rectangular public library is getting new carpet. The area of the room is $(x^3 - 18x^2 + 81x)$ square feet. The length of the room is $(x - 9)$ feet. What is the width?

*20. Justify Give an example to show $\frac{a}{b} \cdot \frac{c}{d} = \frac{ac}{bd}$.

Combining Rational Expressions with Unlike Denominators

LESSON 95

21. Multiple Choice Simplify: $\dfrac{\frac{x^2-9}{x^2-5x+6}}{\frac{x^2+5x+6}{x^2-4}}$.
(92)

A 1 **B** $\dfrac{(x-3)^2}{(x-2)^2}$ **C** -1 **D** $\dfrac{(x-3)}{(x-2)}$

22. Solve and graph the inequality $|x| > 17$.
(91)

23. Measurement When measuring something in centimeters, the accuracy is within 0.1 centimeter. A board is measured to be 15.6 centimeters. The accuracy of the measurement can be represented by the absolute-value inequality $|x - 15.6| \leq 0.1$. Solve the inequality.
(91)

24. Multi-Step Reggie has made a stew that is at a temperature of 100°F, and he either wants to heat it up to 165°F or cool it down to 45°F. To heat up the stew will take at most 45 minutes plus 1 minute for each degree the temperature is raised. To cool down the stew will take at least 10 minutes plus 2 minutes for each degree the temperature is lowered. How much time does Reggie need to allow for changing the food temperature either up or down?
(82)
a. How many degrees up and down does the stew need to go?
b. How much time will it take to cool down or heat up the stew?
c. Is it faster to heat up or cool down the stew?

25. (Olympic Swimming Pool) An Olympic swimming pool must have a width of 25 meters and a length of 50 meters. Find the length of a diagonal of an Olympic swimming pool to the nearest tenth of a meter.
(85)

26. Multi-Step A graphing calculator screen is 128 pixels wide and 240 pixels long. The "pixel coordinates" of three points are shown.
(86)

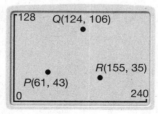

a. Find the distance in pixels between each pair of points.
b. List the line segments in order from shortest to longest.

27. Write Create a polynomial in which each term has a common factor of $4a$, and then factor the expression.
(87)

Combining Rational Expressions with Unlike Denominators

LESSON 95

28. (**Hiking**) In celebration of getting to the top of Beacon Rock in southern Washington, Renata throws her hat up and off the top of Beacon Rock. The height of the hat x seconds after the throw (in meters) can be approximated by the function $y = -5x^2 + 10x + 260$. After how many seconds will the hat be at its maximum height? What is this height?

29. (**Commuting**) Mr. Shakour's round trip commute to and from work totals 30 miles. Because of traffic, his speed on the way home is 5 miles per hour less than what it is on the way to work. Write a simplified expression to represent Mr. Shakour's total commuting time.

30. A teacher randomly picks a shirt and a skirt from her closet to wear to school. Use the table to find the theoretical probability of the teacher choosing an outfit with a blue shirt and khaki skirt.

		Shirts			
		Red	Blue	Blue	White	White
Skirts	Khaki	KR	KB	KB	KW	KW
	Navy	NR	NB	NB	NW	NW
	Black	BR	BB	BB	BW	BW
	White	WR	WB	WB	WW	WW

Name _____ Date _____ Class _____

Graphing Quadratic Functions
LESSON 96

1. Find the zeros of the function shown.
   (89)

2. Add $\dfrac{25}{16x^2y} + \dfrac{xy}{32y^5}$.
   (90)

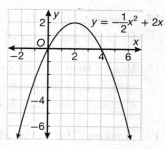

*3. Solve the equation $\dfrac{10|x|}{3} + 18 = 4$ and graph the solution.
(94)

4. Solve $-0.3 + 0.14n = 2.78$.   5. Solve $\dfrac{6}{x-3} = \dfrac{3}{10}$.
(24)   (31)

6. Find the LCD of $\dfrac{6}{x+6} - \dfrac{12}{x^2 + 8x + 12}$.
(95)

7. The table lists the ordered pairs from a relation. Determine whether they form
(25) a function.

Domain (x)	Range (y)
10	15
11	17
8	11
9	13
5	5

8. Simplify $\dfrac{\dfrac{-x^5}{21x+3}}{\dfrac{5x^9}{28x+4}}$.
(92)

*9. Graph the function $y = x^2 - 2x - 8$.
(96)

*10. **Write** Explain how to reflect a point across the axis of symmetry to get a second
(96) point on the parabola.

*11. **Justify** Show that the vertex of $y = 4x^2 - 24x + 9$ is $(3, -27)$.
(96)

*12. **Multiple Choice** Which function has the vertex $(6, -160)$?
(96)
    A $y = 6x^2 - 72x + 56$ **B** $y = 2x^2 - 8x + 48$
    C $y = 3x^2 + 42x - 12$ **D** $y = 5x^2 - 5x + 43$

*13. (**Diving**) A diver moves upward with an initial velocity of 10 feet per second.
(96) How high will he be 0.5 seconds after diving from a 6-foot platform? Use
$h = -16t^2 + vt + s$.

14. **Multiple Choice** Find the LCD of $\dfrac{3}{2x-10}$ and $\dfrac{5x}{2x^2 - 4x - 30}$.
(95)
    A $2(x-5)(x+3)$ **B** $(x-5)(x+3)$
    C $(x+5)(x-3)$ **D** $\dfrac{2}{(x-5)(x+3)}$

Graphing Quadratic Functions
LESSON 96

*15. Geometry One side of a triangle is $\frac{2}{x+2}$ yards and two sides are each $\frac{-5}{3x+6}$ yards. Find the perimeter of the triangle.
(95)

*16. Measurement Carrie measured a distance of $\frac{3x^2}{9x-18}$ yards and Jessie measured a distance of $\frac{4x-5}{x^2-4}$ yards. How much longer is Carrie's measurement than Jessie's?
(95)

17. (Banking) The dollar amount in a student's banking account is represented by the absolute-value inequality $|x - 200| \leq 110$. Solve the inequality and graph the solution.
(91)

18. Generalize Why is a place holder needed for missing variables in a polynomial dividend?
(93)

19. Multiple Choice Simplify $(-5x + 2x^2 - 3) \div (x - 3)$.
(93)
  A $2x - 1$ **B** $2x + 1$ **C** $\frac{x-2}{2x}$ **D** $\frac{x^2-3}{5x}$

*20. (Physics) A family is going to see friends that live in two different towns. They will have to travel 100 miles plus or minus 10 miles to see either of them. They want to spend 2 hours in the car. What are the minimum and maximum rates that they need to go?
(94)

*21. Error Analysis Two students graph the solution to the equation $|2x + 10| = 8$. Which student is correct? Explain the error.
(94)

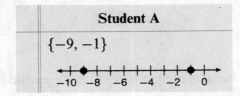

22. (Art) Jeremy's picture frame has an area of $x^2 - 18x + 80$ square inches. He has two square pictures in it, one measuring $\frac{1}{4}x$ inch on each side and the other measuring $\frac{1}{2}x$ inch on each side. Write a simplified expression for the total fraction of the frame covered by pictures.
(90)

23. Verify Divide the rational expression $\frac{3x^2 + 2x}{9y} \div \frac{y+2}{y}$. How can you check your answer?
(88)

24. Multi-Step The volume of a prism is $6x^3 + 14x^2 + 4x$ cm³. What is the length of the greatest dimension?
(87)
  a. Factor out common terms.
  b. Factor completely.
  c. Find the greatest dimension.

Graphing Quadratic Functions

LESSON 96

25. (Baseball) A baseball diamond is a square that is 90 feet long on each side. To use a
(86) coordinate grid to model positions of players on the field, place home plate at (0, 0), first base at (90, 0), second base at (90, 90), and third base at (0, 90). An outfielder located at (150, 80) throws to the first-baseman. How long is the throw? Round your answer to the nearest foot.

26. Multi-Step A square has a side length of a centimeters. A smaller square has a side
(83) length of b centimeters.

  a. If the difference in the areas of the squares is $a^2 - 16$, what is the value of b?

  b. If the area of the larger square is $36b^2 + 60b + 25$, what is the side length in terms of b?

  c. Using your answers to parts **a** and **b**, find the side length and area of the larger square.

27. Determine if the inequality $6x > 7x$ is never, sometimes, or always true. If it is
(81) sometimes true, identify the solution set.

28. Use the graph to find the theoretical probability of choosing each color.
(80)

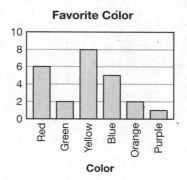

29. Suppose d varies inversely with b and jointly with a and c. Find the constant of
(Inv 8) variation when $a = 4$, $b = 5$, $c = 2$, and $d = 8$. Express the relationship between these quantities. What is d when $a = 9$, $b = 15$, and $c = 6$?

30. Is $(x + 10)(x - 2)$ the correct factorization for $x^2 - 8x - 20$? Explain.
(Inv 9)

Name _____ Date _____ Class _____

Graphing Linear Inequalities LESSON 97

Simplify.

1. $\dfrac{30x^{-2}y^{12}}{6y^{-5}}$
(32)

2. $\sqrt{0.09q^2 r} + q\sqrt{0.04r}$
(69)

3. $\dfrac{16g^4}{2g+3} - \dfrac{81}{2g+3}$
(90)

4. Find the range of the data set that includes the ages of 9 members of a chess club:
(48) 23, 7, 44, 31, 18, 27, 35, 39, 66.

5. Find the product $(4x^2 + 8)(2x - 7)$ using the FOIL method.
(58)

*6. Add $\dfrac{9}{9x-36} + \dfrac{-24}{3x^2 - 48}$.
(95)

*7. Jim ran a total of $\dfrac{x}{x^2 + 2x + 1}$ miles in the gym and $\dfrac{x+2}{x+1}$ miles outside. How
(95) many more miles did he run inside?

8. Find the quotient of $(x^2 - 14x + 49) \div (x - 7)$.
(93)

Solve and graph the inequality.

9. $13 \le 2x + 7 < 15$
(82)

10. $\dfrac{|x|}{6} > 8$
(91)

11. Determine if the inequality $3x - 4x \ge 6 - x + 8$ is never, sometimes, or always true.
(81) If it is sometimes true, identify the solution set.

*12. Determine if the ordered pair (2, 6) is a solution of the inequality $y > 3x - 2$.
(97)

*13. Graph the function. $y = x^2 + 2x - 24$
(96)

*14. **Write** What points on a graph of an inequality satisfy the inequality? Explain.
(97)

*15. **Generalize** How do you know which half-plane to shade for the graph of a linear
(97) inequality?

Graphing Linear Inequalities
LESSON 97

*16. Multiple Choice Which inequality represents the graph on the coordinate plane?
(97)

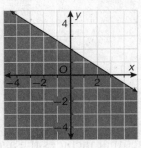

A $y \geq -\frac{2}{3}x + 2$ **B** $y \leq -\frac{2}{3}x + 2$

C $y \leq \frac{2}{3}x + 2$ **D** $y < -\frac{2}{3}x + 2$

17. (Football) Tickets for the school football game cost five dollars for adults and
(97) three dollars for students. In order to buy new helmets, at least $9000 worth of tickets must be sold. Write an inequality that describes the total number of tickets that must be sold in order to buy new helmets.

*18. Error Analysis Two students find the vertex for $y = x^2 - 6x + 19$. Which student is
(96) correct? Explain the error.

Student A	Student B
$\frac{-b}{2a} = \frac{6}{2} = 3$	$\frac{-b}{2a} = \frac{6}{2} = 3$
The vertex is (3, 0).	$3^2 - 6(3) + 19 = 10$
	The vertex is (3, 10).

*19. Geometry The area of a rectangle is 48 square inches. The length is three times
(96) the width. Find the width of the rectangle by finding the positive zero of the function $y = 3w^2 - 48$.

*20. Multi-Step The height y of a golf ball in feet is given by the function $y = -16x^2 + 49x$.
(96) **a.** What is the y-intercept?
b. What does this y-intercept represent?
c. What answer does the equation give for the height of the ball after 5 seconds?
d. What does that height mean?

*21. (Commuting) Jeff traveled $\frac{1}{2x^2 - 4x}$ miles for his job on Monday and $\frac{1}{x^3 - 2x^2}$ miles for his
(95) job on Tuesday. How many more miles did he travel on Tuesday?

Graphing Linear Inequalities

LESSON 97

22. Multiple Choice Add $\dfrac{3y+2}{y+z} + \dfrac{4}{2y+2z}$.
(95)

   A $\dfrac{3y+4}{y+z}$ **B** $\dfrac{y+4}{y+z}$

   C $\dfrac{6y}{4mn}$ **D** $\dfrac{3y-4}{y-z}$

23. Verify Show that -11 is a solution to the inequality $|3x| - 2 = 31$.
(94)

24. Multiple Choice Solve $4|x-8| = 12$.
(94)

   A $\{-5, 5\}$ **B** $\{11, -11\}$

   C $\{5, 11\}$ **D** $\{-20, 20\}$

25. (Bike Riding) Ron rode his bike for $\dfrac{10}{45x^2 + 4x - 1}$ minutes to get to his grandmother's house that was $\dfrac{1}{45x-5} + \dfrac{2x}{25x+5}$ miles away. Find his rate in miles per minute.
(92)

26. (Carpentry) A carpenter uses a measuring tape with an accuracy of $\pm\dfrac{1}{32}$ inches. He measures the height of a bookshelf to be $95\dfrac{5}{8}$ inches. Solve the inequality $\left|x - 95\dfrac{5}{8}\right| \leq \dfrac{1}{32}$ to find the range of the height of the bookshelf.
(91)

27. Generalize How are the number of zeros of a function related to the location of the vertex of the function's parabola?
(89)

28. Probability The probability of winning a certain game is $\dfrac{2x^4y^2}{15xy^3}$. The probability of winning a different game is $\dfrac{5x^2y}{8x^3y^2}$. What is tshe probability of winning both games?
(88)

29. (Rate) An orange juice machine squeezes juice out of $x^2 + 30x$ oranges every hour. How much time in days will it take to squeeze 3000 oranges?
(88)

30. Multi-Step A circle has a radius of x. Another circle has a radius of $3x$.
(84)
   a. Write equations for the areas of both circles.

   b. Graph both functions in the same coordinate plane.

   c. Compare the graphs.

Name _____ Date _____ Class _____

Solving Quadratic Equations by Factoring

LESSON 98

Solve and graph the inequality.

1. $11 < 2(x + 5) < 20$
(82)

2. $|x| + 1.5 \leq 7.6$
(91)

3. Determine whether the polynomial $-121 + 9x^2$ is a perfect-square trinomial or a
(83) difference of two squares. Then factor the polynomial.

4. Graph the function $y = 2x^2 + 8x + 6$.
(96)

5. The number of Apples A and Oranges O grown in a certain fruit orchard can be
(53) modeled by the given expressions where x is the number of years since the trees were planted. Find a model that represents the total number of apples and oranges grown in this orchard.

$$A = 15x^3 + 17x - 20$$
$$O = 20x^3 + 11x - 4$$

6. Write an equation for a line that passes through $(1, 2)$ and is perpendicular
(65) to $y = -\frac{3}{4}x + 2\frac{3}{4}$.

7. Solve the equation $\frac{4|x|}{9} + 3 = 11$ and graph the solution.
(94)

Simplify.

8. $\dfrac{\frac{3x + 6}{7x - 7}}{\frac{5x + 10}{14x - 14}}$
(92)

9. $(5xyz)^2(3x^{-1}y)^2$
(40)

*10. Determine if the ordered pair $(5, 5)$ is a solution of the inequality $y < -5x + 4$.
(97)

*11. **Write** Explain the Zero Product Property in your own words.
(98)

*12. **Justify** What property allows you to use the following step when solving an equation?
(98)

$$(x + 4)(x + 5) = 0$$
$$x + 4 = 0 \quad x + 5 = 0$$

*13. **Multiple Choice** What is the solution set of $0 = (3x - 5)(x + 2)$?
(98)

A $\left\{\frac{5}{3}, -2\right\}$ B $\left\{\frac{5}{3}, 2\right\}$ C $\left\{-\frac{5}{3}, -2\right\}$ D $\left\{-\frac{5}{3}, 2\right\}$

*14. **Ages** A girl is 27 years younger than her mother. Her mother is m years old. The
(98) product of their ages is 324. How old is each person?

Saxon Algebra 1

Solving Quadratic Equations by Factoring

LESSON 98

***15. Multi-Step** Seve plans to go shopping for new jeans and shorts. She plans to spend no more than $70. Each pair of jeans costs $20 and each pair of shorts costs $10.
(97)
  a. Write an inequality that describes this situation.
  b. Graph the inequality.
  c. If Seve wants to spend exactly $70, what is a possible number of each she can spend her money on?

***16. Geometry** The Triangle Inequality Theorem states that the sum of the lengths of any two sides of a triangle is greater than the length of the third side. The sides of a triangle are labeled $4x$ inches, $2y$ inches, and 8 inches. James wrote an inequality that satisfies the Triangle Inequality Theorem. He wrote the inequality $4x + 2y > 8$. Use a graphing calculator to graph the inequality.
(97)

***17.** Solve $(x + 4)(x - 9) = 0$.
(98)

18. Error Analysis Students were asked to write an inequality that results in a dashed horizontal boundary line. Which student is correct? Explain the error.
(97)

Student A	Student B
$y > -3$	$y \geq 4$

19. (Horseback Riding) It took Joe $\frac{2x - 10}{2x^5}$ minutes to ride his horse to Darrell's house that was $\frac{3x^2 - 15x}{3x}$ miles away. Find his rate in miles per minute.
(92)

20. Measurement The area of a triangle can be expressed as $4x^2 - 2x - 6$ square meters. The height of the triangle is $x + 1$ meters. Find the length of the base of the triangle.
(92)

21. (Art) Michael bought a rectangular painting from a local artist. The area of the painting was $(20x + 5 + x^3)$ square inches. The width was $(x - 5)$ inches. What was the length?
(93)

22. Analyze Should you find the LCD when multiplying or dividing rational fractions?
(95)

23. Multi-Step Erika and Casey started a new walking program. They walked $\frac{4x}{3x + 9}$ miles Thursday and $\frac{16}{x^2 + 12x + 27}$ miles Friday.
(95)
  a. What is the total distance that they walked?
  b. If their rate was $\frac{4x}{x + 3}$ miles per hour, how much time did it take them to walk on Thursday and Friday?

***24. (Physics)** A ball is dropped from 100 feet in the air. What is its height after 2 seconds? Use $h = -16t^2 + vt + s$. (Hint: Its initial velocity is 0 feet per second.)
(96)

Solving Quadratic Equations by Factoring

LESSON 98

*25. **Error Analysis** Two students find the zeros of the function $y = x^2 - 8x - 33$. Which
(96) student is correct? Explain the error.

Student A	Student B
$y = x^2 - 8x - 33$	$0 = x^2 - 8x - 33$
$y = (0)^2 - 8(0) - 33$	$0 = (x - 11)(x + 3)$
$y = -33$	$x - 11 = 0$ or $x + 3 = 0$
$(0, -33)$	$x = 11$ or $x = -3$
	11 and -3

26. **Multi-Step** Hideyo has a picture frame that measures 20 centimeters by 26 centimeters.
(85) The frame is 1.5 centimeters wide.
   a. Find the length and width of the picture area.
   b. Find the length of the diagonal of the picture area to the nearest tenth of a centimeter.

27. **Profit** A business sold $x^2 + 6x + 5$ items. The profit for each item sold
(88) is $\frac{x^2}{100x}$ dollars.
   a. What is the profit in terms of x?
   b. What is the profit (in dollars) if $x = 50$?

28. **Multi-Step** Javed has a garden with an area of 36 square feet. The width of his garden
(89) is 9 feet less than the length. What are the dimensions of his garden?
   a. Write a formula to find the dimensions of the garden and describe how you will solve it.
   b. What are the dimensions of the garden?

29. **Generalize** Describe the steps for subtracting $\frac{x-6}{x+5}$ from $\frac{x^2 - x - 30}{x+5}$.
(90)

30. The width of a rectangle is represented by the expression $(4x - 6)$ and the
(Inv 9) length $(8x + 7)$. Would the area of the rectangle be correctly expressed as $32x^2 + 20x - 42$? If not, what is the correct area?

Name _____ Date _____ Class _____

Solving Rational Equations

LESSON 99

Solve.

1. $\dfrac{4}{x} = \dfrac{8}{x+4}$
(99)

2. $(x-13)(x+22) = 0$
(98)

3. (Physics) A student is biking to a friend's house. He bikes at 10 miles per hour. The friend lives 20 miles away, give or take 2 miles. What are the minimum and maximum times it will take him to get there?
(94)

*4. Verify Without graphing, show that the point $(2, -6)$ lies on the graph of $y = x^2 + x - 12$.
(96)

5. Simplify: $\dfrac{\dfrac{4x}{12x-60} + \dfrac{1}{4x-16}}{\dfrac{-2}{9x-20-x^2}}$.
(92)

6. Factor $x + 4x^2 - 5$.
(75)

7. Larry weighed 180 pounds. He has lost 2 pounds a month for x months. Write a linear equation to model his weight after 8 months.
(52)

*8. Write What is an extraneous solution?
(99)

9. Find the LCD of $\dfrac{2x}{2x^2 - 72} - \dfrac{12}{x^2 + 13x + 42}$.
(95)

10. (Population) The function $y = -0.0003x^2 + 0.03x + 1.3$ shows the population of Philadelphia County between the years 1900 and 1990, where x is the number of years after 1900 and y is the population for that year in millions of people. Find the vertex of the parabola that represents the function. What does it represent in terms of the scenario?
(89)

11. Multi-Step The coordinates of three friends' houses on a city map are $P(3, 3)$, $Q(5, 9)$, and $R(11, 3)$. The friends plan to meet at the point that is half-way between Q and the midpoint of $\overline{PR}$.
(86)
a. Find the midpoint of $\overline{PR}$.
b. Find the coordinates of the point where the friends plan to meet.

12. Find the quotient of $(18x^2 - 120 + 6x^5) \div (x - 2)$.
(93)

13. What is the ratio of the volume of a cube with side lengths of 5 to the volume of a cube with side lengths of 3?
(36)

*14. Formulate How can you quickly tell if a possible solution is extraneous using the denominators in the original equation?
(99)

*15. Multiple Choice Which of the following is an extraneous solution to the equation $\dfrac{x^2}{x-1} + \dfrac{4x^2 - 20x}{(x-1)(x-5)} = \dfrac{10}{2x-2}$?
(99)

A $x = -1$

B $x = 0$

C $x = 1$

D $x = 5$

Name _____ Date _____ Class _____

Solving Rational Equations
LESSON 99

*16. **(Housekeeping)** It takes a man 8 hours to clean the house. His friend can clean it
(99) in 6 hours. How long will it take them if they clean together?

*17. **Error Analysis** Two students solve the equation $0 = (x - 5)(x + 11)$. Which student
(98) is correct? Explain the error.

Student A	
$x - 5 = 0$	$x + 11 = 0$
$x = 5$	$x = -11$

Student B	
$x - 5 = 0$	$x + 11 = 0$
$x = -5$	$x = 11$

18. **Geometry** The area of a triangle is 24 square centimeters. The height is four more than
(98) two times the base. Find the base and height of the triangle.

*19. **Multi-Step** The length of a lawn is twice the width. The lawnmower cuts 2-foot strips.
(98) One strip along the length and width has already been cut.

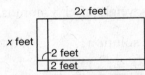

a. Write expressions for the length and width of the area left to be cut.

b. The area left to be cut is 144 square feet. Find the width of the yard.

c. What is the length of the yard?

20. **Generalize** How do you know when there is no solution to an absolute-value
(91) inequality?

21. **(Architecture)** Mia is designing a rectangular city hall. She accidentally spills water on
(93) her newly revised sketches. She is only able to determine the area, which is
$(x^2 - 15x + 56)$ square feet, and the length, which is $(x - 7)$ feet. What is the width?

*22. **Multiple Choice** What are the zeros for the function $y = 4x^2 + 28x - 72$?
(96)
A 0, 4
B 0, −4
C 2, −9
D 2, −76

*23. **(Band)** The school band is performing a music concert. Tickets cost $3 for adults
(97) and $2 for students. In order to cover expenses, at least $200 worth of tickets
must be sold. Write an inequality that describes the graph of this situation.

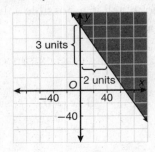

© Saxon. All rights reserved.   380   Saxon Algebra 1

Solving Rational Equations

LESSON 99

24. Measurement Jesse bought a new glass window to go in his front room. The area of
(93) the window is $(2 + 3x^3 - 8x)$ square inches. The width is $(x + 4)$ inches. What is the length?

*25. Error Analysis Students were asked to write an inequality that results in a solid, vertical
(97) boundary line. Which student is correct? Explain the error.

Student A	Student B
$x \leq 6$	$x < -2$

26. Graph the inequality $4x + 5y \geq -7$.
(97)

27. Multi-Step Pedro biked 6 miles on dirt trails and 12 miles on the street. His biking
(90) rate on the dirt trails was 25% of what it was on the street.
   a. Write a simplified expression for Pedro's total biking time.
   b. **Analyze** Explain how finding the simplified expression would change if Pedro's biking rate on the trails was 50% of what it was on the streets.

28. Graph the function $y - 3 = -x^2 + 3$.
(84)

29. A square has a side length s. The square grows larger and has a new area
(83) of $s^2 + 16s + 64$. What is the new side length?

30. Write an equation where j is inversely proportional to m and n and directly
(Inv 8) proportional to p and q.

Name _____ Date _____ Class _____

Solving Quadratic Equations by Graphing — LESSON 100

Solve.

1. $x(2x - 11) = 0$
   (98)

2. $\dfrac{12}{x - 6} = \dfrac{4}{x}$
   (99)

*3. **Generalize** Using the path of a ball thrown into the air as an example, describe in mathematical terms each part of the graph the path of the ball creates.
(100)

*4. **Generalize** What does the graph of a quadratic equation look like when there is no solution? one solution? two solutions?
(100)

5. Given that y varies directly with x, identify the constant of variation such that when $x = 15$, $y = 30$.
(Inv 7)

*6. **Basketball** Ramero shoots a basketball into the air. The ball's movement forms a parabola given by the quadratic equation $h = -16t^2 + 7t + 7$, where h is the height in feet and t is the time in seconds. Find the maximum height of the path the basketball makes and the time t when the basketball hits the ground. Round to the nearest hundredth.
(100)

*7. **Multiple Choice** What is the equation of the axis of symmetry of the parabola defined by $y = \frac{1}{4}(x - 4)^2 + 5$?
(100)
   A $x = 1$ **B** $x = 4$ **C** $x = 5$ **D** $x = -4$

*8. Solve $-7x^2 - 10 = 0$ by graphing.
(100)

*9. Solve $\dfrac{6}{x} = \dfrac{8}{x + 7}$.
(99)

10. A deck of ten cards has 5 red and 5 black cards. Cards are replaced in the deck after each draw. Use an equation to find the probability of drawing a black card twice and rolling a 6 on a number cube.
(80)

11. **Geometry** The altitude of the right triangle divides the hypotenuse into segments of lengths x units and 5 units. To find x, solve the equation $\dfrac{x + 5}{6} = \dfrac{6}{x}$.
(99)

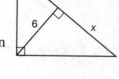

*12. **Multi-Step** Henry starts working a half-hour before Martha. He can complete the job in 4 hours. Martha can complete the same job in 3 hours.
(99)
   a. Let t represent the total time they work together. In terms of t, how long does Henry work?
   b. Use an equation to find how long they work together to complete the job.
   c. How long does Henry work?

13. Find the quotient of $\dfrac{a^2 + 10a - 24}{a - 2}$.
(93)

14. Simplify $\sqrt{49y^5}$.
(61)

© Saxon. All rights reserved.   383   Saxon Algebra 1

Name _____ Date _____ Class _____

Solving Quadratic Equations by Graphing

LESSON 100

15. **(Profit)** An entrepreneur makes $3 profit on each object sold. She would like to make $270 plus or minus $30 total. What is the minimum and maximum number of objects she needs to sell?
(94)

16. **Data Analysis** A student knows there will be 4 tests that determine her semester grade. She wants her average to be an 85, plus or minus 5 points. What is the minimum and maximum number of points she needs to earn during the semester?
(94)

17. Solve the equation $|10x| - 3 = 87$.
(94)

18. **(Exercise)** Tom ran a total of $\frac{7x}{x^2 + 3x - 18}$ miles in August and $\frac{2x + 1}{7x + 42}$ miles in September. How many more miles did he run in August?
(95)

19. Graph the function $y = 5x^2 - 10x + 5$.
(96)

*20. **Verify** A boundary line is a vertical line. The inequality contains a < symbol. Which half-plane should be shaded on the graph?
(97)

21. **Multiple Choice** Which point does not satisfy the inequality $x + 2y < 5$?
(97)
   A (0, 0)   B (2, 1)   C (3, −4)   D (−1, 3)

*22. **(Ages)** A boy is b years old. His father is 23 years older than the boy. The product of their ages is 50. How old is each person?
(98)

*23. **Error Analysis** Two students find the roots of $3x^2 - 6x = 24$. Which student is correct? Explain the error.
(98)

Student A	Student B
$3x^2 - 6x = 24$	$3x^2 - 6x = 24$
$3x(x - 2) = 24$	$3x^2 - 6x - 24 = 0$
$3x = 0 \quad x - 2 = 0$	$3(x^2 - 2x - 8) = 0$
$x = 0 \quad\quad x = 2$	$3(x - 4)(x + 2) = 0$
	$x - 4 = 0 \quad x + 2 = 0$
	$x = 4 \quad\quad x = -2$

24. Does the graph of $y + 2x^2 = 12 + x$ open upward or downward?
(84)

25. Do the side lengths 18, 80, and 82 form a Pythagorean triple?
(85)

26. **Multi-Step** The volume of a prism is $3x^3 + 12x^2 + 9x$. What are the possible dimensions of the prism?
(87)
   a. Factor out common terms.
   b. Factor completely.
   c. Find the dimensions.

Solving Quadratic Equations by Graphing

27. **(Travel)** The Jackson family drove 480 miles on Saturday and 300 miles on Sunday.
(90) Their average rate on Sunday was 10 miles per hour less than their rate was on Saturday. Write a simplified expression that represents their total driving time.

28. **Multi-Step** At the carnival, a man says that he will guess your weight within
(91) 5 pounds.
    a. You weigh 120 pounds. Write an absolute-value inequality to show the range of acceptable guesses.
    b. Solve the inequality to find the actual range of acceptable guesses.

29. **Verify** If the numerator of a rational expression is a polynomial and the denominator
(92) of the rational expression is a different polynomial, will factoring the polynomials always provide a way to simplify the expression? Verify your answer by giving an example.

30. If a 9% decrease from the original price resulted in a new price of $227,500,
(47) what was the original price?

Name _____ Date _____ Class _____

Solving Multi-Step Absolute-Value Inequalities

LESSON 101

*1. Solve and graph the inequality $7|x| - 4 \geq 3$.
(101)

*2. **Error Analysis** Two students solve the inequality $|x - 4| + 2 \leq 6$. Which student is
(101) correct? Explain the error.

Student A	Student B
$|x - 4| + 2 \leq 6$	$|x - 4| + 2 \leq 6$
$|x - 4| \leq 4$	$-6 \leq x - 4 + 2 \leq 6$
$-4 \leq x - 4 \leq 4$	$-6 \leq x - 2 \leq 6$
$0 \leq x \leq 8$	$-4 \leq x \leq 8$

*3. **Write** Describe the three steps needed to solve the inequality $\frac{|x|}{2} + 11 \leq 16$.
(101)

4. Simplify $\dfrac{pt^{-2}}{m^3}\left(\dfrac{p^{-2}wt}{4m^{-1}} + 6t^4w^{-1} - \dfrac{w}{m^{-3}}\right)$.
(39)

*5. **Analyze** Suppose that a, b, and c are all positive integers. Will the solution of
(101) the inequality $-a|x - b| \geq -c$ be a compound inequality that uses AND or a
compound inequality that uses OR?

*6. **(Oven Temperature)** Liam's oven's temperature t varies by no more than 9°F from the
(101) set temperature. Liam sets his oven to 475°F. Write an absolute-value inequality that
models the possible actual temperatures inside the oven. What is the highest possible
temperature?

*7. **Error Analysis** Students were asked if a quadratic equation could have more than one
(100) solution. Which student is correct? Explain the error.

Student A	Student B
yes; A quadratic equation can have two solutions. When a parabola crosses the x-axis twice, there are two solutions.	no; A quadratic equation cannot cross the x-axis more than once. So, there can only be one solution.

*8. **Multi-Step** Shaw hits a tennis ball into the air. Its movement forms a parabola given
(100) by the quadratic equation $h = -16t^2 + 2t + 9$, where h is the height in feet and t is
the time in seconds.
  a. Find the maximum height of the arc the ball makes in its flight. Round to the
     nearest tenth.
  b. Find the time t when the ball hits the ground. Round to the nearest hundredth.
  c. Find the time t when the ball is at its maximum height. Round to the nearest
     hundredth.

9. Find the LCM of $(6w^3 - 48w^5)$ and $(9w - 72w^3)$.
(57)

© Saxon. All rights reserved.   Saxon Algebra 1

Solving Multi-Step Absolute-Value Inequalities

LESSON 101

*10. Geometry A boy spills a cup of juice on the sidewalk. As time increases, the area of the spill changes. The area of the spill is given by the function $A = -2t^2 + 5t + 125$, where A is the area in square feet and t is the time in seconds. Find the time when the area is 60 square feet. Round to the nearest hundredth.
(100)

11. Solve $x^2 + 9 = -6x$ by graphing.
(100)

12. Solve the equation $|8x| + 4 = 28$.
(94)

13. (Traveling) Mia walked $\frac{4}{r-2}$ miles to her neighbors' house on Monday and walked $\frac{r^2}{2-r}$ miles on Tuesday to go see her grandmother. How many miles total did she walk on Monday and Tuesday?
(95)

14. Subtract $\frac{5}{x-3} - \frac{2}{x-2}$.
(95)

15. (Soccer) A soccer ball on the ground is passed with an initial velocity of 62 feet per second. What is its height after 3 seconds? Use $h = -16t^2 + vt + s$.
(96)

16. Measurement A girl is 24 years younger than her mother. The product of their ages is 81. Find the mother's age by finding the positive zero of the function $y = x^2 - 24x - 81$.
(96)

17. Determine if the ordered pair $(-7, 2)$ is a solution of the inequality $y \leq 3$.
(97)

18. Verify Show that $\frac{3}{4}$ is a solution to $(4x - 3)(5x + 7) = 0$.
(98)

19. Multiple Choice What are the roots of the equation $0 = x^2 - 10x - 39$?
(98)
   A 0, 39   **B** 10, 0   **C** 3, -13   **D** 13, -3

*20. Solve and check: $\frac{x}{11} = \frac{6}{x-5}$.
(99)

21. Does the graph of $-8x^2 - 12 = 3 - y$ open upward or downward?
(84)

22. (Office Management) Maria can complete all the copies in 1 hour. It takes Lachelle 2 hours. How long will it take them if they use two identical copiers and work together?
(99)

23. Error Analysis Two students solve $\frac{x-8}{x+2} = \frac{x-6}{3x+6}$. Which student is correct? Explain the error.
(99)

Student A	Student B
$(x-8)(3x+6) = (x+2)(x-6)$ $3x^2 - 18x - 48 = x^2 - 4x - 12$ $2x^2 - 14x - 36 = 0$ $2(x-9)(x+2) = 0$ $\{9\}$	$(x-8)(3x+6) = (x+2)(x-6)$ $3x^2 - 18x - 48 = x^2 - 4x - 12$ $2x^2 - 14x - 36 = 0$ $2(x-9)(x+2) = 0$ $\{-2, 9\}$

Solving Multi-Step Absolute-Value Inequalities

LESSON 101

24. Do the side lengths 3, $3\sqrt{3}$, and 6 form a Pythagorean triple?
(85)

25. Let $P = (-2, 1)$, $Q = (0, 2)$, $R = (1, -2)$, and $S = (-1, -3)$. Use the distance formula to determine whether $PQRS$ is a rhombus.
(86)

26. Multi-Step Find the product of $\frac{5x^2y^2}{3x^3y^3} \cdot \frac{9xy^2}{25xy^3}$ using two different methods.
(88)
  a. Solve the expression by multiplying first and then simplifying.
  b. Solve the expression by simplifying each factor and then multiplying.
  c. Explain which method you prefer.

27. (Road Trip) Carlos tracks the mileage for a road trip on his car's odometer. The total distance is 974.6 miles plus or minus 0.1 miles. Solve and graph the inequality $|x - 974.6| \leq 0.1$.
(91)

28. Multi-Step Amy skipped for $\frac{3x - 6}{9x}$ hours to get to her grandmother's house that was $\frac{2x^2 - 4x}{7x^3}$ miles away.
(92)
  a. Find her rate in miles per hour.
  b. If the rate is divided by $\frac{1}{x^2}$, what is the new rate?

29. How do you write a remainder of 5 for a division problem that has a divisor of $(3x^2 + 7x + 8)$?
(93)

*30. What is the parent quadratic function defined to be? What is the shape of its graph and where is it located on the coordinate system?
(Inv 10)

Name _____ Date _____ Class _____

Solving Quadratic Equations Using Square Roots
LESSON 102

Simplify.

1. $4(2p^{-2}q)^2(3p^3q)^2$
(40)

2. $(7\sqrt{8})^2$
(76)

*3. Error Analysis Two students want to find the length of the sides of a square with an
(102) area that is 720 square meters less than 1161 square meters. Which student is correct?
Explain the error.

Student A	Student B
$x^2 + 720 = 1161$	$x^2 + 720 = 1161$
$-720 \quad -720$	$-720 \quad -720$
$x^2 = 441$	$x^2 = 441$
$\sqrt{x^2} = \pm\sqrt{441}$	$\sqrt{x^2} = \pm\sqrt{441}$
$x = \pm 21$	$x = \pm 21$
The sides of the square are ± 21 m.	The sides of the square are 21 m.

*4. Multi-Step Dominic wants to fence the perimeter of his property. The property is in
(102) the shape of a square. The area of the yard is 12,600 ft², and the area of the house is 1800 ft².

a. Write an equation to find the length of the sides of the property.

b. Solve the equation.

c. How many feet of fencing will Dominic need?

*5. (Banking) Serena places $1000 in an interest-earning account where the interest
(102) compounds annually. After two years, there is $1123.60 in the account. Use the formula $1000(1 + r)^2 = $1123.60 to find the interest rate of the account.

*6. Verify True or False: If $8x^2 - 72 = 0$, then $x = \pm 3$. If the answer is false, provide the
(102) correct answer.

*7. Estimate Find the length of the side of a square with an area of 680 square
(102) kilometers. Round to the nearest thousandth.

8. Solve and graph the inequality $\frac{|x|}{3} + 6 < 13$.
(101)

*9. Coordinate Geometry One side of a rectangle drawn in the coordinate plane has
(101) points whose y-coordinates are 7 and whose x-coordinates are the solutions of the inequality $|x + 1| - 8 \leq -4$. Another side has points whose x-coordinates are -5 and whose y-coordinates are solutions of the inequality $|y - 4| + 6 \leq 9$.

a. Solve the inequality $|x + 1| - 8 \leq -4$.

b. Solve the inequality $|y - 4| + 6 \leq 9$.

c. What are the coordinates of the four vertices of the rectangle?

10. Find the product of $(x - 7)(-7x^2 - x + 7)$ using the vertical method.
(58)

© Saxon. All rights reserved.   Saxon Algebra 1

Solving Quadratic Equations Using Square Roots

LESSON 102

11. Graph the function $y = 4x^2 + 6$.
(96)

12. (Water Balloons) A water balloon is dropped from a third-story window. Its height in feet
(96) is represented by $h = -16t^2 + 30$. How high is the balloon after 1 second?

*13. Multi-Step When the temperature t of the gas argon is within 1.65 degrees of
(101) $-187.65°C$, it will be in a liquid form. This can be modeled by the absolute-value inequality $|t - (-187.65)| < 1.65$.
  a. Solve and graph the inequality $|t - (-187.65)| < 1.65$.
  b. One endpoint of the graph represents the boiling point of argon—the temperature at which argon changes from liquid to gas. The other endpoint represents the melting point—the temperature at which argon turns from solid to liquid. The higher temperature is the boiling point, and the lower temperature is the melting point. What is the boiling point of argon? What is the melting point?

14. Factor.
(Inv 9)
  a. $x^2 + 10x + 25$
  b. $x^2 - 25$

15. Measurement The sides of a triangle are labeled $5x$ inches, $4y$ inches, and
(97) 20 inches. Jonas wrote an inequality that satisfies the Triangle Inequality Theorem: $5x + 4y > 20$. Graph the inequality.

16. (Art Supplies) Tim plans to go shopping for new paper and paint for his students,
(97) and he does not want to spend more than $40. Each pack of paper costs $2, and each set of paints costs $10. Write an inequality that describes this situation, and graph it on a graphing calculator.

17. Solve $x(x + 12) = 0$.
(98)

18. Verify Show that $x = 1$ is an extraneous solution to $\frac{1}{x-1} = \frac{3}{2x-2}$.
(99)

19. Multiple Choice Solve $\frac{2}{x-3} = \frac{x}{9}$.
(99)
  A $\{3, 6\}$ **B** $\{-3, 6\}$
  C $\{3, -6\}$ **D** $\{6\}$

20. Add $\frac{m}{m^2 - 4} + \frac{2}{3m + 6}$.
(95)

*21. Error Analysis Students were asked to write a quadratic equation that had no solution.
(100) Which student is correct? Explain the error.

Student A	Student B
$f(x) = x^2 - 3x + 12$	$f(x) = x^2 + 11x + 11$

Solving Quadratic Equations Using Square Roots

LESSON 102

*22. **(Rocket)** Malachi shot a rocket for his science project. The path of the rocket's
(100) movement formed a parabola given by the quadratic equation $h = -16t^2 + 4t + 10$,
where h is the height in feet and t is the time in seconds. Find the maximum height of
the path the rocket makes and the time t when the rocket hits the ground. Round to
the nearest hundredth.

*23. Solve $x^2 + 12x + 40 = 0$ by graphing.
(100)

24. Find the midpoint of the line segment with the endpoints $(13, -3)$ and $(-7, -3)$.
(86)

25. Factor $-3y^3 - 9yz + 5y^2 + 15z$.
(87)

26. **Multi-Step** Mr. Tranh's lawn has an area of 144 square feet. The length of his yard is
(89) 7 feet more than the width. What are the dimensions of his yard?
   a. Write a formula to find the dimensions of the yard and describe how you will
      solve it.
   b. What are the dimensions of the yard?

27. Do the side lengths 3, 7, and 8 form a Pythagorean triple?
(85)

28. **(Running)** It took Wayne $\frac{x}{2x^2 + x - 15}$ minutes to run to the gym that was
(92) $\frac{9x}{4x - 10} + \frac{5x^2}{3x + 9}$ miles away. Find his rate in miles per minute.

29. **Multi-Step** Raj is measuring the area of his rectangular living room. He determined
(93) that the area is $-64x + x^3 - 2x^2 + 128$ square feet. The width is $\frac{(x^2 - 64)}{(x + 8)}$ feet.
   a. Simplify the expression for the width of the living room.
   b. Find the expression for the length.

30. **Generalize** What is the difference between solving an absolute-value equation with
(94) operations on the outside and solving absolute-value equations with operations
on the inside?

Name _____ Date _____ Class _____

Dividing Radical Expressions
LESSON 103

***1.** Simplify $\dfrac{35}{\sqrt{7}}$.
(103)

2. Solve $\dfrac{8}{x-1} = \dfrac{x}{7}$.
(99)

***3. Error Analysis** Two students simplified the following expression. Which student is correct? Explain the error.
(103)

Student A	Student B
$\dfrac{1}{3+\sqrt{2}}$	$\dfrac{1}{3+\sqrt{2}}$
$\dfrac{1}{3+\sqrt{2}} \cdot \dfrac{3-\sqrt{2}}{3-\sqrt{2}}$	$\dfrac{1}{3+\sqrt{2}} \cdot \dfrac{\sqrt{2}}{\sqrt{2}}$
$\dfrac{3-\sqrt{2}}{7}$	$\dfrac{\sqrt{2}}{3\sqrt{2}+2}$

***4. Skydiving** A 150-pound skydiver reaches terminal velocity after free-falling for a number of seconds. The formula for the terminal velocity V of a skydiver (in feet per second) can be estimated by the formula $V = \sqrt{\dfrac{2W}{0.0063}}$, where W equals the weight of the skydiver in pounds. Write a rational expression for the terminal velocity of the skydiver.
(103)

5. What is 400% of 40? Use a proportion to solve.
(42)

***6. Write** Is $\dfrac{2\sqrt{3}}{\sqrt{2}}$ in simplest form? Explain.
(103)

***7. Predict** If $2 \div \sqrt{2}$ is $\sqrt{2}$, and $3 \div \sqrt{3}$ is $\sqrt{3}$, what is a good prediction of what the quotient of $239 \div \sqrt{239}$ might be?
(103)

8. Multi-Step The area of a square is $9x^2$. The length of one of its sides plus 32 is 47.
(102)
a. What is the length of one of its sides?
b. What is the area of the square?
c. What is x?

***9. Time and Distance** A stone is dropped from a height of 450 feet. Use the equation $25t^2 - 450 = 0$ to find how many seconds it takes for the stone to hit the ground.
(102)

10. Solve $\begin{matrix} 4x + 2y = 22 \\ 6x - 5y = 9 \end{matrix}$ by substitution.
(59)

*11. Verify True or False: $5x^2 + 125 = 0$; $x = \pm 5$. If the answer is false, provide the correct answer.
(102)

© Saxon. All rights reserved.   Saxon Algebra 1

Dividing Radical Expreassions

LESSON 103

Solve and graph the inequality.

12. $-6|x| + 20 \geq 2$
(101)

13. $14x + 2y > 6$
(97)

*14. Error Analysis Two students solve the inequality $-12|x| - 15 > -39$. Which student is
(101) correct? Explain the error.

Student A	Student B
$-12\|x\| - 15 > -39$	$-12\|x\| - 15 > -39$
$-12\|x\| > -24$	$12\|x\| + 15 < 39$
$\|x\| > 2$	$12\|x\| < 24$
$x < -2$ OR $x > 2$	$\|x\| < 2$
	$-2 < x < 2$

*15. Geometry Find the length of the line segment that is the graph of the
(101) inequality $\left|\frac{x}{7} + 6\right| - 5 \leq 4$.

*16. (Tennis) The diameter d of a tennis ball should vary no more than $\frac{1}{16}$ inch
(101) from $2\frac{5}{16}$ inches. Write and solve an absolute-value inequality that models
the acceptable diameters. What is the greatest acceptable diameter?

17. Graph the function $y = 10x^2 - 20$.
(96)

18. (Soccer) The height h in meters of a kicked soccer ball is represented by the function
(98) $h = -5t^2 + 20t$, where t stands for the number of seconds after the ball is kicked.
When is the ball on the ground?

19. Data Analysis A teacher graphed the test grades. He found that the distribution formed
(98) a parabola. Solve the equation $0 = x^2 - 170x + 7000$ to find its roots.

20. Write What are the other names for the x-intercepts of a function?
(100)

21. Multiple Choice What is the equation of the parabola that passes through the points
(100) $(0, 2)$, $(-2, 6)$, and $(6, 14)$?

A $y = x^2 - x + 2$ **B** $y = -\frac{1}{2}x^2 - x + 2$

C $y = \frac{1}{2}x^2 + x - 2$ **D** $y = \frac{1}{2}x^2 - x + 2$

22. Factor $2x^2y + 4xy - 7xyz - 14yz$.
(87)

23. Multiply $\frac{90}{24a} \cdot \frac{6a^2b^2}{25b}$.
(88)

24. Do the side lengths 15, 36, and 39 form a Pythagorean triple?
(85)

Dividing Radical Expressions

LESSON 103

25. Multi-Step A plane left the airport and traveled west, with a tailwind, at a cruising speed of 230 miles per hour for 300 miles. After dropping passengers off, the plane traveled east at the same cruising speed, but into a headwind, for 220 miles before landing for fuel.
   a. Justify Write an expression for the time going west and another expression for the time going east. Explain how you came up with these expressions.
   b. Add the expressions and simplify.
   c. What does the simplified expression represent?

26. Gardening Jasmine has a rectangular garden with an area of $(x^2 - 14x + 45)$ square feet and a length of $(x - 5)$ feet. What is the width of her garden?

27. Multi-Step A bike rental company charges $8 for each bike rental plus $10 for each hour it is rented. A couple has budgeted $66 for both of them to rent bikes. They hope that the total cost is within $10 of their budget.
   a. Write an absolute-value equation for the minimum and maximum number of hours the couple can ride bikes.
   b. What is the minimum and maximum number of hours the couple can ride bikes?

28. Write How do you find $\frac{4y - 5}{6}$ as the difference of two rational expressions?

29. Can $x^2 + x + 1$ be factored? Explain.

30. If a quadratic function has been vertically stretched, does that mean the parabola is wider or narrower than the parent quadratic function, $f(x) = x^2$?

Solving Quadratic Equations by Completing the Square

LESSON 104

*1. Find the missing term of the perfect-square trinomial: $c^2 + 100c + \underline{}$.
(104)

*2. Find the missing term of the perfect-square trinomial: $y^2 - 26y + \underline{}$.
(104)

*3. Multiple Choice What is the missing value for the perfect-square trinomial?
(104)
$$x^2 - 30x + \underline{}$$

  A -225 **B** -15 **C** 15 **D** 225

*4. Justify Solve $6x^2 - 12x - 18 = 0$ by completing the square. Justify each step in your solution. Then check the answer(s).
(104)

*5. (Design) The diagram shows the cutout for an open box. The height of the box is 3 inches. The length is 5 inches greater than the width. The area of the base of the box is 24 square inches. What are the dimensions of the box?
(104)

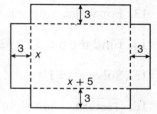

6. Simplify $\dfrac{\sqrt{3}}{\sqrt{11}}$.
(103)

7. A deck has 6 green cards and 2 yellow cards in it. What is the probability of drawing a green card, keeping it, and then drawing a yellow card?
(33)

*8. Multi-Step A circle has an area of 6 square meters. Find the radius of the circle. Use $\frac{22}{7}$ for π. (Hint: Area of a circle = πr^2)
(103)
  a. Write the formula for finding the radius of the circle.
  b. Write the equation for finding the radius after substituting in $\frac{22}{7}$ for π.
  c. What is the radius of the circle?

*9. Coordinate Geometry A right triangle is plotted at points $A\left(\dfrac{\sqrt{5}}{3}, \dfrac{3\sqrt{3}}{4}\right)$, $B\left(\dfrac{\sqrt{5}}{3}, \dfrac{\sqrt{3}}{4}\right)$, and $C\left(\dfrac{2\sqrt{5}}{3}, \dfrac{\sqrt{3}}{4}\right)$, and line segment AC forms the hypotenuse of the triangle. What is the length of the hypotenuse of triangle ABC?
(86)

10. (Printing) A photographer has printing paper that is 8 inches by 10 inches with a half-inch margin on the left and right side and a one-inch margin on the top and bottom. He can print out six square images. What are the dimensions of the image?
(102)

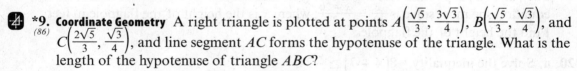

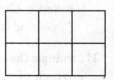

*11. Geometry The volume of a cylindrical container is 339.12 cubic meters. The formula representing the volume of the container is $(18\pi)r^2 = 339.12$. Find r, the radius of the container. Use 3.14 for π.
(102)

Solving Quadratic Equations by Completing the Square

LESSON 104

12. Error Analysis Two students were asked to find the LCD of
(95) $\dfrac{4x^2}{x^2 + 15x + 56} + \dfrac{7x + 1}{-3x + 21}$. Which student is correct? Explain the error.

Student A	Student B
$\dfrac{4x^2}{x^2 + 15x + 56} + \dfrac{7x + 1}{-3x + 21}$	$\dfrac{4x^2}{x^2 + 15x + 56} + \dfrac{7x + 1}{-3x + 21}$
$\dfrac{4x^2}{(x + 7)(x + 8)} + \dfrac{7x + 1}{-3(x - 7)}$	$\dfrac{4x^2}{(x + 7)(x + 8)} + \dfrac{7x + 1}{-3(x + 7)}$
LCD $= -3(x - 7)(x + 7)(x + 8)$	LCD $= -3(x + 7)(x + 8)$

13. For which values is the rational expression $\dfrac{x - 6}{x}$ undefined?
(43)

14. Find the roots of $32x - 3x = 24 - 4x^2$.
(98)

15. Solve $x^2 = 100$.
(102)

16. (**Football**) The height of a punted ball at time t is represented by the function
(98) $-32t^2 + 12t + 2 = h$, where t stands for the number of seconds after the ball is kicked. When does the ball land on the ground?

17. (**Masonry**) Pedro can build a brick fence in 10 hours. His partner can build the
(99) same brick fence in 12 hours. How long would it take them to do the masonry work together?

18. Solve $x^2 + 81 = 18x$ by graphing.
(100)

19. Measurement A student uses indirect measurement to find the height of a flagpole.
(99) She writes a proportion relating the heights and lengths of the shadows. The equation she must solve is $\dfrac{x}{10} = \dfrac{x - 20}{2}$; where x is the height of the flagpole in feet. Find the height of the flagpole.

*20. a. Solve the inequality $-8|x + 7| \geq -24$.
(101)
  b. Verify Choose two x-values in the solution set you found in part **a**. Verify that each x-value satisfies the original inequality.

*21. Multiple Choice Suppose a number n is a solution of the inequality $|5x - 2| < 9$.
(101) Which of the following inequalities does not have n as a solution?
  A $5x - 2 > -9$ **B** $5x - 2 > 9$ **C** $-9 < 5x - 2$ **D** $5x - 2 < 9$

Solving Quadratic Equations by Completing the Square

LESSON 104

22. Find the zeros of the function shown.
(89)

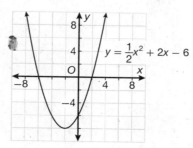

23. Find the quotient of $\dfrac{49x^2 + 21xy}{5x^2} \div \dfrac{14x}{25xy^2}$.
(88)

24. Find the product $(6y - 3)(6y + 3)$.
(60)

25. Factor $4x^4 - 64$ completely.
(Inv 9)

26. Multi-Step When budgeting to purchase a new car, a student is willing to spend $3000 plus or minus $200.
(91)

    a. Write an absolute-value inequality to show the range of prices the student is willing to consider.

    b. Solve and find the range of the actual price the student might pay.

27. Find the midpoint of the line segment with the endpoints $(-4, 3)$ and $(2, 4)$.
(86)

28. **Cell Phone** A student budgets $25 for his cell phone each month. He pays $10 for the service and $0.05 per minute. He knows that his budget can be off by $5 in either direction. What is the maximum and minimum number of minutes he can talk each month?
(94)

29. Justify $\dfrac{a}{x} + \dfrac{a}{b} \neq \dfrac{2a}{x + b}$
(95)

30. Generalize When does a quadratic function only have one zero?
(96)

Recognizing and Extending Geometric Sequences

LESSON 105

***1.** Find the common ratio of the geometric sequence $-80, 20, -5, 1\frac{1}{4}, \ldots$.
(105)

***2.** **Multiple Choice** Which rule can you use to find the nth term of the sequence
(105) $4, 6, 9, 13.5, \ldots$?

A $A(n) = 4(1.5)^n$ **B** $A(n) = 4(1.5)^{n-1}$

C $A(n) = 4(2)^{n-1}$ **D** $A(n) = 3(1.5)^n$

***3.** (Depreciation) Harper buys a car in 2007 for $20,000. Each year, the car decreases
(105) in value by 18%. How much will the car be worth in 2012? Round to the nearest cent.

***4.** **Write** The third term of a sequence is 0. The first two terms are not 0. Can this
(105) be a geometric sequence? Explain.

5. Write a recursive formula for the arithmetic sequence with $a_1 = \frac{1}{2}$ and common
(34) difference $d = \frac{1}{2}$. Then find the first four terms of the sequence.

***6.** **Analyze** Write two possible rules for the nth term of the geometric sequence with a
(105) first term of 5 and a third term of 605.

7. Find the missing term of the perfect-square trinomial: $x^2 + 18x + \underline{}$.
(104)

8. **Error Analysis** Dominic stated that the missing value for completing the square
(104) of $g^2 + 28g + \underline{}$ is 14. Is this correct? Explain.

***9.** **Multi-Step** A design for a rectangular flower bed is shown.
(104) The total area of the flower bed is 880 square feet.

a. Write an equation to represent the problem.

b. Write the quadratic equation in the form $x^2 + bx = c$.

c. What is the width of the interior of the flower bed?

d. What is the area of the border?

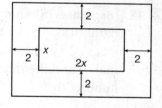

Simplify.

10. $\dfrac{11x + 22}{22x^2 + 44x}$
(43)

11. $\dfrac{\sqrt{63}}{\sqrt{18}}$
(103)

***12.** **Geometry** The base of a right triangle is 14 units longer than its height. The
(104) hypotenuse is 26 units. What are the base and height measurements of the triangle?

Recognizing and Extending Geometric Sequences

LESSON 105

*13. **Error Analysis** Two students simplified the given expression. Which student is correct? Explain the error.
(103)

Student A	Student B
$\dfrac{5}{\sqrt{8}}$	$\dfrac{5}{\sqrt{8}}$
$\dfrac{5}{2\sqrt{2}} \cdot \dfrac{1}{\sqrt{2}}$	$\dfrac{5}{2\sqrt{2}} \cdot \dfrac{\sqrt{2}}{\sqrt{2}}$
$\dfrac{5}{4}$	$\dfrac{5\sqrt{2}}{4}$

*14. (**Architecture**) In the city of Rotterdam, Netherlands, architect Piet Blom designed a group of cube-shaped houses that each sit upon its vertex. If the surface area of each cube measures $337\frac{1}{2}$ square meters, write a rational expression representing the edge length of the cube. (Hint: edge length = $\sqrt{\dfrac{A}{6}}$)
(103)

15. Find the roots: $14x^2 - 2x = 3 - 21x$.
(98)

16. (**Construction**) It takes a woman 3 hours to build a doghouse. Her husband can build it in 4 hours. How long will it take them if they build the doghouse together?
(99)

17. Use mental math to find the product of 29^2.
(60)

18. (**Horseshoes**) Shannon plays a game of horseshoes. The horseshoe's movement forms a parabola given by the quadratic equation $h = -16t^2 + 6t + 6$ where h is the height in feet and t is the time in seconds. Find the maximum height of the path the horseshoe makes and the time t when the horseshoe hits the ground. Round to the nearest hundredth.
(100)

19. **Measurement** A puddle of water creates a shape on the ground. As time increases, the area of the puddle changes. The area of the puddle is given by the function $A = 3t^2 + 8t - 70$, where A is the area in square feet and t is the time in seconds. Find the time when the area is 55 square feet. Round to the nearest hundredth.
(100)

20. Solve and graph the inequality $2|x| - 12 > -5$.
(101)

21. **Verify** True or False: $4x^2 - 64 = 0$; $x = 4$. Verify that the answer is true. If the answer is false, provide the correct answer.
(102)

Recognizing and Extending Geometric Sequences

LESSON 105

*22. Error Analysis Student A and Student B want to find the length of the sides of a square
(102) with an area 460 square meters less than 685 square meters. Which student is correct? Explain the error.

Student A	Student B
$x^2 + 460 = 685$	$x^2 + 460 = 685$
$-460 \quad -460$	$+460 \quad +460$
$x^2 = 225$	$x^2 = 1145$
$\sqrt{x^2} = \sqrt{225}$	$\sqrt{x^2} = \sqrt{1145}$
$x = \pm 15$	$x \approx \pm 3.838$
The sides of the square are 15 m.	The sides of the square are about 3.838 m.

23. Give the coordinates of the parabola's vertex. Then
(89) give the minimum or maximum value.

24. Add $\dfrac{d}{d-10} + \dfrac{10}{10-d}$.
(90)

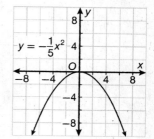

25. Solve and check: $\dfrac{18}{2x} - 4 = \dfrac{15}{3x}$.
(99)

26. Multi-Step Louis walked for $\dfrac{6x^2 - 24x}{6x}$ minutes to get to a
(92) grocery store that was $\dfrac{4x - 16}{x^3}$ miles away.

a. Find his rate in miles per minute.

b. If the rate is divided by $\frac{1}{x}$, what is the new rate?

27. (Football) The school football team is going to a camp that is $\dfrac{8x^2}{x^2 - 11x + 18}$ miles
(95) away. The team traveled $\dfrac{2}{8x - 72}$ miles on the first day. How many miles are left to travel?

28. Find the midpoint of the line segment with the endpoints $(-5, 0)$ and $(1, 14)$.
(86)

29. Multi-Step A ball is thrown into the air from the top of a cliff at an initial velocity of
(96) 32 feet per second. (Use $h = -16t^2 + vt + 0$.)

a. How high is the ball after 2 seconds?

b. What does this height represent?

c. After 3 seconds, the ball is -48 feet. What does this height represent?

30. Justify List the inequality symbols that result in graphs with dashed boundary lines
(97) and list the inequality symbols that result in graphs with solid boundary lines.

Solving Radical Equations — LESSON 106

1. Solve $x^2 = 64$.
   (102)

2. Factor $x^2 - 9x + 20$.
   (Inv 9)

3. Translate the inequality $3z + 4 < 10$ into a sentence.
   (45)

*4. **Multiple Choice** Which of the following radical equations will require the use of division to isolate the radical?
   (106)
   A $\sqrt{x} - 12 = 2$ B $\sqrt{x} + 12 = 13$

   C $\dfrac{\sqrt{x}}{7} = 5$ D $14\sqrt{x} = 70$

*5. **Verify** Solve $\sqrt{x-1} = \sqrt{3x+2}$. Check your answer.
   (106)

*6. **Justify** Solve $\dfrac{\sqrt{x}}{4} = 32$. Justify your answer.
   (106)

*7. Find the common ratio of the geometric sequence $18, -9, 4\tfrac{1}{2}, -2\tfrac{1}{4}, \ldots$
   (105)

*8. Find the 6th term in the geometric series that has a common ratio of 2 and an initial term of 5.
   (105)

9. **Multi-Step** Leila drops a ball from a height of 1 meter. The height of each bounce is 75% of the previous height.
   (105)
   a. What is the ball's height after the first bounce?
   b. What rule can be used to find the ball's height after n bounces?
   c. What is the height of the sixth bounce? Round your answer to the nearest hundredth.

10. **Geometry** Each unit square in the figure represents 5 square feet. If the pattern continues, what will the area of the ninth figure be?
    (105)

*11. (**Botany**) The growth of an ivy plant in feet can be described by $2\sqrt{x} - 4$. How many days x will it take for the ivy to reach a length of 20 feet?
    (106)

12. Solve $\begin{array}{l} -5x + 4y = -37 \\ 3x - 6y = 33 \end{array}$
    (63)

*13. (**Fractals**) Fractals are geometric patterns that repeat themselves at smaller scales. The pattern shows fractals of equilateral triangles. How many unshaded triangles will be in the sixth figure?
    (105)

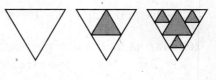

*14. Solve $x^2 + 9x = 4.75$ by completing the square.
    (104)

Solving Radical Equations

LESSON 106

15. Error Analysis Two students started solving the equation $2x^2 + 20x = -18$ as shown below. Which student is correct? Explain the error.
(104)

Student A	Student B
$2x^2 + 20x = -18$	$2x^2 + 20x = -18$
$x^2 + 10x = -9$	$x^2 + 10x = -18$
$x^2 + 10x + 25 = -9 + 25$	$x^2 + 10x + 25 = -18 + 25$
$(x + 5)^2 = 16$	$(x + 5)^2 = 7$

*16. (Business) The marketing group for a cosmetics company determined that the expression $u^2 - 0.8u$ represents the profit for every 1000 units u of mascara sold. How many units need to be sold to have a profit margin of $0.33?
(104)

Solve each equation. Check your answer.

17. $\dfrac{60}{4x} + \dfrac{45}{5x} = 3$
(99)

*18. $\sqrt{x} = 9$
(106)

19. (Egg Toss) Tyrese and Jameka were playing an egg-toss game. The egg's movement through the air formed a parabola given by the quadratic equation $h = -16t^2 + 9t + 4$, where h is the height in feet and t is the time in seconds. Find the maximum height of the path the egg makes and the time t when the egg hits the ground. Round to the nearest hundredth.
(100)

20. Solve $x^2 - 16 = 0$ by graphing.
(100)

21. (Tennis) The weight w of a tennis ball should vary no more than $\dfrac{1}{12}$ ounce from $2\dfrac{1}{12}$ ounces. Write an absolute-value inequality that models the acceptable weights. What is the least acceptable weight?
(101)

22. Multiple Choice Which is the simplest form of $\dfrac{18\sqrt{7}}{3\sqrt{28}}$?
(103)

A $\dfrac{3}{2}$ **B** 3 **C** $\dfrac{6\sqrt{7}}{\sqrt{28}}$ **D** $\dfrac{6\sqrt{7}}{7}$

23. Write Anton wants to estimate the quotient of $\dfrac{\sqrt{145}}{2\sqrt{9}}$. How should he do this?
(103)

24. Subtract $\dfrac{2r}{r-4} - \dfrac{6}{12-3r}$.
(90)

25. Solve and graph $|x - 16| \leq 12$.
(91)

26. Martha built a new playroom. She determined that the rectangular reading area is $(9x^2 + 44x - 5)$ square feet. The width is $(x + 5)$ feet. What is the length?
(93)

27. (Volleyball) A server's hand is 3 feet above the floor when it hits the volleyball. After the volleyball is hit, it has an initial velocity of 23 feet per second. What is its height after 1 second? Use $h = -16t^2 + vt + s$.
(96)

Solving Radical Equations

28. **Multi-Step** Tickets for the Valley High School production of *Romeo and Juliet* are $5
(97) for adults and $4 for students. In order to cover expenses, at least $2500 worth of tickets must be sold.

  a. Write an inequality that describes this situation.

  b. Graph the inequality.

  c. If 200 adult and 400 student tickets are sold, will the expenses be covered?

29. **Generalize** Consider the equation $(x - 5)(x + 8) = 0$. How can you quickly tell what
(98) the roots are?

30. The graph of $f(x) = x^2 + bx + 3$ has an axis of symmetry $x = 4$. What is the value
(Inv 10) of b?

Graphing Absolute-Value Functions

LESSON 107

*1. **Estimate** Without graphing the function, which direction would the function
(107) $f(x) = |x| - 6$ shift the parent function?

2. Solve $\sqrt{2x} = 14$. Check your answer.
(106)

3. Write $5y - 29 = -14x$ in standard form.
(35)

*4. **Multiple Choice** What absolute-value function is shown by
(107) the graph?
  A $f(x) = 2|x|$ **B** $f(x) = 0.5|x|$
  C $f(x) = -5|x|$ **D** $f(x) = -0.5|x|$

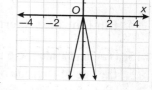

5. Translate the inequality $3b + \frac{2}{5} \geq 1\frac{3}{5}$ into a sentence.
(45)

*6. **(Boating)** The path of a sailboat is represented by the function
(107) $f(x) = \left|\frac{3}{5}x - 30\right| + 30$. At what point does the sailboat tack (turn)?

*7. **Write** Why does the graph of an absolute-value function not extend past
(107) the vertex?

8. Solve the system of linear equations: $\quad 4y = -3x - 4$
(63) $\qquad\qquad\qquad\qquad\qquad\qquad\qquad 4x + 6 = -5y$

*9. **Geometry** The perimeter of the square is 20 centimeters. Solve for x.
(106)

*10. **Error Analysis** Two students found the solution for a radical equation. Which student is
(106) correct? Explain the error.

Student A	Student B
$\sqrt{x} + 7 = 14$	$\sqrt{x} + 7 = 14$
$\sqrt{x} = 7$	$\sqrt{x} + 49 = 196$
$x = 49$	$\sqrt{x} = 147$
	$x = 21,609$

11. Jason built a new deck with an area of $(-20x + 100 + x^2)$ square feet. The width is
(93) $(x - 10)$ feet. What is the length?

Graphing Absolute-Value Functions

LESSON 107

*12. Multi-Step A triangular brace is constructed in the shape of a right triangle. The
(106) two legs of the brace are $\sqrt{x+5}$ and $\sqrt{x}$ units long.
  a. What expression could be used to solve for the length, l, of the third side of the brace?
  b. Simplify the equation so it does not contain any radicals.
  c. Find the value of x for which the length of the third side of the brace is equal to 10.

13. Coordinate Geometry Find the coordinates of the point(s) at which the graphs of
(106) $y = x$ and $y = \sqrt{x}$ intersect.

14. Find the next 3 terms of the sequence 125, 25, 5, 1.
(105)

*15. Carbon Dating Scientists can use the ratio of radioactive carbon-14 to carbon-12 to find
(105) the age of organic objects. Carbon-14 has a half-life of about 5730 years, which means that after 5730 years, half the original amount remains. Carbon dating can date objects to about 50,000 years ago, or about 9 half-lives. About what percent of the original amount of carbon-14 remains in objects about 50,000 years old?

*16. Error Analysis Two students find the 5th term in a geometric series that has a common
(105) ratio of $\frac{1}{2}$ and a first term of 6. Which student is correct? Explain the error.

Student A	Student B
$A(n) = ar^{n-1}$	$A(n) = ar^{n-1}$
$= 6 \cdot \left(\frac{1}{2}\right)^4$	$= 6 \cdot \frac{1}{2} \cdot 4$
$= \frac{3}{8}$	$= 12$

17. Solve by graphing on a graphing calculator. Round to the nearest tenth.
(100)
$$-11x^2 + x = -4$$

Solve and graph each inequality.

18. $|x - 4| + 15 \geq 21$ **19.** $|x| + 45 \leq 34$
(101)   (91)

20. Football NCAA rules require that the circumference c of a football, measured around its
(101) widest part, 21 inches, to vary by no more than 0.25 inches. Write and solve an absolute-value inequality that models the acceptable circumferences. What is the least acceptable circumference?

21. Area of a Pool Maria wants to increase the radius of a pool by
(102) 3 meters. The new area of the pool is 200.96 square meters.
  a. Write a formula to find the original radius of the pool.
  b. Solve the formula.
  c. What will the new diameter of the pool be?

412   Saxon Algebra 1

Graphing Absolute-Value Functions

LESSON 107

*22. Graph the function $f(x) = |x| + 3$.
(107)

23. Multiple Choice Solve $-3x^2 + 24x = 36$.
(104)
   A $x = -8$ or 0 **B** $x = -6$ or 2 **C** $x = 2$ or 6 **D** $x = 0$ or 8

24. Analyze Determine what values of c would make the equation $x^2 - 50x = c$ have no solution.
(104)

Simplify.

25. $\dfrac{\dfrac{4x}{2x+12} + \dfrac{x}{3x+18}}{\dfrac{8x^2}{x^2+8x+12}}$
(92)

26. $\sqrt{\dfrac{20}{3}}$
(103)

27. Multi-Step A businessman makes $50 profit on each item sold. He would like to make $950 plus or minus $100 total each week.
(94)
   a. Write an absolute-value equation for the minimum and maximum profit he desires.
   b. What is the minimum and maximum number of items he needs to sell each week?

28. (**School Dance**) Tickets for the school dance are $4 for middle school students and $6 for high school students. In order to cover expenses, at least $600 worth of tickets must be sold. Write an inequality that models this situation and graph it.
(97)

29. Multi-Step A painting is 5 inches by 4 inches. The frame around it is x inches wide.
(98)
   a. Write expressions for the length and width of the picture with the frame.
   b. The total area of the picture and frame is 42 square inches. What is the width of the frame?

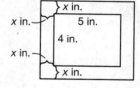

30. Justify Explain how to transform $\dfrac{x}{x-3} = \dfrac{4}{x}$ to $x^2 = 4x - 12$.
(99)

Name _____ Date _____ Class _____

Identifying and Graphing Exponential Functions

LESSON 108

***1.** Evaluate the function $f(x) = 2(5)^x$ for $x = -2, 0,$ and 2.
(108)

2. Graph the function $f(x) = |x - 2|$.
(107)

***3. Justify** Why is $f(x) = 4(1)^x$ not an exponential function?
(108)

***4. Multiple Choice** Which could be the function graphed?
(108)

A $y = -\left(\dfrac{1}{2}\right)^x$ **B** $y = \left(\dfrac{1}{2}\right)^x$

C $y = -(2)^x$ **D** $y = 2^x$

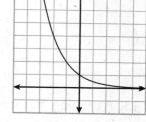

***5.** (**Population**) The exponential function $y = 20.85(1.0212)^x$
(108) can model the approximate population of Texas from 2000 to 2006, where x is the number of years after 2000 and y represents millions of people. Assuming the model does not change, what is the difference in expected populations for 2010 and 2020?

***6. Verify** Show that the set $\{(3, -4), (2, -1), (5, -64), (4, -16)\}$ is an exponential
(108) function when $b = 4$.

7. Name the corresponding sides and angles if $\triangle RST \sim \triangle NVQ$.
(36)

***8. Multi-Step** Graph the parent function $f(x) = |x|$. Translate the function down
(107) by 2. Then reflect the function across the x-axis. What is the new function?

9. Is the graph an absolute-value function? Explain.
(107)

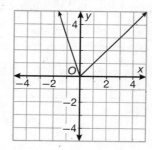

10. Evaluate $\sqrt[3]{x}$ when $x = (-4)^3$.
(46)

***11. Geometry** Describe why the function $f(x) = |x|$ is in the shape of a "V".
(107)

12. Error Analysis Two students found the solution to a radical equation. Which student is
(106) correct? Explain the error.

Student A	Student B
$\sqrt{x + 3} = 6$	$\sqrt{x + 3} = 6$
$x + 9 = 36$	$x + 3 = 36$
$x = 27$	$x = 33$

Saxon Algebra 1

Identifying and Graphing Exponential Functions

LESSON 108

13. Solve $\sqrt{x} - 2 = 8$. Check your answer.
(106)

*14. Write the equation of the function graphed.
(107)

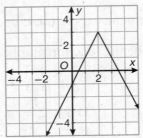

15. Assuming that y varies inversely as x, what is y when $x = 8$, if $y = 55$ when
(64) $x = 11.6$? $y = 79.75$

16. (**Meteorology**) In the mountains snow will accumulate quickly in winter. If the average
(106) accumulation can be described using the expression $12\sqrt{x}$, find the value of x when the accumulation is equal to 108 inches?

17. Solve and graph the inequality $\frac{|x|}{8} - 10 < -9$.
(101)

Solve.

18. $x^2 = -9$
(102)

19. $12|x + 9| - 11 = 1$
(94)

20. (**Building**) Tom's house has two square rooms. He knocks down a wall separating
(102) the rooms. The area of the new room is 338 square feet. What were the dimensions of the original rooms?

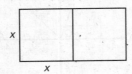

21. Simplify $\dfrac{\frac{24a^2b}{7c^2}}{\frac{8ab^2}{49c^2}}$.
(92)

22. Find the missing term of the perfect-square trinomial: $x^2 + 7x + \underline{}$.
(104)

*23. Multiple Choice What is the common ratio of the geometric sequence $-\frac{5}{8}, -\frac{5}{16}$,
(105) $-\frac{5}{32}, -\frac{5}{64}, \ldots$?

A -2 **B** $-\frac{1}{2}$ **C** $\frac{1}{2}$ **D** 2

Identifying and Graphing Exponential Functions

LESSON 108

*24. **(Landscaping)** Li is designing a triangular flower bed in one corner of her rectangular
₍₁₀₃₎ yard. She plans on making one leg of the triangle $1\frac{11}{12}$ meters long and the other leg
$2\frac{5}{12}$ meters long. She wants to know how much edging material she needs to buy to
place along the hypotenuse of the triangle. Write a rational expression to show how
much material Li needs to buy.

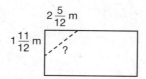

25. Analyze Is the sequence $-72, -57.6, 46.08, 36.864, \ldots$ geometric? Explain.
₍₁₀₅₎

26. Find the quotient of $(36x + 12x^2 + 15) \div (2x + 1)$.
₍₉₃₎

27. Multi-Step Amber drove $\frac{7x^2}{x^2 - 49}$ miles on Monday and $\frac{x-1}{4x + 28}$ miles on Tuesday while
₍₉₅₎ delivering pizzas.
   a. What is the total distance she drove?
   b. If her rate was $\frac{7}{7x + 49}$ miles per hour, how much time did it take her to deliver
   pizzas on Monday and Tuesday?

28. (Construction) A box needs to be built so that its rectangular top has a length that is
₍₉₈₎ 3 more inches than the width, and so that its area is 88 square inches. Find the
length and the width.

29. Multi-Step Sherry can enter all weekly data into the computer in 16 hours. When she
₍₉₉₎ works with Kim, they complete the data entry in 9 hours 36 minutes.
   a. Convert 9 hours 36 minutes to hours.
   b. Write an equation to find how long it would take Kim to enter the same data.
   c. How long would it take Kim to enter the data alone?

30. Analyze If the y-coordinate of the ordered pair represents the maximum height of the
₍₁₀₀₎ path of a ball thrown into the air, what does the x-coordinate represent?

Name _____ Date _____ Class _____

Graphing Systems of Linear Inequalities

LESSON 109

*1. Multiple Choice Which system is represented in the graph?
(109)

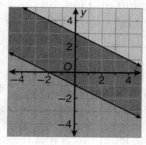

A $y \leq -0.5x + 3$
   $y \geq -0.5x - 1$

B $y \leq -0.5x + 3$
   $y \leq -0.5x - 1$

C $y \geq -0.5x + 3$
   $y \geq -0.5x - 1$

D $y \geq -0.5x + 3$
   $y \leq -0.5x - 1$

*2. (**Sports**) The requirements for a major league baseball are shown in
(109) the graph. Write the system of inequalities that matches the graph.

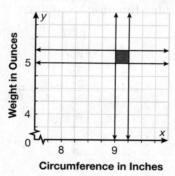

3. Graph the function $f(x) = -3|x|$.
(107)

*4. Write Explain how to represent the solution set of $\begin{array}{l} y \leq -3x + 4 \\ y < 2x - 1 \end{array}$.
(109)

*5. Verify Graph the solution set of $\begin{array}{l} y \geq -x \\ y \leq 2x \end{array}$ to verify that $(1, -2)$ is not a
(109) solution of the system.

*6. Evaluate the function $f(x) = 3\left(\dfrac{1}{3}\right)^x$ for $x = -2, 0,$ and 2.
(108)

7. If the original price was increased 44% to a new price of $900, what was the
(47) original price?

Simplify.

8. $10\sqrt{8x^2y^3} - 5y\sqrt{98x^2y}$
(69)

9. $\sqrt{\dfrac{24y^8}{6x^3}}$
(103)

Saxon Algebra 1

Graphing Systems of Linear Inequalities

LESSON 109

10. Error Analysis Student A said that the following set satisfies an exponential function
(108) because there is a common ratio of 3 among the y-values. Student B said that this is not so. Which student is correct? Explain the error.
$$\{(3, 1), (5, 3), (6, 9), (7, 27)\}$$

*11. Multi-Step Niall has a baseball card whose value, in dollars, x years after he acquired
(108) it, is represented by the function $f(x) = 4.8(1.25)^x$. If Niall bought the card in the year 2000, how much more is it worth in 2010 than it was in 2005?

*12. Geometry Mr. Flores gives the length of a rectangle, in inches, as $f(x) = 16\left(\frac{1}{2}\right)^x$, where
(108) x is the number of times he cuts the length in half. What is the length of the rectangle after Mr. Flores has cut it in half 4 times? 6 times? 0 times?

*13. Probability For the function $f(x) = 7(5)^x$, what is the probability that for a randomly
(108) chosen x-value from the domain of $\{0, 1, 2, 3, 4, 5\}$, $f(x)$ is a number between 100 and 1000?

14. Is the graph an absolute-value function? Explain.
(107)

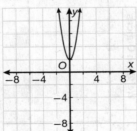

*15. Graph the system $\begin{array}{l} y > \frac{1}{4}x + 3 \\ y > -\frac{1}{4}x + 3 \end{array}$.
(109)

16. Baseball An outfielder catches a ball 120 feet from the pitcher's mound and throws it
(107) to home. If $d = |90t - 120|$ represents the ball's distance from the pitcher's mound, how would the graph change if the outfielder caught the ball 100 feet from the pitcher's mound?

17. Renovations Nadia is using 48 tiles to cover a floor. The tiles come in 6-inch,
(102) 12-inch, and 13-inch sizes. If the total area of the floor is 6912 square inches, which tile size will fit best?

18. Projectile Motion The equation for the time in seconds (t) it takes an object to strike
(104) the ground is $-4.9t^2 - 53.9t = -127.4$. When will the object strike the ground?

19. Find the next 3 terms of the sequence 5, 4.5, 4.05, 3.645, ...
(105)

Graphing Systems of Linear Inequalities

LESSON 109

*20. **Multiple Choice** Which of the following radical equations has no solution?
(106)
 A $\sqrt{x-3} = x - 9$ **C** $\sqrt{x} + 7 = -2$
 B $13\sqrt{x} = 65$ **D** $\sqrt{x+10} = \sqrt{2x+8}$

21. **Write** Why is it important to isolate the radical in a radical equation?
(106)

22. Jim's rectangular home gym has an area of $(x^2 - 144)$ square feet. The length is $(x - 12)$ feet. What is the width?
(93)

Solve.

23. $4|x + 2| - 9 = 19$ **24.** $x^2 = -49$ **25.** $2\left|\dfrac{x}{4} - 6\right| = 8$
(94)   (102)   (94)

26. **Multi-Step** A pitcher throws a softball. The height in feet is represented by the function $h = -16t^2 + 47t + 5$.
(96)
 a. How high is the ball after 1 second?
 b. How high is the ball when it is released?
 c. What is the initial velocity of the ball?

27. (**Gardening**) It takes a boy 2 hours to pull all the weeds in the garden. It takes his sister 4 hours. How long will it take them if they pull weeds together?
(99)

28. **Multi-Step** Andrew hits a golf ball into the air. Its movement forms a parabola given by the quadratic equation $h = -16t^2 + 31t + 7$, where h is the height in feet and t is the time in seconds.
(100)
 a. Find the time t when the ball is at its maximum height. Round to the nearest hundredth.
 b. Find the time t when the ball hits the ground. Round to the nearest hundredth.
 c. Find the maximum height of the arc the ball makes in its flight. Round to the nearest hundredth.

29. **Write** Describe the similarities and differences between solving the inequality $2|x| + 1 < 7$ and solving the inequality $|2x + 1| < 7$.
(101)

30. If the area of a rectangle is represented by the expression $3x^2 + 22x - 45$ and the width by the expression $(x + 9)$, what would the length be?
(Inv 9)

Name _____ Date _____ Class _____

Using the Quadratic Formula — LESSON 110

Use the quadratic formula to solve for x. Check the solutions.

*1. $x^2 - 2x - 35 = 0$ *2. $x^2 - 10x + 25 = 0$
(110)   (110)

*3. **Multi-Step** Determine why $16h^2 + 25 = 40h$ has only 1 solution using the quadratic
(110) formula.
   a. Rearrange the equation into the $ax^2 + bx + c = 0$ form.
   b. What is different about $b^2 - 4ac$?
   c. Generalize When will the equation $ax^2 + bx + c = 0$ have only 1 solution?

4. Compare: $12{,}000 \bigcirc 1.2 \times 10^3$.
(37)

5. Find the zeros of the function. $y = x^2 + 12x + 36$
(96)

6. Describe the graph of an indirect variation when the constant of variation is positive.
(Inv 7)

7. Identify the outlier or outliers in the data set.
(48)
   number of cars for sixteen households: 3, 2, 2, 1, 2, 3, 6, 2, 1, 1, 3, 2, 2, 2, 1, 3

*8. **Predict** Use mental math to predict whether the quadratic formula is necessary to
(110) solve $3b^2 + 15b - 20 = 0$. Solve.

*9. (**Soccer**) A 1.5-meter-tall soccer player bounces a soccer ball off his head at a velocity
(110) of 7 meters per second upward. Use the formula $h = -4.9t^2 + v_0 t + h_0$ to estimate how
many seconds it will take the ball to hit the ground.

*10. **Error Analysis** For the system of inequalities graphed, Student A said that
(109) $(1, -4)$ is a solution of the system and Student B said that $(4, 2)$ is a solution
of the system. Which student is correct? Explain the error.

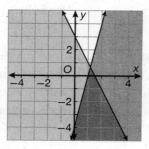

11. Graph the system $\begin{array}{l} y \leq 2 \\ x \geq 2 \end{array}$.
(109)

*12. **Multi-Step** A student group is planning on washing cars in an effort to raise at least $300.
(109) They want to charge $5 for a basic wash, which will take about 10 minutes, and $15 for
a detailed wash, which will take about 30 minutes. They have the car-wash lot rented
for 8 hours. Write and graph a system of linear inequalities to describe this situation.
Explain your findings.

Using the Quadratic Formula

LESSON 110

13. Geometry Suppose the perimeter of a rectangle must be less than 50 units and the width must be greater than 5 units. Graph a system of linear inequalities to describe this situation. Give one set of possible dimensions for the rectangle.
(109)

14. Evaluate the function $f(x) = -3(6)^x$ for $x = -2, 0,$ and 2.
(108)

15. Error Analysis Which student correctly evaluated $f(x) = 2(3)^x$ for $x = 2$? Explain the error.
(108)

Student A	Student B
$f(x) = 2(3)^x$	$f(x) = 2(3)^x$
$= 6^x$	$= 2(3)^2$
$= 6^2 = 36$	$= 2(9) = 18$

*16. Chemistry Amaro uses $f(x) = 10\left(\frac{1}{2}\right)^x$ to give the amount remaining from 10 grams of a radioactive substance after x number of half-lives. Which graph represents this function?
(108)

Graph A
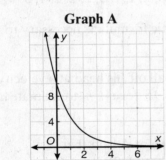

Graph B
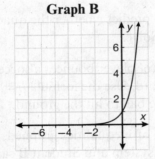

Graph C
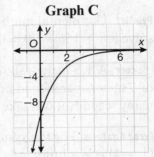

17. Simplify $\dfrac{\sqrt{15xy}}{3\sqrt{10xy^3}}$.
(103)

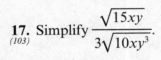

18. Subtract $\dfrac{5x^2}{10x - 30} - \dfrac{2x - 5}{x^2 - 9}$.
(95)

19. Astronomy Astronomers can use the formula $T = \sqrt{d^3}$ to find the time T it takes a planet to orbit the Sun (in earth years), knowing the distance d of the planet from the Sun (in astronomical units, AU). If Mars is about $\frac{3}{2}$ AU from the Sun, about how long does it take Mars to orbit the Sun in earth years? Give your answer as a rational expression.
(103)

Using the Quadratic Formula

LESSON 110

20. **Multiple Choice** What is the absolute-value function of the graph?
(107)

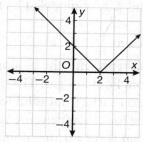

A $f(x) = |x + 2|$
B $f(x) = |x - 2|$
C $f(x) = |x| + 2$
D $f(x) = |x| - 2$

21. Solve $p^2 + 13p = -50$ by completing the square.
(104)

*22. (**Compound Interest**) The formula for a fund that compounds interest is
(105)
$A_n = P\left(1 + \frac{r}{n}\right)^{nt}$, where A is the balance, P is the initial amount deposited, r is the annual interest rate, t is the number of years, and n is the number of times the interest is compounded per year. Gretchen deposits $1500 into an account that pays 4.5% interest compounded annually. Write the first 4 terms of the sequence representing Gretchen's balance after t years. Round to the nearest cent.

23. Solve $\sqrt{x + 11} = 16$. Check your answer.
(106)

*24. **Analyze** Are the graphs for $f(x) = 5|x|$ and $f(x) = |5x|$ the same? Explain.
(107)

25. Solve the equation $9\left|\frac{x}{2} - 6\right| = 27$.   **26.** Factor $x^2 + 42 + 13x$.
(94)   (72)

27. **Multi-Step** Lisa plans to shop for books and magazines and she plans to spend no
(97) more than $32. Each book costs $14 and each magazine costs $4.
  a. Write an inequality that describes this situation.
  b. Graph the inequality.
  c. If Lisa wants to spend exactly $32, what is a possible number of each she can spend her money on?

28. (**Volleyball**) Diego hits a volleyball into the air. The ball's movement forms a
(100) parabola given by the quadratic equation $h = -16t^2 + 3t + 14$ where h is the height in feet and t is the time in seconds. Find the maximum height of the path the volleyball makes and the time when the volleyball hits the ground. Round to the nearest hundredth.

Using the Quadratic Formula

LESSON 110

29. Multi-Step When the temperature (t) of the gas neon is within 1.25° of −247.35°C
(101) it will be in a liquid form. This can be modeled by the absolute-value inequality
$|t - (-247.35)| < 1.25$.
  a. Solve and graph the inequality $|t - (-247.35)| < 1.25$.
  b. One endpoint of the graph represents the boiling point of neon, the temperature at which neon changes from liquid to gas. The other endpoint represents the melting point, at which neon turns from solid to liquid. The higher temperature is the boiling point and the lower temperature is the melting point. What is the boiling point of neon? What is the melting point?

30. Measurement The following formula represents the area of circle A:
(102) $\pi r^2 - 165.05 \text{ m}^2 = 0$. What is the approximate measurement, in meters, of the radius r? Use 3.14 for π.

Solving Problems Involving Permutations

LESSON 111

*1. Draw a tree diagram to represent the possible outcomes of flipping a coin three times.
(111)

*2. **Multiple Choice** Evaluate 10!.
(111)
A 3,628,800   B 362,880   C 55   D 9

*3. (**Video Rental**) For movie night, you want to rent one drama, one comedy, and one science fiction movie. The video store has 5 new releases for drama, 6 new releases for comedy, and 3 new releases for science fiction. How many possible movie combinations are there?
(111)

4. Simplify the rational expression $\frac{3d}{2x^3} - \frac{5d}{2x^3}$ if possible.
(51)

Find the zeros of each function.

5. $y = x^2 - 8x + 16$
(96)

6. $y = 3x^2 + 36x - 39$
(96)

*7. **Model** Draw a tree diagram to determine the number of possible outcomes of earning an A, B, or C in history, English, and math classes.
(111)

*8. **Justify** Explain how to find the number of outfits possible if you have 5 shirts and 4 pairs of pants to choose from.
(111)

*9. Use the quadratic formula to solve $c^2 + 16c - 36 = 0$. Check the solutions.
(110)

*10. **Estimate** Find the best whole number estimate for the solutions to $70 - 52x = -x^2$.
(110)

11. Find and correct the error the student made in graphing $\begin{array}{l} y - 2.5 > 0 \\ y - 4 < -2x \end{array}$.
(109)

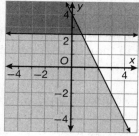

12. Graph the system $\begin{array}{l} 4x - 2y < 6 \\ y + 1 \geq 2x \end{array}$.
(109)

13. Solve $r^2 - 24r = -144$ by completing the square.
(104)

Solving Problems Involving Permutations

LESSON 111

*14. Error Analysis Two students used the quadratic formula to solve a quadratic equation. Which student is correct? Explain the error.
(110)

Student A	Student B
$8a = -10a^2 + 1$	$8a = -10a^2 + 1$
$8a - 10a^2 + 1 = 0$	$10a^2 + 8a - 1 = 0$
$x = \dfrac{-(-10) \pm \sqrt{-10^2 - 4(8)(1)}}{2(8)}$	$x = \dfrac{-8 \pm \sqrt{8^2 - 4(10)(-1)}}{2(10)}$
$x = \dfrac{10 \pm \sqrt{100 + 32}}{16}$	$x = \dfrac{-8 \pm \sqrt{64 + 40}}{20}$
$x = \dfrac{10 \pm \sqrt{132}}{16}$	$x = \dfrac{-8 \pm \sqrt{104}}{20}$
$x = \dfrac{10 \pm 2\sqrt{33}}{16}$	$x = \dfrac{-8 \pm 2\sqrt{26}}{20}$
$x = \dfrac{5 \pm \sqrt{33}}{8}$	$x = \dfrac{-4 \pm \sqrt{26}}{10}$

*15. Space Shuttle The external tank of the space shuttle separates after 8.5 minutes at
(110) a velocity of 28,067 kilometers per hour. Can the formula $-4.9t^2 + v_0 t + y_0 = 0$ be used to find the distance above earth? Explain.

16. Measurement The length of a piece of wood must measure between 15 and
(109) 17 centimeters and the width must measure between 9 and 11 centimeters. Write a system of linear inequalities to represent the possible dimensions of the wood piece, in inches, given that 1 inch is equal to 2.54 centimeters.

17. Business The total profit on a particular skateboard is represented as $p^2 - 7p$ where
(104) p is the number of units sold in thousands. How many units need to be sold to have a profit of $23,750? Round to the nearest hundred.

18. Find the next 3 terms of the sequence $\dfrac{1}{2187}, \dfrac{1}{729}, \dfrac{1}{243}, \dfrac{1}{81}, \ldots$.
(105)

19. Chemistry Oxygen evaporation from a body of water increases with the temperature.
(106) This process of oxygen depletion can be modeled by the expression $\dfrac{\sqrt{x}}{6}$ where x is the temperature in C°. What value of x corresponds to an evaporation of 9 cubic feet of oxygen?

20. Graph the function $f(x) = 3|x|$.
(107)

21. Multiple Choice Which function is not an exponential function?
(108)
 A $y = 4(3)^x$ **B** $y = -4(3)^x$ **C** $y = 4^3 x$ **D** $y = 4\left(\dfrac{1}{3}\right)^x$

22. Analyze For an exponential function with $a = 5$ and $b = 3$, why is it necessary to put
(108) parentheses around the 3 when writing the function rule?

Solving Problems Involving Permutations

23. Geometry The diagram shows a right triangle with a hypotenuse that is an irrational number. What set of numbers would include the hypotenuse?
(1)

24. Evaluate $x^2 - 8x + 15$ and its factors for $x = -2$.
(72)

25. Oven Temperature The actual temperature (t) of Jeannine's oven varies by no more than 9°F from the set temperature. Jeannine sets her oven to 350°F. Write an absolute-value inequality that models the possible actual temperatures inside the oven. What is the lowest possible temperature?
(101)

26. Multi-Step The length of a picture is 2 inches greater than its width. A 3-inch-wide border is added to the bottom of the picture for a scrapbook page.
(98)
  a. Write expressions for the width and length of the picture with the border.
  b. The area of the picture with the border is 110 square inches. Find the length and width of the original photo.

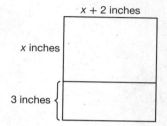

27. Multi-Step Jasmine wants to plant tulips around the perimeter of her property. The property is the shape of a square. The area of the yard is 21,000 square feet and the area of the house is 1500 square feet.
(102)
  a. Write a formula to find the length of the sides of the property.
  b. Solve for x.
  c. Jasmine changes her mind and decides to buy enough bulbs to plant them 6 inches apart along just one edge of the property. How many bulbs will she need if she starts at the first corner and goes to the second corner?

28. Justify Explain how to simplify $\dfrac{6}{\sqrt{5}-7}$.
(103)

29. Subtract $\dfrac{2x^2}{x^2-49} - \dfrac{x-7}{x^2-6x-7}$.
(95)

*30. If $f(x) = \dfrac{1}{3}^x$ and $g(x) = 3^x$, which function represents exponential growth and which function represents exponential decay?
(Inv. 11)

Graphing and Solving Systems of Linear and Quadratic Equations

LESSON 112

***1.** Solve this system by graphing: $\begin{array}{l} y = -x^2 + 12 \\ y = -x + 6 \end{array}$
(112)

***2.** **Multiple Choice** Which system of equations has no solution?
(112)

A $\begin{array}{l} y = x^2 + 2 \\ y = 3 \end{array}$
B $\begin{array}{l} y = x^2 - 2 \\ y = 3 \end{array}$
C $\begin{array}{l} y = -x^2 + 2 \\ y = 3 \end{array}$
D $\begin{array}{l} y = -x^2 - 2 \\ y = -3 \end{array}$

3. Simplify the rational expression $c^{-2}f^{-5} + \frac{6}{c^2 f^5}$, if possible.
(51)

4. Write a compound inequality that represents all real numbers that are greater than -4 and less than 8.
(73)

***5.** (Architecture) In a European castle, a room with an arched ceiling is covered by a slanted roof. The ceiling is modeled by the equation $y = -x^2 + 4$ and the roofline by the equation $y = -2x + 5$. Assume that the dimensions are in meters. What are the coordinates for the point of intersection of the roof with the ceiling assuming that the vertex of the parabola is (0, 4)?
(112)

***6.** **Analyze** A system of three equations consists of the quadratic equation $y = x^2$ and two linear equations that do not describe the same line. What is the maximum number of ordered pairs in the solution set? Explain.
(112)

7. A six-sided number cube is rolled three times. How many outcomes are possible?
(111)

***8.** **Error Analysis** Two students are finding the value of $_6P_6$. Which student is correct? Explain the error.
(111)

Student A	Student B
$_6P_6 = \dfrac{6!}{(6-6)!}$ $= \dfrac{6!}{0!}$ $=$ undefined	$_6P_6 = \dfrac{6!}{(6-6)!}$ $= \dfrac{6!}{0!}$ $= \dfrac{720}{1} = 720$

9. **Geometry** A triangle can be classified according to its sides or according to its angles.
(111) There are three side length categories—equilateral, isosceles, and scalene—and three angle categories—acute, obtuse, and right.
   a. How many possibilities are there for classifying triangles according to both sides and angles?
   b. How many of these triangles are not possible? Which ones are they?

10. **Multi-Step** There are 7 runners on the track team. Runners will be selected randomly
(111) for the first, second, third, and final positions on the 4-member relay team.
   a. How many different relay teams can be formed?
   b. What is the probability that a runner at random is chosen to be on the relay team?

Graphing and Solving Systems of Linear and Quadratic Equations

LESSON 112

11. Use the quadratic formula to solve $x^2 - 60 + 17x = 0$. Check the solutions.
(110)

12. **Multiple Choice** What are the solutions to $2a^2 + 20a - 30 = 0$?
(110)
   A $20 \pm 4\sqrt{10}$
   B $-20 \pm \sqrt{10}$
   C $-5 \pm 2\sqrt{10}$
   D $-5 \pm \sqrt{10}$

*13. **Measurement** A rectangle has sides of length x feet and $2x + 2$ feet with an area
(110) of 24 square feet. Cassandra uses the quadratic formula and finds that x equals 3 and -4. She determines that this means the sides of the rectangle are -4 by -6 or 3 by 8. Why is she incorrect?

14. **Construction** Suzanne would like to place a fence around her rectangular yard, which
(110) has a perimeter of 200 feet. The fencing for the front length of the house will cost $5 per foot and the fencing for the side and back of the yard will cost $3 per foot. Her total cost is $720. What are the dimensions of her property?

15. Find the next 3 terms of the sequence $-0.032, 0.16, -0.8, 4, \ldots$.
(105)

16. **Paper Folding** Solange folds a piece of paper, making two rectangles. When she folds it
(105) again, she makes 4 rectangles. Each fold doubles the number of rectangles. A sequence describing this process is $2, 4, 8, \ldots$. If someone folds a piece of paper is 12 times, how many rectangles did the 12 folds form?

17. Solve the equation $\frac{\sqrt{x}}{6} = 12$. Check your answer.
(106)

18. **Population** The exponential function $y = 11.35(1.00183)^x$ can model the approximate
(108) population of Ohio from 2000 to 2006, where x is the number of years after 2000 and y represents millions of people. What was the population in 2003?

19. Evaluate the function $f(x) = -2(4)^x$ for $x = -2, 0,$ and 2.
(108)

20. **Multiple Choice** Which system has no solutions?
(109)
   A $\begin{array}{c} y < 2 \\ y < 1 \end{array}$
   B $\begin{array}{c} y > 2 \\ y > 1 \end{array}$
   C $\begin{array}{c} y < 2 \\ y > 1 \end{array}$
   D $\begin{array}{c} y > 2 \\ y < 1 \end{array}$

21. **Analyze** What inequality symbols should go into the boxes so that the solution set lies
(109) between the lines and does not include the boundary points?

$$y \,\square\, \tfrac{3}{5}x + 7$$

$$y \,\square\, \tfrac{3}{5}x + 1$$

22. Find the zeros of the function $y = x^2 - 6x - 72$.
(96)

23. Graph the inequality $4x - y \leq -5$.
(97)

Graphing and Solving Systems of Linear and Quadratic Equations

LESSON 112

24. Multi-Step A girl takes 4 hours to complete a job. Her mother can complete the same job in 3 hours. Her little sister takes 6 hours to complete it.
(99)
   a. Write an equation representing how long it takes the three of them to complete the job working together.
   b. How long will it take to complete the job, in hours, if all three famiy members work together?
   c. How many minutes is that?

25. (Biking) Dustin and Roberto leave their house at the same time. Dustin rides his bike 49 feet east. Roberto rides his bike 81 feet south. Use the formula $(49)^2 + (81)^2 = x^2$ to find the distance between Dustin and Roberto.
(102)

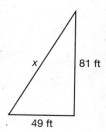

*26. Formulate In the system $\begin{matrix} y = x^2 - 3 \\ y = a \end{matrix}$, a is a real number. What is the minimum value of a so that the system will have two solutions?
(112)

27. Multi-Step A race-car driver is driving at a rate of $\sqrt{10{,}800}$ miles per hour. How long does it take the driver to go 85 miles? Give the answer as a rational expression in simplest form. (Hint: distance = rate times time)
(103)
   a. Write the equation to find the driver's travel time using the given values.
   b. Find the solution.

28. Write Tell how to remove any coefficients of the x^2-term in a quadratic equation before completing the square.
(104)

*29. Describe the transformation of $f(x) = -x^2 + 2$ from the parent quadratic function.
(Inv 10)

*30. Charlotte invested $1000 in an account that doubles her balance every 7 years. Does this situation model exponential growth or decay? Express the function that represents this situation. After 42 years, how many times will her balance have doubled? What will that balance be after 42 years?
(Inv 11)

Interpreting the Discriminant — LESSON 113

*1. Find the value of the discriminant of the equation $3x^2 - x + 2 = 0$.
(113)

2. The new rectangular basketball court at the high school has a width of $9x^2 + x + 36$
(53) and a length of $4x^2 + 2x + 2$. What is the perimeter of the new court?

3. Solve $6|z - 3| = 18$.
(74)

4. Find $8!$.
(111)

*5. Multiple Choice Which is a possible value for the discriminant of the equation
(113) graphed?

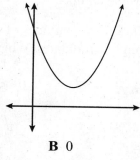

A -5 B 0

C 3 D 5

*6. Model Draw the graph of a quadratic equation that has a discriminant that is greater
(113) than zero.

*7. Generalize Describe the values of the discriminant that indicate two real solutions.
(113)

*8. Solve this system $\begin{array}{l} y = -\dfrac{x^2}{2} + 8 \\ y = -2x + 10 \end{array}$ by graphing.
(112)

*9. Error Analysis Two students are solving the system of equations $\begin{array}{l} y = x^2 + 3 \\ y = -3x + 1 \end{array}$
(112) by substitution. Which student is correct? Explain the error.

Student A	Student B
$y = x^2 + 3 \quad y = -3x + 1$ $x^2 + 3 = -3x + 1$ $x^2 + 3 + 3x - 1 = 0$ $x^2 + 3x + 2 = 0$ $(x + 2)(x + 1) = 0$ So, $x = -2, x = -1$, and the solutions are $(-2, 7)$ and $(-1, 4)$.	$y = x^2 + 3 \quad y = -3x + 1$ $x^2 + 3 - 3x + 1 = 0$ $x^2 - 3x + 4 = 0$ no solution

Interpreting the Discriminant
LESSON 113

10. **Geometry** For safety reasons, a guy wire must connect the top of a utility pole to the ground at a particular angle. The utility pole is located at the base of a hill described by the equation $y = -\frac{x^2}{25} + 2x$. The equation for the correct angle of the wire is $y = -x + 14$. At what altitude on the hill should the ground stake be located? (Assume all dimensions are in feet.)
(112)

*11. In designing a necklace, a goldsmith places a gold wire on a workbench so that the wire takes on the shape of a parabola described by the equation $y = \frac{x^2}{2}$. The goldsmith then lays a straight wire across the first so that the second follows the equation $y = \frac{x}{6} + 6$. Use a graphing calculator to determine the coordinates for the points of intersection. Round answers to the nearest whole number.
(112)

*12. **Error Analysis** Two students are finding the number of ways to choose a president and a vice president from a list of eight candidates. Which student is correct? Explain the error.
(111)

Student A	Student B
$_8P_2 = \frac{8!}{(8-2)!}$ $= \frac{8!}{6!}$ $= 56$	$_8P_2 = \frac{8!}{2!}$ $= 20{,}160$

13. (**Dining**) A restaurant offers a choice of 3 sandwiches, 3 chips, and 5 soft drinks. How many different meal combinations are offered?
(111)

14. **Probability** A CD has 9 tracks. The CD player is set to play the songs randomly so that each song plays only once. What is the probability that the first 3 songs are the first 3 tracks in order?
(111)

15. Solve the equation $\sqrt{x} + 2 = 8$. Check your answer.
(106)

16. (**Architecture**) An architect is designing a structure that merges two different right triangles along the hypotenuse of each triangle. The hypotenuse of one triangle is $\sqrt{x+2}$ units long and the hypotenuse of the second is $\sqrt{2x-4}$. At what value of x are the two lengths equal?
(106)

17. Graph the function $f(x) = |x + 4|$.
(107)

*18. **Multi-Step** A plot of land is 143 square feet with dimensions of x and $x + 2$. What is the perimeter of the plot of land?
(110)
  a. Use the quadratic formula to find the dimensions of the plot of land.
  b. What is the perimeter of the plot of land?

Interpreting the Discriminant

LESSON 113

19. Multi-Step Emmanuel throws a football into the air. Its movement forms a parabola given by the quadratic equation $h = -16t^2 + 14t + 50$, where h is the height in feet and t is the time in seconds.
  a. Find the time t when the ball is at its maximum height. Round to the nearest hundredth.
  b. Find the time t when the ball hits the ground. Round to the nearest hundredth.
  c. Find the maximum height of the arc the ball makes in its flight. Round to the nearest tenth.

20. (**Firefighting**) A forest ranger is stationed at the Delilah Lookout fire tower in the Sequoia National Forest in California. The distance d (in miles) he can see to the horizon can be estimated by the formula $d = \sqrt{\frac{3h}{2}}$, where h is the height of the observer's eyes (in feet) above sea level. If Delilah Lookout is located at an elevation of 5176 feet above sea level, write a radical expression that shows the distance the ranger can see to the horizon.

21. Graph the system: $\begin{array}{l} y \geq -\frac{3}{5}x + 3 \\ y \geq \frac{3}{4}x + 3 \end{array}$

22. Analyze Compare $-4.9t^2 + v_0 t + y_0 = 0$ and $-4.9t^2 + v_0 t = 0$.

23. Write the inequality that is graphed on the coordinate plane.

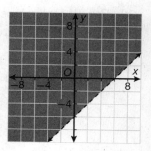

*24. (**Projectile Motion**) A projectile is shot up in the air from the ground with an initial velocity of 84 feet per second. Using $y = -16t^2 + 84t$, write an equation to model the situation and use the discriminant to determine if the projectile will reach a height of 200 feet.

25. Find the roots of $36x = 9x^2 + 36$.

Interpreting the Discriminant
LESSON 113

26. **Finance** The amount of money Ricardo has after x years of investing $100 at his local bank is $f(x) = 100(1.065)^x$. Which graph could represent this function?
(108)

A 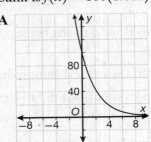 B 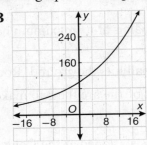 C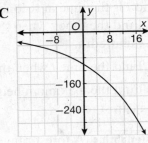

27. **Multi-Step** The time in minutes t it takes for a projectile to strike the ground is described by the equation $-4.9t^2 - 29.4t + 34.3 = 0$.
(104)
  a. Write the quadratic equation in the form $x^2 + bx = c$.
  b. Find the real-number solutions by completing the square.
  c. At what time does the object strike the ground? Explain your answer.

28. **Verify** The fifth term of a geometric sequence is -1. The first is -81. Randy thinks the common ratio is $\frac{1}{3}$. Robin says it could be $-\frac{1}{3}$. Could both be correct? Explain.
(105)

29. If a quadratic function has been vertically compressed, does that mean the parabola is wider or narrower than the parent quadratic function $f(x) = x^2$?
(Inv 10)

30. For all real values of the domain, describe the relationship between the graphs of an exponential growth and an exponential decay function.
(Inv 11)

Name _____ Date _____ Class _____

Graphing Square-Root Functions — LESSON 114

Solve.

1. $|z + 5| + 11 = 10$
(74)

2. $10x^2 = 70x$
(98)

3. $24x = 32x^2$
(98)

4. $\dfrac{5}{x+1} - \dfrac{2}{x} = \dfrac{5}{10x}$
(99)

*5. **Multiple Choice** Evaluate the equation $y = \sqrt{x + 6} - 1$ for $x = 2$.
(114)
 A $\sqrt{2}$ **B** $\sqrt{7}$ **C** $2\sqrt{2} - 1$ **D** no solution

*6. (**Oceanography**) A good approximation of the speed of a wave in deep ocean water is
(114) given by the equation $y = \sqrt{10d}$. In this equation, y is the wave's speed in meters per second and d is the ocean's depth in meters. What is the speed of a wave if the depth is 400 meters? Round to the nearest whole number.

*7. **Analyze** Given the function $f(x) = \sqrt{\dfrac{4x}{3} - 1}$, for what values of x will $f(x)$ be greater
(114) than 5? Show your work.

*8. **Analyze** Explain how to graph $f(x) = \sqrt{x - 2} + 3$ in terms of its parent function.
(114)

9. Find the value of the discriminant of the equation $2x^2 - 5x - 4 = 0$.
(113)

*10. **Error Analysis** Two students are using the discriminant to find the number of real
(113) solutions to the equation $5x^2 - 3x = 2$. Which student is correct? Explain the error.

Student A	Student B
$5x^2 - 3x = 2$ $\\$ $b^2 - 4ac = (-3)^2 - 4(5)(2)$ $\\$ $= 9 - 40$ $\\$ $= -31$ $\\$ As the discriminant is negative, there are no x-intercepts.	$5x^2 - 3x = 2$ $\\$ $5x^2 - 3x - 2 = 0$ $\\$ $b^2 - 4ac = (-3)^2 - 4(5)(-2)$ $\\$ $= 9 + 40$ $\\$ $= 49$ $\\$ As the discriminant is positive, there are two x-intercepts.

*11. **Geometry** The length of a rectangle is $x + 12$ inches and the width is $x + 8$ inches.
(113) Is there a value for x that makes the area of the rectangle 50 square inches? Explain your reasoning.

*12. **Multi-Step** The equation $288 = (3 + x)(6 - x)$ can be used to determine if the base of
(113) a rectangular box with a length of $(3 + x)$ inches and a width of $(6 - x)$ inches can have an area of 288 square inches.
 a. Write the equation setting it equal to zero.
 b. Use the equation to find the values of a, b, and c.
 c. Find the value of the discriminant.
 d. Can a box with these dimensions be made? Explain.

13. Solve this system by graphing: $\begin{array}{l} y = x^2 + 3 \\ y = -2x + 3 \end{array}$
(112)

Graphing Square-Root Functions

LESSON 114

***14. Error Analysis** Two students are solving the system of equations $\begin{array}{l} y = x^2 + 4x \\ y = -4 \end{array}$ by substitution. Which student is correct? Explain the error.
(112)

Student A	Student B
$y = x^2 + 4x \quad y = -4$ $x^2 + 4x = -4$ $(x^2 + 4x) - 4 = 0$ $x^2 + 4x - 4 = 0$ no solution	$y = x^2 + 4x \quad y = -4$ $x^2 + 4x = -4$ $(x^2 + 4x) + 4 = 0$ $x^2 + 4x + 4 = 0$ $(x + 2)(x + 2) = 0$ So, $x = -2$, and the solution is $(-2, -4)$.

*15. (Physics) Miguel is standing at the base of a ramp. He tosses a ball into the air. The path of the ball is described by the equation $y = -x^2 + 7x$. The equation $y = x$ represents the ramp. At what altitude does the ball strike the ramp? Assume that dimensions are in feet.
(112)

16. Measurement On what scale would the distance between the x-coordinates in the solution set of the system $\begin{array}{l} y = \frac{x^2}{2} \\ y = 4x - 6 \end{array}$ be 8 centimeters?
(112)

17. Graph the function $f(x) = |x| - 2$.
(107)

18. (Temperature) The temperature outside yesterday was 65°. Today the temperature changed by $|5°|$. Give the possible temperatures outside today.
(107)

19. Determine if the set of ordered pairs $\{(6, 3), (4, 2), (2, 1), (8, 4)\}$ satisfies an exponential function.
(108)

20. (Engineering) A small bridge has a weight limit of 8000 pounds. A photographer wants to photograph at least 5 vehicles on the bridge. The cars weigh about 1800 pounds each and the motorcycles weigh about 600 pounds each. There must be at least one car and four motorcycles in the photo. Graph a system of linear equations to describe the situation. Give two combinations of cars and motorcycles that are solutions.
(109)

21. Use the quadratic formula to solve $x^2 = 19x - 60$. Check the solutions.
(110)

22. Multiple Choice There are three numbers in a locker combination: 19, 22, and 28. How many different ways can the numbers be arranged?
(111)
   A 3   **B** 6   **C** 12   **D** 24

*23. Write Explain what types of situations apply to permutations.
(111)

Graphing Square-Root Functions

LESSON 114

24. Multi-Step In a bowling lane, the distance (d) from the foul line to the center of the
(101) Number 1 pin should be 60 feet and should vary from this length by no more
than $\frac{1}{2}$ inch.

   a. Convert 60 feet to inches.

   b. Write and solve an absolute-value inequality that models the acceptable distances
   from the foul line to the center of the Number 1 pin.

   c. If the diameter of the base of the Number 1 pin is $4\frac{1}{8}$ inches, what is the shortest
   possible distance between the foul line and the front of the Number 1 pin?

25. Evaluate $y = \sqrt{2x} + 3$ for $x = 8$.
(114)

26. (**Property**) Mr. Kinsey's property is in the shape of a right triangle. The legal description
(104) states that the property has an area of 900 yd² and that the base of the property is
30 yards longer than the height. What are the actual dimensions of the property?

27. Multi-Step A company gives its employees a 4% raise at the beginning of every year.
(105) This year, Jordan earns $32,000.

   a. Write a rule that can be used to find Jordan's salary after n years.

   b. How many years will it take for Jordan to earn $40,000?

   c. What will Jordan's salary be in 12 years? Round to the nearest cent.

28. Analyze Write the radical equation $\sqrt{x+3} = 2x$ so that the equation has no radical
(106) and is equal to zero.

29. Has the graph of the parent quadratic function been stretched or compressed to
(Inv 10) produce the graph $f(x) = 4x^2 + 2$?

30. Describe the similarity and difference between the graphs of $f(x) = 3^x$ and
(Inv 11) $g(x) = 4 \cdot 3^x$.

Name _____ Date _____ Class _____

Graphing Cubic Functions — LESSON 115

Solve and check.

1. (99) $\dfrac{x-2}{x+7} = \dfrac{x-6}{3x+21}$

2. (99) $\dfrac{x-4}{x+1} = \dfrac{x+5}{2x+2}$

*3. (115) Graph the cubic function $y = \frac{1}{3}x^3$. Use it to solve the equation $0 = \frac{1}{3}x^3$.

*4. (115) **Multiple Choice** Which equation represents a cubic function?
  A $y = 3x - 4y$
  B $y = 6x^2 + 2$
  C $y = x^3 - 4x + 1$
  D $y = 10x^4 + 3x^2 - 5$

*5. (115) (**Capacity**) The volume of a box is represented by the equation $V = x^3 - 4$. Use a table or graph to find the value of x that corresponds to a volume of 23 cubic units.

*6. (115) (**Games**) The volume of a whiffle ball is represented by the equation $V = \frac{4}{3}\pi r^3$. Use a graphing calculator to graph the equation and then use the graph to estimate the volume of air in a ball with a radius of 2 inches.

*7. (115) **Write** Describe the characteristics of the graph of a cubic function.

*8. (115) **Formulate** Write an example of a cubic function.

9. (114) Evaluate $y = \sqrt{4x} - 5$ for $x = 3$. Round to the nearest tenth.

*10. (114) **Error Analysis** Two students are evaluating the equation $y = \sqrt{2x - 5} + 2$ for $x = 6$. Which student is correct? Explain the error.

Student A	Student B
$y = \sqrt{2x - 5} + 2$	$y = \sqrt{2x - 5} + 2$
$y = \sqrt{2 \cdot 6 - 5} + 2$	$y = \sqrt{2 \cdot 6 - 5} + 2$
$y = \sqrt{2} + 2$	$y = \sqrt{12 - 5} + 2$
	$y = \sqrt{7} + 2$

11. (114) **Multi-Step** An apple fell from a tree limb. The function $t = 0.45\sqrt{x}$ represents how long it takes an object to fall from a height of x meters.
  a. Graph the function. (Hint: Increment the x-axis by 1 and the y-axis by 0.1, and if a graphing calculator is not used, then use the following values for x: 0, 4, 9, and 16.)
  b. Use the graph to estimate how long it took the apple to fall if the limb was 12 meters above the ground.

12. (113) Use the discriminant to find the number of real solutions of the equation $6x^2 + 2x - 1 = 0$.

*13. (113) **Error Analysis** Two students are using the discriminant to find the number of real solutions to the equation $2x^2 + 3x - 4 = 0$. Which student is correct? Explain the error.

Graphing Cubic Functions

LESSON 115

Student A	Student B
$2x^2 + 3x - 4 = 0$	$2x^2 + 3x - 4 = 0$
$b^2 - 4ac = 3^2 - 4(2)(4)$	$b^2 - 4ac = 3^2 - 4(2)(-4)$
$= 9 - 32$	$= 9 + 32$
$= -23$	$= 41$
As the discriminant is negative, there are no x-intercepts.	As the discriminant is positive, there are two x-intercepts.

14. Gardening The length of a garden is $6 + x$ meters and the width is $10 - x$ meters. Write an equation to model the area of the garden, and use the discriminant to determine if there is a value for x that will allow the area of the garden to be 50 square meters.
(113)

15. Measurement The length of a fence is $15 - x$ feet and the width is $12 + x$ feet. Can the fence enclose an area of 200 square feet? Explain.
(113)

16. Determine if the set of ordered pairs $\left\{\left(-3, \frac{1}{8}\right), \left(-1, \frac{1}{2}\right), \left(-2, \frac{1}{4}\right), \left(-4, \frac{1}{16}\right)\right\}$ satisfies an exponential function.
(108)

17. Graph the system $\begin{array}{l} 21x + 7y \geq -14 \\ \frac{1}{2}y \leq -x + 2 \end{array}$.
(109)

18. Geometry If the area of the triangle is 48 square units, what are the lengths of the base and the height to the nearest whole number?
(110)

*19. Graph the cubic function $y = 3x^3$. Use it to solve the equation $0 = 3x^3$.
(115)

20. A new 3-digit area code is being created for new telephone numbers. If the first digit must be even but not 0, the second digit is 0 or 1, and the third digit can be any number except 0, how many new area codes are possible?
(111)

21. Multiple Choice Which system of equations has the solution $(-1, 1)$?
(112)
A $y = x^2$
   $y = x + 6$
B $y = x^2$
   $y = 6$
C $y = x^2$
   $y = -2x - 1$
D $y = x^2$
   $y = -x + 6$

Graphing Cubic Functions

LESSON 115

*22. **Analyze** A system of three equations consists of a quadratic, given by $y = x^2 - 3$, and
(112) two linear equations. One linear equation intersects the parabola at two points. If the
second linear equation is parallel to the first, how many solutions does the system
have? Explain.

23. (**Accessories**) Candida has plans to shop for hair bows and does not plan on spending
(97) more than $20. Each big bow costs $5 and each small bow costs $2. Write an
inequality and graph it to describe the situation.

24. Solve $-x^2 + 2 = -7x$ by using a graphing calculator. Round to the nearest tenth.
(100)

25. **Multiple Choice** Solve $x^2 + 7 = -42$.
(102)
    A 7   **B** ±7   **C** no solution   **D** $\pm 7\sqrt{1}$

26. (**Phone Chains**) In order to relay information quickly, staff at a school use a phone
(105) chain. The superintendent first notifies 3 people of a snow day. In the second set of
calls, these 3 people each call 2 people. Each person called then calls 2 other people.
How many sets of calls need to be made to notify 96 people?

27. **Multi-Step** A square frame is to be made so that its side length is $\sqrt{x + 1}$.
(106)
    a. What is the perimeter of the square?

    b. For what value of x will the perimeter of the frame be equal to 8 units?

28. **Generalize** Look at the function $f(x) = -0.5|x|$. How can you find the direction of
(107) the "V" without graphing it?

29. (**Football**) The distance d from the goal post in feet of a football during a field goal
(107) kick is represented by the function $d = |60t - 90|$ where t is the time in seconds.
If the ball were kicked at 80 feet per second how would the graph change?

30. Write, in order, the function that grows the slowest to the one that grows the fastest:
(Inv 11) exponential, linear, quadratic.

Solving Simple and Compound Interest Problems

LESSON 116

*1. $900 is invested at 3% simple interest for 5 years. How much interest is earned?
(116)

*2. **Write** Explain the difference between simple and compound interest.
(116)

*3. **Formulate** The graph shows the value of a money market account that pays compound interest. How much principal was originally invested?
(116)

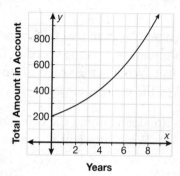

4. **Population** The exponential function $y = 3.45(1.00617)^x$ can model the
(108) approximate population of Oklahoma from 2000 to 2006, where x is the number of years after 2000 and y represents millions of people. Assuming the model does not change, predict when the population will reach 4 million?

*5. **Multiple Choice** $600 is invested at 11% simple interest. What is the value of
(116) the investment after 14 years?
  A $924   B $1524   C $2586.26   D $92,400

*6. **Mutual Funds** Over the past 20 years, a mutual fund averages paying
(116) 10% interest compounded annually. If a woman had invested $3000 originally, how much would her account be worth now?

*7. Graph the cubic function $y = -3x^3$. Use the graph to find the roots of
(115) the equation.

8. **Error Analysis** Two students write the equation "y equals x cubed plus five."
(115) Which student is correct? Explain the error.

Student A	Student B
$y = x^3 + 5$	$y = x^2 + 5$

9. **Geometry** The formula for the volume of a cube is $V = s^3$. Graph the
(115) equation and find the volume of the cube if the side length is 2 units.

Solving Simple and Compound Interest Problems

LESSON 116

*10. **Multi-Step** The volume of a packing container is given by the
(115) function $y = x^3 + 5$.
   a. Make a table of values for the equation.
   b. Graph the equation.
   c. Find the volume when x is 3 feet.

11. Evaluate $y = 3\sqrt{7x + 2} - 7$ for $x = 2$.
(114)

12. (**Physics**) The speed at which an object in free fall drops is modeled by the
(114) equation $y = 8\sqrt{x}$. In this equation, y is the speed in feet per second and
x is the distance fallen in feet. What is the speed of an apple after it falls a
distance of 8 feet? Round to the nearest tenth.

*13. **Error Analysis** Two students are determining the domain and range of the
(114) function $f(x) = \sqrt{x - 5} + 1$. Which student is correct? Explain the error.

Student A	Student B
$f(x) = \sqrt{x - 5} + 1$	$f(x) = \sqrt{x - 5} + 1$
$x - 4 \geq 0$	$x - 5 \geq 0$
$x \geq 4; y \geq 0$	$x \geq 5; y \geq 1$

14. **Measurement** The function $s = \sqrt{A}$ gives the side length of a square with area A.
(114) What is the side length of a square that has an area of 625 square feet?

15. Graph the system $\begin{array}{l} y \geq \frac{2}{5}x - 4 \\ y \leq 0 \end{array}$
(109)

16. Use the quadratic formula to solve $46 + 16x = -x^2$. Find approximate answers
(110) to four decimal places.

17. (**Sports**) The American League Central Division in Major League Baseball
(116) has 5 teams. How many different ways are there for the teams to finish first
through fifth?

18. Solve this system by graphing: $\begin{array}{l} y = 2x^2 - 6x + 1 \\ y = -x - 4 \end{array}$
(112)

19. **Multiple Choice** How many x-intercepts does the equation $y = 4x^2 + 8x - 2$ have?
(113)
   A 0   **B** 1   **C** 2   **D** 3

20. **Write** Explain what the discriminant tells about the graph of a quadratic equation.
(111)

21. Solve $4x^2 + 8 = -6x$ by using a graphing calculator. Round to the nearest tenth.
(100)

© Saxon. All rights reserved.   448   Saxon Algebra 1

Solving Simple and Compound Interest Problems

LESSON 116

22. Solve and graph the inequality $|4x - 3| + 1 > 10$.
(101)

23. **(Structural Engineering)** The water pressure p on a dam is a function of the depth
(106) of the water x behind the dam: $p = 4905\sqrt{x}$. For what value of x is the pressure equal to 44,145?

24. Multi-Step Graph the function $f(x) = |x| - 4$, and then translate the function to the
(107) left by 2. What is the vertex of this new function?

*25. $4500 is borrowed at 3.5% simple interest. The total amount of interest paid is
(116) $1260. For how many years was the money borrowed?

*26. **(Credit Cards)** A man uses a credit card to make a $1200 purchase. The credit
(116) card charges 22% annual interest compounded monthly and requires no payments for the first year. At the end of one year, how much will he owe?

27. Justify Why is $f(x) = 4(-2)^x$ not an exponential function?
(108)

28. Multi-Step Study the numbers in the sequence.
(103)

$$3, \sqrt{3}, 1, \frac{\sqrt{3}}{3}, \frac{1}{3}, \ldots$$

a. Find the pattern.

b. What is the next term in the sequence?

29. If $f(x) = 3x^2 - 12x + 2$, where is the axis of symmetry located? Give the x- and
(Inv 10) y-coordinates of the vertex.

30. Identify which function is linear, quadratic, exponential growth, and exponential
(Inv 11) decay: $f(x) = \left(\frac{1}{5}\right)^x$, $g(x) = x^2$, $h(x) = 5^x$, and $j(x) = 5x$.

Using Trigonometric Ratios

LESSON 117

***1.** Find sin A, cos A, and tan A
(117)

***2.** Find sin A, cos A, and tan A.
(117)

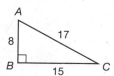

3. Write an equation for a direct variation that includes the point, (24, 3).
(56)

***4.** If $\angle A = 77°$, find sin A, cos A, and tan A to the nearest ten-thousandth.
(117)

***5. Error Analysis** Two students are finding the measure of $\angle A$. Which student is correct? Explain the error.
(117)

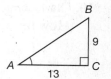

Student A	Student B
$\tan A = \dfrac{9}{13}$	$\tan A = \dfrac{13}{9}$
$\tan^{-1}(\tan A) = \tan^{-1}\left(\dfrac{9}{13}\right)$	$\tan^{-1}(\tan A) = \tan^{-1}\left(\dfrac{13}{9}\right)$
$A \approx 34.7°$	$A \approx 55.3°$

***6. Geometry** In a right isosceles triangle, the acute angles are congruent. Find the measures of the acute angles. Then use the sine or cosine ratio to find the length of a leg of a right isosceles triangle to the nearest hundredth if the hypotenuse is 5 centimeters.
(117)

***7. Multi-Step** You are standing on the roof of a 70-foot-tall building looking across at another building. Use the picture to answer the questions.
(117)

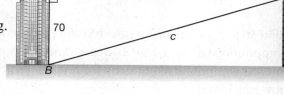

a. Find the distance from the bottom of the building where you are standing to the top of the other building.

b. Find the measure of $\angle A$.

Using Trigonometric Ratios
LESSON 117

*8. **(Nature)** A tree casts a shadow of 25 feet along the ground. The angle from the ground to the top of the tree is 45°. How tall is the tree?

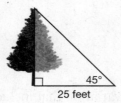

9. If $1100 is borrowed for 2 years at 9% simple interest, how much interest is paid?

*10. **(Navigation)** A submarine begins diving from the water's surface at an angle of 7°. How far below the water's surface is the submarine after it has traveled 3.4 miles?

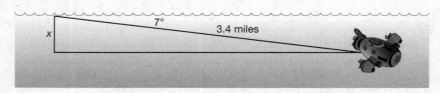

*11. Generalize Explain the meaning of opposite leg and adjacent leg to an acute angle in a right triangle.

12. Error Analysis A $1500 investment earns 8% simple interest. Two students find the value of the account after 25 years. Which student is correct? Explain the error.

Student A	Student B
$I = Prt$	$I = Prt$
$I = 1500 \cdot 0.08 \cdot 25$	$I = 1500(0.08)(25)$
$I = 3000$	$I = 3000$
$3000	$3000 + 1500 = 4500$
	$4500

13. Multi-Step A boy plans to invest $100 in an account that pays 10% interest compounded annually for 10 years. Another option is an account that earns 20% interest compounded annually for 5 years. Which will earn him more money, and how much more?

14. Graph the cubic function $y = x^3 + 3$. Use the graph to evaluate the equation for $x = 0$.

Using Trigonometric Ratios

LESSON 117

15. Error Analysis Two students draw a graph of the equation $y = 2x^3$. Which student is correct? Explain the error.
(115)

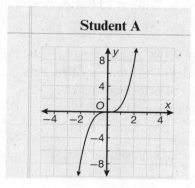

Student A

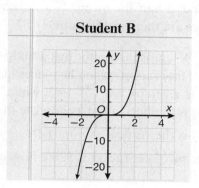

Student B

16. (Packaging) A rectangular box has a volume of $V = x^3 + 3$ cubic units. Use a table or graph to find the value of x that corresponds to a volume of 30 cubic units.
(115)

17. Use the quadratic formula to solve $2x^2 + 9 = 9x$. Check the solutions.
(1)

18. Find $_{12}P_3$.
(111)

19. Solve $x^2 + 5 = 9$.
(102)

20. (Astronomy) Near a planet, a satellite follows a trajectory described by the equation $y = \frac{x^2}{8} + \frac{7}{4}$. The trajectory is intercepted by a radio signal represented by the line $y = -\frac{9x}{8}$. At what coordinates will the radio signal intersect the trajectory?
(112)

21. Use the discriminant to find the number of real solutions of the equation $x^2 + 2x - 2 = 0$.
(113)

22. Multiple Choice What is the domain of the function $f(x) = 2\sqrt{x+6} - 1$?
(114)
  A $x \geq -5$ **B** $x \geq -6$ **C** $x \geq 6$ **D** $x \geq 0$

23. Write Describe the graph of $f(x) = \sqrt{x+4}$ in terms of its parent function.
(114)

24. Solve and graph the inequality $3|8x+2| < 12$.
(101)

25. Multi-Step The sum of the squares of two consecutive odd numbers is 74.
(104)
  a. Write expressions for two consecutive odd numbers.
  b. Write an equation to represent the problem.
  c. What is the possible solution(s)?

*26. If $\angle A = 81°$, find $\csc A$, $\sec A$, and $\cot A$ to the nearest ten-thousandth.
(117)

27. Multi-Step Anita figures that the value of her car, in thousands of dollars, can be approximated by $f(x) = 15\left(\frac{4}{5}\right)^x$, where x is the number of years after the car's manufacture. Evaluate the function for $x = 0, 1,$ and 2 and then sketch the function.
(108)

Using Trigonometric Ratios

LESSON 117

28. Justify Explain why $\begin{array}{l} y > 4 \\ y < 4 \end{array}$ has no solutions but $\begin{array}{l} y \geq 4 \\ y \leq 4 \end{array}$ does.
(109)

29. What does a half-life mean? If a substance's half-life is 25 hours, how many half-lives are there in 150 hours?
(Inv 11)

30. (Tennis) A tennis instructor has a budget of $2000 to buy new rackets. He will receive 2 free rackets when he places his order. The number of rackets, y, that he can get is given by $y = \frac{2000}{x} + 2$, where x is the price per racket.
(78)

  a. What is the horizontal asymptote of this rational function?

  b. What is the vertical asymptote?

  c. If the price per racket is $200, how many rackets will he receive?

Name _____ Date _____ Class _____

Solving Problems Involving Combinations

LESSON 118

***1. Photography** A photographer wants to take a picture of a group of 4 students from a class of 15. How many different pictures could she take?
(118)

Calculate each combination.

***2.** $_{11}C_4$
(118)

3. $_9C_7$
(118)

4. $_{12}C_5$
(118)

***5. Write** Explain the difference between permutations and combinations.
(118)

***6. Verify** Show that $_8C_3 = \dfrac{_8P_3}{\text{number of ways to order 3 items}}$.
(118)

***7. Multiple Choice** Find $_5C_3$.
(118)
    A 10   **B** 12   **C** 20   **D** 60

***8. Nutrition** Teenage girls need 3 servings of dairy products per day. How many combinations can a girl make from 10 different dairy products?
(118)

9. Find the LCM of $(15x - 10)$ and $(3x - 2)$.
(57)

***10. Geometry** How many straight line segments can be formed by connecting any 2 of 8 points?
(118)

***11. Multi-Step** A student chooses 9 stuffed animals to give to a charity. He had a total of 25 stuffed animals.
(118)
    a. How many combinations of 9 animals could be chosen?
    b. If one stuffed animal has already been chosen, then what is the probability that he chooses his 8 favorite animals.

***12.** Find the six trigonometric ratios for $\angle A$.
(117)

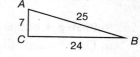

13. Multi-Step A kite is caught in the top of a 12-foot tree and a string 20 feet long is stretched out to the ground.
(117)
    a. Find the distance along the ground from the base of the tree to the end of the string.
    b. Find $\angle A$, the angle the string makes with the ground.

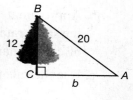

14. Coordinate Geometry A right triangle has coordinate $A(2, 1)$, $B(8, 9)$, and $C(8, 1)$. Plot the points and find the measure of acute angle A.
(117)

15. If $9200 is borrowed for 3 years at 5% simple interest, how much money will be owed after 3 years?
(116)

16. Bonds A woman invests $20,000 in a bond that pays 6% interest compounded annually. How much interest will she earn in 5 years?
(116)

Solving Problems Involving Combinations

LESSON 118

***17. Error Analysis** A CD earns 4% interest compounded quarterly. A woman deposits $10,000 for 20 years. Two students find the value of her account. Which student is correct? Explain the error.
(116)

Student A	Student B
$A = 10,000(1 + 0.01)^{80}$	$A = 10,000(1 + 0.04)^{20}$
$A = 10,000(1.01^{80})$	$A = 10,000(1.04^{20})$
$A = 22,167.15$	$A = 21,911.23$
$22,167.15	$21,911.23

18. Find $\dfrac{10!}{5!}$.
(111)

19. Simplify $\dfrac{4}{\sqrt{3}-2}$.
(103)

20. Solve this system by substitution: $\begin{array}{l} y = x^2 - 5 \\ y = 4x \end{array}$
(112)

21. (Football) A punter kicks a football straight up in the air from 2 feet off the ground with an initial velocity of 75 feet per second. Using $y = -16t^2 + 75t + 2$, write an equation to model the situation and use the discriminant to determine if the ball will reach a height of 45 feet.
(113)

22. Evaluate $y = \sqrt{\dfrac{3}{x} + 2}$ for $x = 6$. Round to the nearest tenth.
(114)

23. Multiple Choice What is the solution to the equation $x^3 - 27 = 0$?
(115)
  A 3   **B** 9   **C** 27   **D** 0

24. Model What is the equation for the parent function of a cubic equation?
(115)

25. Find the solution of $x^2 = 45$. Round the answer to three decimal places.
(102)

26. Multi-Step Colin is experimenting with a type of cell that multiplies 5 times each day.
(105)
  a. On Monday, there are 500 cells. If all cells survive, how many are there on Friday?
  b. Approximately one fourth of the cells die off each day. If the cells still multiply at the same rate, what rule represents a geometric sequence to represent the number of cells remaining each day?
  c. Use the rule to find the number of cells on Friday if there are 500 on Monday.

27. Multi-Step Tickets to a school play are sold to teachers for $3 each and to students for $0.50 each. The drama class hopes to earn at least $200 from the ticket sales. The theater seats 250 people.
(109)
  a. Write and graph a system of linear inequalities to describe the situation.
  b. If 15 teachers buy tickets, is it possible for the drama class to meet their goal? Explain.

Solving Problems Involving Combinations

LESSON 118

28. **Estimate** Estimate to the nearest whole number the value of the zeros for the equation $2v^2 + 20v = 21$.
(110)

29. Factor completely $-88z^3 - 2r^2z^3 - 30rz^3$.
(79)

30. Radioactive glucose is used in cancer detection. It has a half-life of 100 minutes. How much of a 320 mg sample remains in the body after 10 hours? First determine how many half-lives there are in a 10-hour period if the half-life is 100 minutes.
(Inv 11)

Graphing and Comparing Linear, Quadratic, and Exponential Functions

LESSON 119

***1. Error Analysis** Students were asked to write an equation that has a quadratic parent function. Which student is correct? Explain the error.

Student A	Student B
$f(x) = 2(x-1)^2 - 4$	$f(x) = 2x + 5$

***2.** Identify the function family.

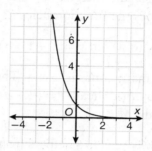

***3.** Identify the function family.

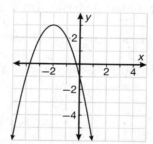

*4. Write How can a parent function be used to graph a family of functions?

*5. Interior Decorating A type of carpet sells for $12 per square foot. Installation is an additional $500. A function can be written to determine the price for installing carpet in a square room with floor x feet in length. Identify the function family and the parent function.

6. Error Analysis Two students find $_{18}C_5$. Which student is correct? Explain the error.

Student A	Student B
$_{18}C_5 = \dfrac{18!}{5!(18-5)!}$ $= 8568$	$_{18}C_5 = \dfrac{18!}{(18-5)!}$ $= 1{,}028{,}160$

*7. Multi-Step A restaurant charges $16 for a large pizza plus $1.50 per topping.
 a. Which type of function best describes the cost of a pizza?
 b. Write a function that describes the cost of a pizza with x toppings.

Graphing and Comparing Linear, Quadratic, and Exponential Functions

LESSON 119

8. (Jewelry) A jewelry store sells necklaces for the cost of the chain plus the cost of the beads. Chains cost $20 each, and beads cost $7.50 each. To which function family does the equation that describes the cost of a necklace with x beads belong?
(119)

*9. Multiple Choice Which of the following describes the graph of $y = 5^x$?
(119)
  A quadratic   **B** linear   **C** exponential   **D** none of these

*10. Geometry A triangle has base b and height $b - 4$. If its area is written as a function, to which function family does the function belong?
(119)

11. Identify the function family to which $y = 2 - 1100x$ belongs.
(119)

*12. Write A fair coin is flipped many times in a row. The probability of all flips resulting in heads is given by $P(\text{all heads}) = \left(\frac{1}{2}\right)^x$, where x is the number of flips. What type of function is this?
(119)

13. Multi-Step A teacher randomly selects 4 helpers from her class of 22.
(118)
  a. How many ways can 4 helpers be selected?
  b. What is the probability that Shawn, Tonia, Torie, and Reid are all chosen?

*14. If $\angle A = 14°$, find $\sin A$, $\cos A$, and $\tan A$ to the nearest ten-thousandth.
(117)

15. Error Analysis Two students are finding the value of x in the figure. Which student is correct? Explain the error.
(117)

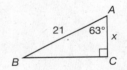

Student A	Student B
$\sin 63° = \dfrac{x}{21}$	$\cos 63° = \dfrac{x}{21}$
$21 \cdot \sin 63° = x$	$21 \cdot \cos 63° = x$
$18.71 \approx x$	$9.53 \approx x$

*16. (Aviation) An airplane begins making its descent at an angle of 11° with the horizontal. If the plane is at an altitude of 8000 feet and will remain at this angle throughout its descent, how far away is the plane from its landing point?
(117)

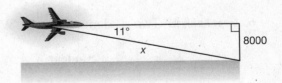

17. Solve this system by graphing: $\begin{array}{l} y = x^2 - 2 \\ y = 2x - 5 \end{array}$
(112)

18. Use the discriminant to find the number of real solutions of the equation $9x^2 - 24x + 16 = 0$.
(113)

Graphing and Comparing Linear, Quadratic, and Exponential Functions

LESSON 119

19. (Business) A baseball-card seller models the costs for attending a sale as the function $y = \$250 + \$0.25\sqrt{n + 30}$. In this function, n represents the number of cards sold and 30 is the number of cards given out free as prizes. What is the domain of the function? Explain.
(114)

20. Graph the cubic function $y = x^3 - 2$. Use the graph to evaluate the equation for $x = 0$.
(115)

21. Verify Show that a $500 investment earning 6% interest compounded annually for 3 years will earn more than earning 6% simple interest for 3 years.
(116)

22. Multiple Choice $1000 is invested in an account paying 5% interest compounded quarterly. What is the value of the account after 20 years?
(116)
    A $1282.04   **B** $2000   **C** $2653.30   **D** $2701.48

23. Find $_8C_3$.
(118)

24. Simplify $\dfrac{6}{\sqrt{7} - 3\sqrt{5}}$.
(103)

25. Solve $-2m^2 - 12m = 10$ by completing the square.
(104)

26. Multi-Step The expression $-\sqrt{x - 4}$, where x is the time in seconds, represents the change in temperature as water cools when ice cubes are placed in a glass of water.
(106)
    a. For what value of x is the temperature change equal to -4?
    b. What must be done to isolate the radical when solving for x?
    c. Graph the radical equation $y = -\sqrt{x - 4}$.

27. Identify the function family to which $y = 100^{(-3x)}$ belongs.
(119)

28. Write Explain why you would not use the quadratic formula for the equation $x^2 - x - 2 = 0$?
(110)

29. Verify Use the formula for permutations to verify that $_6P_2 = 30$.
(111)

30. Given the function $f(x) = ax^2 + bx + c$, describe the general shapes of the graphs of $f(x)$ when:
(Inv 10)
    a. $a \neq 0$ and $b \neq 0$;
    b. $a = 0$ and $b \neq 0$;
    c. $a \neq 0$ and $b = 0$;
    d. $a = 0$ and $b = 0$.

Name _____ Date _____ Class _____

Using Geometric Formulas to Find the Probability of an Event

LESSON 120

***1.** Find the probability of landing in the shaded area.
(120)

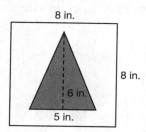

2. Write Explain what geometric probability is.
(120)

3. Generalize A system of equations consists of a quadratic and a linear equation. If the graphs of the two equations do not intersect, what can you conclude about the solution to the system?
(112)

***4. Verify** Show that the probability of landing in the shaded region is $\frac{1}{2}$.
(120)

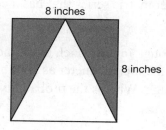

5. Find $_7C_2$ 21
(118)

6. (**Manufacturing**) A number cube is made with side lengths of 5 centimeters. Use the function $V = s^3$ to find the volume of plastic that is contained in the number cube.
(115)

***7. Multiple Choice** A parachutist will land in a rectangular field with a circular landing area as her target. What is the probability that she will land on target?
(120)

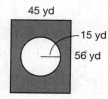

A ≈ 0.09   **B** ≈ 0.28   **C** ≈ 0.72   **D** ≈ 0.92

8. (**Jewelry**) To make a friendship bracelet, 8 beads are used. How many different combinations of beads could be on the bracelet if there are 20 different beads?
(118)

Saxon. All rights reserved.   463   Saxon Algebra 1

Using Geometric Formulas to Find the Probability of an Event

LESSON 120

*9. (**Puzzles**) A children's stacking puzzle teaches shapes. A child randomly points to the puzzle. What is the probability that the child's finger lands on the shaded part of the square?
(120)

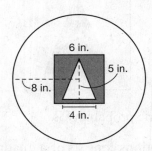

10. Model Draw right $\triangle ABC$ with right angle C so that $\sin A = \frac{3}{5}$ and $\cos A = \frac{4}{5}$.
(117)

11. What is the domain of $f(x) = 3\sqrt{x} - 5$?
(114)

12. Find the 4th term in the geometric series that has a common ratio of -1.1 and a first term of 7.
(105)

*13. (**Baking**) After rolling the dough into a 9-inch by 12-inch rectangle, a boy cuts as many biscuits with a 3-inch diameter as possible. His little sister comes and touches the dough. What is the probability that she did not touch a biscuit?
(120)

*14. Error Analysis Two students find the probability of landing in the shaded region. Which student is correct? Explain the error.
(120)

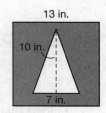

Student A	Student B
$A = \frac{1}{2}(10)(7) = 35$	$A = \frac{1}{2}(10)(7) = 35$
$A = 13^2 = 169$	$A = 13^2 = 169$
$P(shaded) = \frac{35}{169}$	$P(shaded) = 1 - \frac{35}{169} = \frac{134}{169}$

*15. Justify Find the probability of landing in the shaded region and explain your steps.
(120)

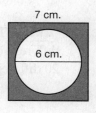

Using Geometric Formulas to Find the Probability of an Event

LESSON 120

*16. Geometry A spinner is divided into equal sectors. The red sector is 120°. What is the probability of not landing on a red sector?
(120)

17. Find the probability of landing in the shaded square.
(120)

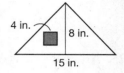

*18. Multi-Step A game uses a large square game board that is divided into congruent smaller squares. The probability of landing on one of those smaller squares is $\frac{1}{4}$. Each of the smaller squares has area 49 square millimeters.
(120)
   a. What is the probability of not landing on that square the next time?
   b. What is the area of the larger square?

19. Error Analysis Students were asked to sketch an example of a linear function. Which student is correct? Explain the error.
(119)

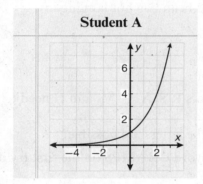
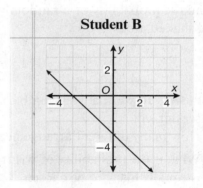

*20. Multi-Step The graph shows the height in feet, h, of a tennis ball hit into the air with initial velocity of 30 feet per second at time t seconds.
(119)
   a. What type of function is this? How do you know?
   b. Use the graph to approximate the maximum height that the ball reaches. Round to the nearest foot.

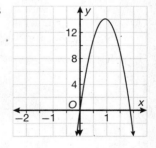

21. Identify the function family: $y = 4x^2 + 2$.
(119)

22. Error Analysis Two students are asked to list all the possible combinations choosing 2 letters from A, B, and C. Which student is correct? Explain the error.
(118)

Student A	Student B
AB, AC, BC	AB, AC, BA, BC, CA, CB

23. Use the discriminant of the equation $-x^2 + 2x + 1 = 0$ to find the number of real solutions.
(113)

465

Saxon Algebra 1

Using Geometric Formulas to Find the Probability of an Event

LESSON 120

24. Find the probability of a randomly tossed dart landing in the shaded area.
(120)

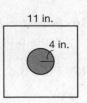

25. An investment of $2600 is made for 7 years at 8% simple interest. How much will be in the account after 7 years?
(116)

26. Multi-Step Graph the function $f(x) = |x + 2|$, then translate the function up by 3. What function does this graph now represent?
(107)

27. Multiple Choice Find $\angle A$ if $\cos A = \frac{5}{7}$.
(117)
   A 44.4°   **B** 45.6°   **C** 0.71°   **D** 1.00°

28. Solve $3x^2 + 9x = 5.25$ by completing the square.
(104)

29. Find the probability of a randomly tossed bean bag not landing in the shaded area.
(120)

30. Multi-Step To work a word jumble the letters P, S, A, C, M, and H need to be unscrambled.
(111)
   a. How many different arrangements of the letters are possible?
   b. What is the probability of writing the word CHAMPS on the first try if each letter is equally likely to be chosen?